THE ART OF CHEMISTRY

THE ART OF CHEMISTRY
Myths, Medicines, and Materials

ARTHUR GREENBERG
Department of Chemistry
College of Engineering and Physical Sciences
University of New Hampshire
Durham, New Hampshire

A JOHN WILEY & SONS, INC., PUBLICATION

Published by John Wiley & Sons, Inc., Hoboken, New Jersey.
Published simultaneously in Canada.

Library of Congress Cataloging-in-Publication Data:

Greenberg, Arthur.
The art of chemistry : myths, medicines, and materials / Arthur Greenberg.
p. ; cm.
Includes index.
ISBN 0-471-07180-3 (cloth : alk. paper)
1. Chemistry—History. 2. Alchemy. 3. Medicine—History.
[DNLM: 1. Chemistry—history. 2. Alchemy. 3. History of Medicine.
] I. Title.
QD11 .G735 2003
540'.9—dc21 2002009950

Printed in the United States of America.

10 9 8 7 6 5 4 3 2 1

First and foremost, this book is dedicated to my wife Susan and to our children David and Rachel, who have been loving, supportive, and understanding in the face of my many human foibles.

I also dedicate this book to my parents Murray and Bella and my three siblings Dee Dee (Ilene), Kenny, and Roberta, who have also been loving and supportive.

CONTENTS

PREFACE

The physician and writer Oliver Sacks has written that "Chemistry has perhaps the most intricate, most fascinating, and certainly most romantic history of all of the sciences."* How does one convey the surprise, pleasure, and excitement of discovery in early chemistry to the college or high school student or, for that matter, the teacher? There is also a receptive public who do indeed wish to learn more about science. So why burden them with some outdated theories alongside those most current? The answer is that it is vital to help nonscientists understand how science works. First and foremost, the practice of science is an intensely human endeavor. Although alchemy is treated today by most (non–"New Age") people as an exercise in *naivité* if not downright fraud, it was in fact a fundamentally human attempt to understand Nature's unity and to try and express it metaphorically. The transformation of these myths, superstitions, and applications to the arts and medicine into our modern science provides the forward motion of this book. But a recurring theme is our very human need to visualize and try to understand the fundamental nature of matter. Another goal is to understand early experimental chemistry at a time when the "guts" of the apparatus were fully visible in contrast to the "black boxes" that today are so ubiquitous. (It is indeed arguable that those who perform DNA sequencing in an automated "black box" that spews out alphabetical sequences of nucleotides may have forgotten that they are actually doing chemistry.)

The Art of Chemistry is very similar in style to my earlier book, *A Chemical History Tour,* published in 2000. The book's foundation is the wonderful artwork employed over the centuries to illustrate chemical apparatus as well as our various metaphors for the nature and structure of matter. It attempts to entertain as well as inform. In the present book, 188 figures are employed to illuminate 72 essays. I have attempted to make these essays accessible to a broad audience, including chemists and chemistry teachers, other scientists and teachers, engineers and physicians, as well as nonscientists who find science interesting and enjoy artwork. This book is not an orderly history of chemistry but rather another idiosyncratic tour including many historical "sites" unvisited in our first tour as well as a few revisits uncovering new insights. The essays are organized in eight sections in roughly chronological order. The first section focuses on the imagery of the spiritual and mythological roots of chemistry—gods and goddesses, winged dragons, witches, the phoenix, of course (was the Japanese film icon *Rodan* a

*Statement on back cover of Greenberg, *A Chemical History Tour,* Wiley, 2000.

phoenix?), passionate birds of prey, the feared basilisk (spitting cobra or *Godzilla*?), and the ouroboros—a metaphor for the conservation of matter and perhaps Kekulé's true inspiration for the structure of benzene. The second and third sections of the book treat the technological aspects of early chemistry. In addition to beautiful plates of sixteenth- and seventeenth-century stills and other apparatus, there is a rather too graphic image of antimony's power as a purgative—both emetic *and* laxative. The fourth section focuses on the period between the mid-1600s and mid-1700s when chemistry began to emerge as a science. In addition to Boyle, Hooke, and Mayow, who almost solve the riddle of combustion and respiration, we have the business machinations surrounding the discovery of phosphorus—first corner the market, *then* decide what the new element is good for. We think of Becher as the *ur*-father of the first true theory of chemistry: phlogiston. However, he was perhaps the foremost mercantilist of his era as well as the economic advisor to Leopold I, Emperor of the Holy Roman Empire. The longest section of *The Art of Chemistry* is devoted to the chemical revolution that occurred during the last half of the eighteenth century. It is not commonly appreciated that while Lavoisier was surely the father of modern chemistry, he was also one of the most influential economists of the eighteenth century. Benjamin Franklin's early contributions to chemistry are visited briefly in this section. In Section VI, Dalton's atomic theory is introduced as the culmination of the chemical revolution. Five essays in this section are devoted to chemistry in America at the beginning of the nineteenth century. One of my favorite figures in this book is an early American (ca. 1790) laboratory apparatus for synthesizing sulfuric acid that combines elements of the farm (clay crocks) and the blacksmith shop (bellows). The major themes in Section VII are the development of specializations in chemistry, exemplified by organic chemistry, and the consolidation introduced by the Periodic Law. Like *A Chemical History Tour,* coverage lightens during the late nineteenth century and is very sparse during the twentieth and twenty-first centuries. The explosive exponential growth of the chemical literature would make balanced and appropriately weighted coverage impossible. Furthermore, we are immersed in a sea of textbooks and monographs treating this modern material. Thus, the final section (VIII) treats some modern topics in a very light manner, although it endeavors to give readers a peek at the future—nanotechnology and self-organization, which are both triumphs of our ability to understand the chemical structure of matter at its most fundamental level.

I have concluded *The Art of Chemistry* with an Epilogue consisting of two brief, more personal essays. One of these is about a friend from adolescent years, Robert Silberglied, a quirky and ingenious butterfly collector and mischief-maker, who became a world-renowned entomology Professor at Harvard before he died at an early age in an airplane crash. The second is a brief essay whimsically visiting my own chemistry genealogy. Although these may appear to be exercises in self-indulgence and self-aggrandizement, they are not meant to be. The purpose is to give the reader a taste for our scientific culture—the early signs of "a natural scientist" and the interest in our personal scientific roots and desire to connect with the past.

But beyond artwork I have attempted to include excerpts from plays and novels and even take a trip into occult realms. Thus, we have some fun with one of Chaucer's *Canterbury Tales*. Dmitri Mendeleev and the great composer Alexander Borodin, both chemists in their mid-twenties, took a leisurely trip to the groundbreaking chemistry conference at Karlsruhe in 1860, stopping frequently to indulge their tastes for music. What an interesting premise for a film. While many readers are familiar with Primo Levi's autobiographical book *The Periodic Table*, how many are familiar with Lewis' *White Lightning* (1923), a 354-page novel consisting of 92 chapters for the chemical elements in order of atomic number? The broader cultural perspective of chemistry has been well served by the play *Oxygen*, written by the distinguished chemists Carl Djerassi and Roald Hoffmann. In addition to including a very brief excerpt from this play, I have also included a brief excerpt of Peter Weiss's 1966 play *Marat/Sade*. There is even a *faux*-Thurber short story inspired by the doodling of a high school student on the title page of her late-nineteenth-century textbook. While I have retained some of the Rabelaisian earthiness of the Renaissance and injected a bit of satire, the ultimate purpose of this book is a serious one- to provide education and enjoyment.

ACKNOWLEDGMENTS

The Art of Chemistry is a sequel to *A Chemical History Tour* published by John Wiley and Sons in 2000. I remain grateful to Professor Roald Hoffmann, who was so supportive of the earlier book; Dr. Barbara Goldman, who accepted and encouraged the project; and Dr. Darla Henderson and Ms. Amy Romano, who helped to bring the book to fruition at Wiley. I am indeed fortunate that Dr. Henderson and Ms. Romano continued to be supportive and fun to work with during the present project.

It is a very happy and fortunate circumstance that the readers of this manuscript have included my biological father, Mr. Murray Greenberg; my "chemistry father," Professor Pierre Laszlo; my closest chemistry friend, Professor Joel F. Liebman; my wife, Ms. Susan J. Greenberg; and my bibliophile friend, Dr. Roy G. Neville, all of whom were educated as chemists. As usual, Joel provided numerous suggestions and stimulated at least one essay. The fortunate opportunity to read drafts of the play *Oxygen*, by Professors Carl Djerassi and Roald Hoffmann, provided both insight and impetus for revisiting the chemical revolution. Professor Dudley Herschbach provided some wonderful information and insight that stimulated my essay on Benjamin Franklin. Almost a decade ago, Professor Robert Vrijenhoek made me aware of the different conclusions about taxonomy that are sometimes drawn when morphological and chemical (genomic) techniques are compared; he supplied figures and very helpful discussions for one of my essays. I am also grateful for the help provided to me by Dr. Arnold Thackray and Ms. Elizabeth Swan of the Chemical Heritage Foundation so appropriately located in Philadelphia. It is also a very great pleasure to acknowledge stimulating discussions with Dr. Roy G. Neville. The Roy G. Neville Historical Chemical Library, in California, provided copies of a number of exceedingly rare figures and plates from its extensive collection. There are many others who have contributed to this book, and they are acknowledged wherever appropriate. I am also very grateful to Ms. Alice Greenleaf (and also to Ms. Nan Collins), University of New Hampshire, for carefully scanning and editing the numerous figures from my own book collection used in this book. These figures were also employed in a number of classes and invited lectures presented during the period in which this book was written.

COLOR PLATE CAPTIONS

FIGURE 1 ▪ Fifteenth-century Tibetan painted mandala in which gods and goddesses represent the dualities that are the origin of the four ancient elements (the square) The circle represents completeness, the cycle of life, and (even) the conservation of matter. (Courtesy Rossi & Rossi, London, and *Asianart.com*.)

FIGURE 3 ▪ Eighteenth-century painting by a Johann Winckler, exhibiting Rosicrucian influences, in which the four friars represent water, air, fire, and earth in a correctly ordered square. The red tincture, in the possession of each friar, is an embodiment of the Philosopher's Stone—the mysterious agent of projection and transmutation.

FIGURE 7 ▪ Painting in a fifteenth-century manuscript depicting the eleventh-century Persian physician Avicenna (Abu Ali-al Hussin ibn Abdallah ibn Sina, 980–1037) in an apothecary shop (© Archivo iconografico, S.A./CORBIS).

FIGURE 32 ▪ Title page (hand-colored) from *Liber de Arte Distillandi* (Heironymous Brunschwig, 1512) depicting a double-still in which the central tower contains cooling water that is continuously replenished. Come to think of it, this apparatus evokes an image of the caduceus: two serpents entwined about the staff of Hermes where the loops symbolize "couplings." (Courtesy Chemical Heritage Foundation.)

FIGURE 92 ▪ Oil-on-porcelain painting by artist L. Sturm, very likely the porcelain painter Ludwig Sturm (source: Dr. Alfred Bader). Although the painting is titled "The Alchemist," rational chemistry is occurring. The key to the picture is the pan of charcoal. It is likely that a metal oxide is being reduced to the metal by charcoal in the red-hot crucible. (I am grateful to the Art Museum of the State University of New York at Binghamton for permission to use this image.)

FIGURE 117 ▪ Madame Lavoisier was instructed in painting by the famous artist Jacques Louis David. This is a photo of the oil portrait she painted of her close friend Benjamin Franklin. (Courtesy of a relative of Benjamin Franklin.)

FIGURE 139 ▪ Color plate from the American edition (1854) of *Notions Générale de Chimie* by Théophile Jules Pelouze and Edmond Fremy. Depicted in *1* is Lavoisier's classic mercury oxidation experiment involving reflux of mercury in air; *2* illustrates the oxidation of an iron wire using a flame in pure oxygen; *3* depicts strong heating of manganese dioxide to release oxygen—an experiment first performed by Scheele.

FIGURE 140 ▪ Color plate from the 1854 American Edition of *Notions Générale de Chimie* by Pelouze and Fremy. Depicted in *4* is the combustion of phosphorus in air leaving unreacted nitrogen; *5* shows the generation of hydrogen through reaction of zinc and sulfuric acid—work first published by Henry Cavendish in 1776. In 6, we see Lavoisier's decomposition of water using an iron wire in a porcelain tube heated red hot in a furnace (see Figure 114).

FIGURE 141 ▪ Color plate from the 1854 American edition of *Notions Générale de Chimie* by Pelouze and Fremy. Depicted in *7* is a synthesis of water involving generation of hydrogen and its combustion in air; 8 shows a clever self-controlling hydrogen gas generator; 9 illustrates a laboratory-scale distillation apparatus appropriate for students.

FIGURE 142 ▪ Color plate from the 1854 American edition of *Notions Générale de Chimie* by Pelouze and Fremy. Depicted in *10* is an industrial-scale still consisting of a copper boiler covered with a hood; *11* displays the very modem-looking water-cooled distillation apparatus designed by Gay-Lussac.

FIGURE 143 ▪ Color plates from the 1854 American edition of *Notions Générale de Chimie* by Pelouze and Fremy. In *12*, phosphorus, suspended by a wire, burns with blinding brightness in pure oxygen; diamonds (*13*) are composed of pure carbon just like the humble mineral plumbago (graphite), another carbon allotrope, and totally different from ruby and other gemstones; heating saltpetre (KNO_3) and sulfuric acid produces nitric acid (*14*) but on an industrial scale (*15*) it is cheaper to use niter ($NaNO_3$).

FIGURE 144 ▪ A wonderfully surrealistic rendition of stalactites (suspended from the ceiling, in case you forgot) and stalagmites composed of limestone ($CaCO_3$) formed by contact between the CO_2 dissolved in groundwater and lime (CaO) in the soil. This image seems to anticipate the artistic style of René Magritte some 45 years before his birth. (From the 1854 American edition of *Notions Générale de Chimie* by Pelouze and Fremy.)

FIGURE 156 ▪ Early advertising cards for Ozone Soap. While we strongly doubt that even the minutest traces of ozone were ever to be found in Ozone Soap, it was still a wonderfully evocative trade name.

FIGURE 164 ▪ An early twentieth-century caricature in *Vanity Fair* of William Ramsay pointing with fatherly pride to his chemical family—the rare gases.

FIGURE 176 ▪ Collectors' cards portraying famous chemists issued in 1938 for *La Cigarette Oriental de Belgique*. Although Topps issued bubblegum trading cards in the early 1950s that included Marie Curie and Louis Pasteur, there seems to be no current market for a "Stars of Chemistry" bubblegum trading card series. *Quel domage!* (I am grateful to Jamie and Steve Berman for this information.)

FIGURE 177 ▪ Collectors' cards issued during the 1920s and 1930s depicting chemistry laboratories and famous chemists. The figure of Gay-Lussac published for *Chocolat Poulain* ("Taste And Compare! Quality Without Rival") is particularly well done. The card at the lower left is a version of the famed drawing of Justus Liebig's laboratory housed at the University of Giessen. The chemist in the front-center, dreamily applying the mortar and pestle, is my *chemical* great-great-great-great-great-grandfather, Adolph Strecker.

FIGURE 178 ▪ Justus Liebig was one of the fundamental pioneers in biochemistry (animal, plant, and food chemistry). He held strong views about the value of "meat juice" in the diet and lent his name to commercial endeavors. (Today there is a company that sells Linus Pauling vitamin C tablets.) The Justus Liebig Company sponsored these Belgian cards printed during the 1930s that tout their line of food items. (I am grateful to Jamie and Steve Berman for this information.)

FIGURE 182 ▪ Linus Pauling continued to be "way ahead of the curve" as he coauthored with long-time artist friend Roger Hayward The *Architecture of Molecules* in 1964. Before the era of molecular graphics, the depiction of the heme molecule, one of 57 color drawings in this lovely book, exemplified their enlightened attempt to convey the beauty of chemistry to the public. From *The Architecture of Molecules* by Linus Pauling and Roger Hayward. © 1964 W.H. Freeman and Company. Used with permission.

FIGURE 187 ▪ Butterflies painted to prove the role of ultraviolet light in the mating behavior of butterflies. This plate is from the final paper presented by Professor Robert E. Silberglied, a boyhood friend of the author and a "most unforgettable character."

THE ART OF CHEMISTRY

SECTION I
SPIRITUAL AND MYTHOLOGICAL ROOTS

EASTERN AND WESTERN ESOTERICA

How do we make sense of our very brief lives, our earthly domain, and the universe surrounding us? We humans have an instinct for symbolism that urges us to represent the tangible and intangible with thoughts, words, pictures, and music. Cave paintings predate our modern era by tens of thousands of years. My son David, barely two years old, called ice cream "um-num"—and it was an apt word-symbol from a toddler who did not yet know that it was frozen cream but certainly knew what tasted and felt good. We are also born with sexual instincts vital to our survival as a species, and we inherently recognize the duality of opposites and respond to sexual symbolism.[1,2] These instincts led to allegories and metaphors for understanding the nature and transformations of matter at least 2500 years ago in central and eastern Asia, biblical lands of the Middle East, and ancient Greece.[3]

Figure 1 shows a mandala painted in central Tibet during the fifteenth century.[4] Mandalas have their origins in Tantrist Hindhuism and Buddhism and are representations for contemplation of the universe. Mandalas are often made by Buddhist monks from painted sand over the course of many days, contemplated, and then returned to the sea—a symbolic act of enriching one's earthly life with thought rather than accumulation of worldly wealth, which is illusory and transient. Central to the Eastern mandala is the circle that represents unity and completeness. Circles also separate domains such as heaven and earth. Circle imagery can also be likened to very ancient ideas about the conservation of matter. Later in this book we encounter the ouroboros, a serpent that forms a circle by devouring its own tail even as it regenerates itself. The act of returning the sand of a mandala to the sea implies both conservation of matter and the cycle of life. Typically, a circle in a mandala will encompass a square surrounded by four gates representing the four cardinal directions (north—actually on the right here, south, east, and west). These four gates lead to an inner sphere inhabited by four gods. On the outside of this sphere are four goddesses in a complementary fourfold array (NE, SE, SW, NW). This male-female duality is also represented by female and male (mother–father) figures outside of the larger circles in positions of sexual embrace. The four elements, commonly coded for by a square, really represent pairs of opposing properties—hot versus cold, dry versus wet. Thus, fire is hot and dry and water is cold and wet. The center of this mandala depicts the boddhisatva ("enlightenment being") Vajrapani holding the thunderbolt and grasping a serpent.[4] The center of a mandala can be likened to the fifth ancient element—the ether.

Adam McLean has an interesting view concerning Western alchemical mandalas.[5] He analyzes 30 images of esoteric alchemy in European texts mostly from the seventeenth century. Although many of these have the circular and square forms similar to Figure 1, others share only the essence but not the form of

FIGURE 1. ■ Fifteenth-century Tibetan painted mandala in which gods and goddesses represent the dualities that are the origin of the four ancient elements (the square). The circle represents completeness, the cycle of life, and (*even*) the conservation of matter. See color plates. (Courtesy Rossi & Rossi, London and *Asianart.com*).

Eastern mandalas. For example, McLean considers the emblem from Libavius' 1606 *Alchymiae* (see Figure 22 in a later essay) to be a mandala.[6] In this light Figure 2, drawn by artist Rita L. Schumaker,[7] also contains the fundamental essence of a mandala. Earth, water, and air are clearly depicted by circular realms while fire penetrates these realms. Dualities are depicted by dark and light doves as well as dragons. Seeds of growth in the earth imply the multiplication of metals. In the next essay we will encounter even more tangible metaphors for the four ancient elements.

FIGURE 2. ■ Pencil drawing, in the style of a mandala, by Ms. Rita L. Schumaker depicting dualities (entwined dragons, dark and light doves) as well as the four ancient elements (water, air, fire, and earth).

---

1. N. Schwartz-Salant, *Encountering Jung on Alchemy*, Princeton University Press, Princeton, 1995.
2. I. MacPhail, *Alchemy and the Occult*, Yale University Library, New Haven, 1968, pp. xv–xxxii (essay by A. Jaffe).
3. B. Pullman, *The Atom in the History of Human Thought*, Oxford University Press, New York, 1998.
4. A.M. Rossi, F. Rossi, and J.C. Singer, *Selections 1994*, Rossi & Rossi, Ltd., London, 1994. I am grateful to Ms. Anna Maria Rossi, Rossi & Rossi, London, and Ian Alsop, Editor, *Asianart.com*, Santa Fe, New Mexico for permission to employ theVajrapani mandala image.
5. A. McLean, *The Alchemical Mandala*, Phanes Press, Grand Rapids, 1989.
6. McLean, op. cit., pp. 62–69.

7. The author thanks Ms. Rita L. Schumaker, Charlotte, North Carolina, for this original drawing and for discussion of its themes.

THE FOUR ANCIENT ELEMENTS

A 1747 oil-on-wood painting signed by a Johann Winckler[1] (Figure 3) joyously employs alchemical, spiritual, and religious symbolism characteristic of Rosicrucian beliefs. Most prominent are the four abbots whose activities symbolize earth, fire, air, and water. They are arrayed in the appropriate order of contrary properties—cold versus hot; wet versus dry. Thus, water is wet and cold, fire is dry and hot; air (think steam) is hot and wet, and earth is cold and dry:

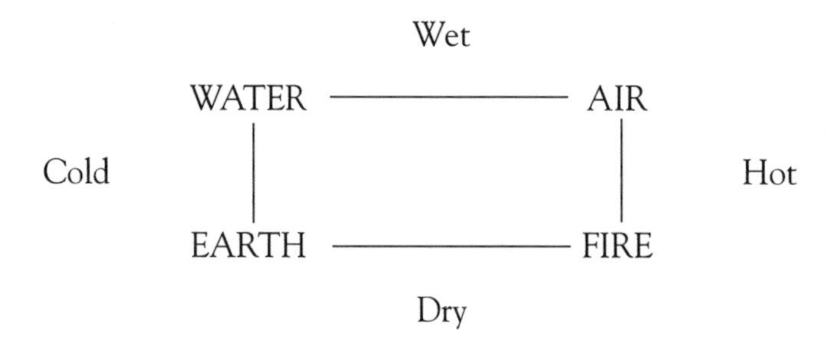

The Cupid (or Mercurious) figure was said by the psychologist Carl Jung to represent "the archer who, chemically, dissolves the gold, and morally, pierces the soul with the dart of passion."[2] "Christian Rosencreutz in *The Chymical Wedding* is pricked with a dart by Cupid after stumbling upon the naked Venus."[2] The four abbots and the Venus figure each possess a vessel containing the Red Tincture, which represents the transmuting agent or Philosopher's Stone[3] or a preliminary stage of the Stone.[4] The castles may represent . . . well . . . castles. Or . . . they may symbolize the athanor or philosopher's furnace, which holds the hermetically sealed philosopher's egg.[5] The pair of doves represent the *albedo*, the white color that follows the *nigredo*, or the initial black color of The Great Work. Initially, metals and other substances are heated to form a black mass. Subsequent heating may calcine this mass to produce a white calx. Now, if that long-tailed bird attached to an abbot by a string is a peacock, we see represented the third color change of The Great Work, the rainbow hues. The fourth and final color is the ruby of the Red Tincture—four cucurbits—full and one goblets-worth in this painting. The phoenix also represents this final ruby red color but no phoenix is seen rising (or expiring) in the painting. No crows are in evidence either, so let's assume that the coals or the ashes in the athanor represent the *nigredo*.

Rosicrucians combine religious, occult, and alchemical beliefs.[6] Although the earliest writings date to the beginning of the seventeenth century, the origins of Rosicrucianism are commonly attributed to a Christian Rosenkreutz ("rosy cross"), allegedly born in 1378. Some consider the early sixteenth-century physician and alchemist Paracelsus to be the true founder. The alchemist Michael Maier appears to have been a Rosicrucian.[7]

FIGURE 3. ■ Eighteenth-century painting by a Johann Winckler, exhibiting Rosicrucian influences, in which the four friars represent water, air, fire, and earth in a correctly ordered square. The red tincture (see color plates) in the possession of each friar, is an embodiment of the Philosopher's Stone—the mysterious agent of projection and transmutation.

The sign in the lower right of this painting may be translated as follows:

1. I search in the water here.
2. The air should give me
3. I search in the earth
4. The fires should become for me
5. Something here, you fools, here in the water, air and earths.
   In the fire, shall you busily search.
6. All here suddenly becomes.

---

1. I am not certain about the identity of the artist. One possibility is Johann Heinrich Winckler (1703–1770).
2. L. Abraham, *A Dictionary of Alchemical Imagery*, Cambridge University Press, Cambridge, UK, 1998, p. 51.
3. J. Read, *Prelude To Chemistry*, The Macmillan Co., New York, 1937, p. 12; p. 148.
4. Abraham, op. cit., p. 169.
5. Abraham, op. cit., pp. 31–32.
6. *The New Encyclopedia Britannica*, Encyclopedia Britannica, Inc., Chicago, 1986, Vol. 10, p. 188.
7. Read, op. cit., pp. 230–232.

MYSTICAL AND MAJESTIC NUMBERS

Certain simple numbers have for centuries been attributed great symbolic significance[1] in alchemy.[2] *One* connotes God or Allah in monotheistic religions as well as the *prima materia*—the origin of all matter. *Two* signifies the male and female principles (opposites—mercury and sulfur) that are present in all things. *Three*—the *tria prima* (mercury, sulfur, and salt—spirit, soul, and body) represents Paracelsus' (1493–1541) extension of the male and female principles; it also represents the Holy Trinity (Father, Son, and Holy Ghost). *Four,* as noted in the first two essays, is the number of ancient elements (earth, water, air, and fire), each one a composite of two opposing qualities or properties—hot/cold and dry/wet. There are also four seasons and four cardinal directions. Aristotle introduced a *fifth* element, the *quinta essencia,* also known as the *quintessence*, representing the heavens or the celestial ether. *Seven* metals known to the ancients (silver, gold, iron, copper, mercury or quicksilver, tin and lead) matched the number of visible "planets" (Moon, Sun, Mars, Venus, Mercury, Jupiter, and Saturn) as well as days of the week. *Twelve* signs of the Zodiac, equivalent to the twelve months in a year, were equated by the English alchemist George Ripley, Canon of Bridlington, to twelve "Gates" or operations en route to the Philosopher's Stone.[2] These were as follows:[2,3]

1. Calcination (action of fire on minerals in air)	Aries, the Ram
2. Congelation (a thickening by cooling)	Taurus, the Bull
3. Fixation (trapping a volatile as a solid or liquid)	Gemini, the Twins
4. Solution (dissolutions or reactions of substances)	Cancer, the Crab

5. Digestion (heat continuously applied; no boiling)	Leo, the Lion
6. Distillation (ascent and descent of a liquid)	Virgo, the Virgin
7. Sublimation (ascent and descent of a solid)	Libra, the Scales
8. Separation (isolation of insoluble liquids, solids)	Scorpio, the Scorpion
9. Ceration (bringing hard material into a soft state)	Sagittarius, the Archer
10. Fermentation (animation of a substance with air)	Capricornis, the Goat
11. Multiplication (increasing potency of the Stone)	Aquarius, the Water Carrier
12. Projection (mysterious action of the Stone)	Pisces, the Fishes

Basil Valentine ("Valiant King"), reputed to be born in 1394, is a perplexing literary forgery.[4] There appears to be fairly widespread agreement that he was in reality a certain sixteenth/seventeenth-century salt boiler and publisher named Johann Thõlde who "edited" Basil Valentine's works.[4] Whoever Basil Valentine was, he was quite knowledgeable about the chemistry of his times,[4] and he described Twelve Keys or operations defining The Great Work. Figures 4 and 5 are from Manget's 1702 *Biblotheca Chemica Curiosa*.[5]

The First Key (*prima clavis*) signifies the chemical wedding and represents the creation of the primitive materials for the stone.[6–8] The wolf represents antimony sulfide, useful for separating gold from other metals. [A related figure was included in Michael Maier's 1618 book *Atalanta Fugiens* (see Figure 17) and is discussed later.] The old man may represent Saturn (lead) for separating sulfur. The Second Key represents watery separation, and the Third Key depicts the dragon as the *Prima Materia* and suggests a circular cycle of volatilization and fixation.[8] The Fourth Key symbolizes putrefaction, a heating, with fire that we now recognize as roasting of metal ores entailing a rather mixed and messy oxidation of sulfide ores to a dark mass. The Fifth Key is considered to represent a solution process (but this can, of course, signify chemical reactions). The Sixth Key represents conjunction—chemical marriage between sophic sulfur (the King) and sophic mercury (the Queen). The Seventh Key is a kind of alchemical mandala[9] representing the four earthly elements, the ether and the three Paracelsan principles. The Eighth Key is a resurrection scene symbolized by planting of seed. As John Read has noted,[2] if putrefaction (the Fourth Key) is associated with oxidation, the reverse process of regeneration or resurrection of metals corresponds to reduction (*reducere*—"to lead back" or "restore") and the return of their souls. The Ninth Key alludes to the three principles, the four ancient elements, and the consecutive colors of The Great Work in order of ascension—the crow (black, putrefaction), the swan (white, calcinations), the peacock (yellow or rainbow hues), and the red phoenix (symbolizing the Red Tincture or Philosopher's Stone). The Tenth Key represents the *tria prima* and has been discussed elsewhere.[6–8] The Hebrew may represent a verse in Psalms but with a kind of Kabbala letter-substitution code.[10] The Eleventh Key symbolizes the process of multiplication, and the Twelfth and final Key symbolizes calcinations, the mysterious fire in the wine barrel, and the fixation of the volatile is symbolized by the Lion (sulfur) eating the snake (mercury).[6–8]

FIGURE 4. ■ And the magical number is . . . 12! Here are the first six keys (*clavis* = key) of the equally famous and fictional Basil Valentine ("Valiant King"). The twelve keys represent 12 alchemical operations leading to the Philosopher's Stone (as well as signs of the Zodiac and months in one year). This figure is from Manget's 1702 *Bibliotheca Chemica Curiosa* (courtesy Chemical Heritage Foundation).

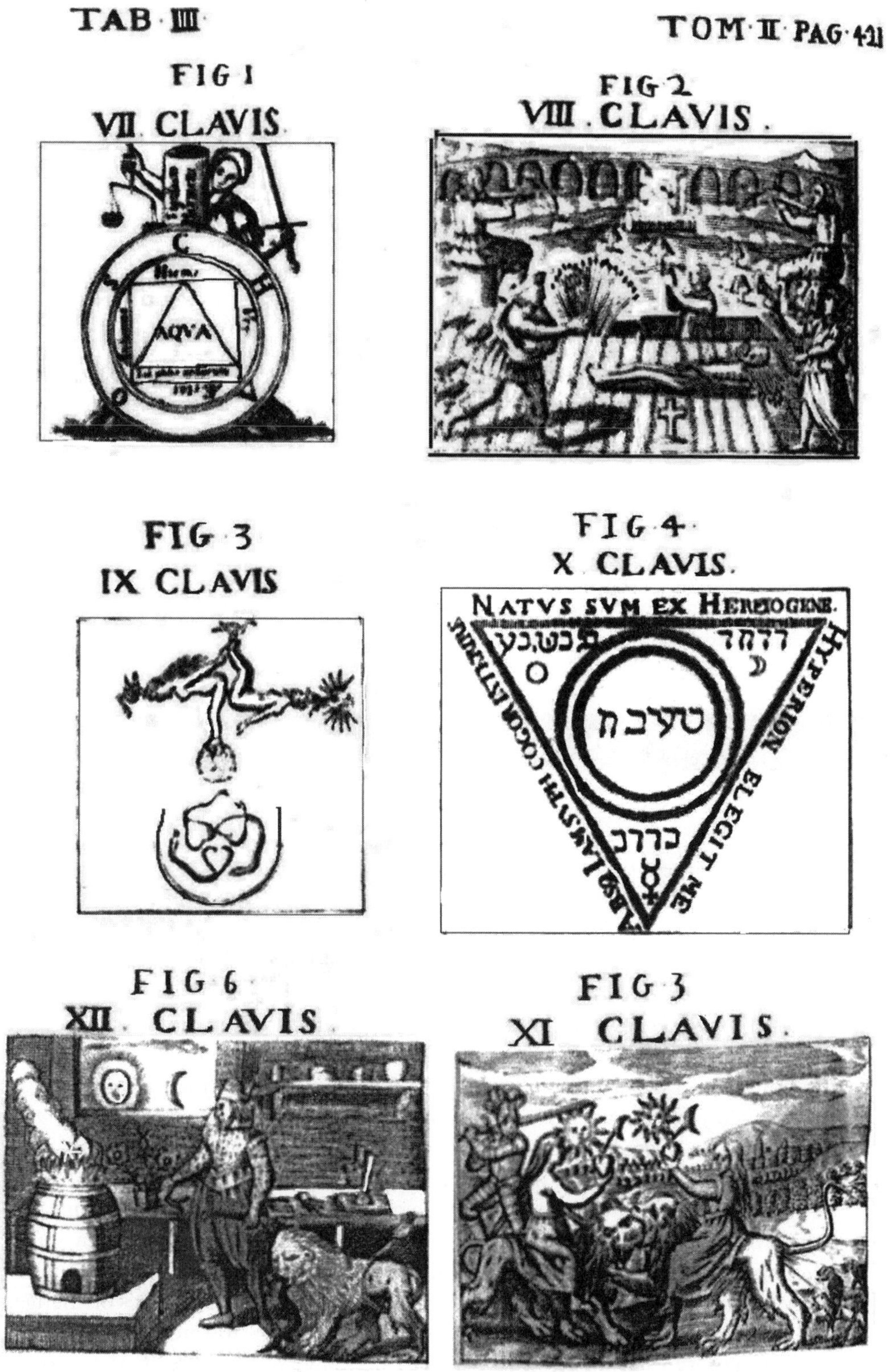

TAB · III · TOM · II · PAG · 421

FIG 1
VII. CLAVIS.

FIG 2
VIII. CLAVIS.

FIG 3
IX CLAVIS

FIG 4
X CLAVIS.

NATVS SVM EX HERMOGENE.
HYPERION ELEGIT ME
SINE IANVA COCOPHENIS

FIG 6
XII. CLAVIS.

FIG 5
XI CLAVIS.

FIGURE 5. ■ The final six keys of Basil Valentine in Manget's 1702 *Bibliotheca Chemica Curiosa* (courtesy Chemical Heritage Foundation). See the text for discussion of the symbolism in the 12 keys.

1. Many faiths and philosophies include their own mystical numerological systems. For example, the Hebrew Kabbala, first appearing in the twelfth century, had its roots more than a millennium earlier. The Creation was a process involving 10 divine numbers and the 22 letters of the Hebrew alphabet. Together these two numbers provided the 32 paths to wisdom (see *The New Encyclopedia Britannica*, Encyclopedia Britannica, Inc., Chicago, 1986, Vol. 6, p. 671). A truly religious fan likewise intrinsically understands the Kabbala of Baseball: 3 (strikes for an out; outs for a half inning); 4 (balls for a walk); 7 (games in the World Series); 9 (innings; field positions); 13 (uniform number to avoid); 44 (the all-time home-run champ Hank Aaron's uniform number—a uniform number to seek); 60, 61, 70, 73 (home-run records established in turn by Babe Ruth, Roger Maris, Mark McGwire, and Barry Bonds); and 1955 (only year in the Modern Era in which Brooklyn won the World Series).
2. J. Read, *From Alchemy to Chemistry*, Dover Publications, Inc., New York, 1995, pp. 32–35.
3. J. Read, *Prelude To Chemistry*, The Macmillan Co., New York, 1937, pp. 139–142.
4. J.R. Partington, *A History of Chemistry*, Macmillan and Co. Ltd., London, 1961, pp. 183–203.
5. J.J. Manget, *Bibliotheca Chemica Curiosa, seu Rerum ad Alchemicum pertinentium Thesaurus Instructissimus* . . . , Sumpt. Chouet, G. de Tournes, Cramer, Perachon, Ritter, & S. de Tournes, Geneva, 1702. I am grateful to Ms. Elizabeth Swan, Chemical Heritage Foundation, for supplying images of these plates.
6. Read (1937), op. cit., pp. 196–208.
7. A. Greenberg, *A Chemical History Tour*, John Wiley and Sons, New York, 2000, pp. 26–31.
8. S.K. De Rola, *The Golden Game—Alchemical Engravings of the Seventeenth Century*, Thames and Hudson Ltd., London, 1988, pp. 120–126.
9. A. McLean, *The Alchemical Mandala*, Phanes Press, Grand Rapids, MI, 1989.
10. Personal communication from Professor Laura Duhan Kaplan to the author.

NATURAL MAGICK: METAMORPHOSES OF WEREWOLVES AND METALS

A twelfth-century ethnography of Ireland, written by a Gerald of Wales, describes an encounter by a Priest with a talking he-wolf in the wild, who entreaties him to give the Eucharist to his dying mate.[1] The Priest does so and the discussion of whether this action is sacrilegious depends on the nature of the creature—is it a true hybrid (like a griffin), a man in wolf's clothing, or a total change in identity? How did a man *change* to become a wolf? What humanity remained? Was the man "wolfish" before any change occurred?[2] Historian Caroline Walker Bynum treats the concepts of *identity* and *change* and posits that, in the final decades of the twelfth century, a new conceptual image took hold in European culture—the metamorphosis—gradual rather than sudden change.[1] For example, Bynum contrasts the ancient New Testament story of the sudden conversion of Saul from a persecutor of Christians to Paul, a disciple of Christ, with the twelfth-century version of his slow, evolutionary, and reasoned conversion on the road to Damascus.[1] Metamorphosis presents a more dynamic and complex story than sudden miraculous change or the mere appearance of a static hybrid. Ancient stories of change, including werewolf folklore, also took on new meanings toward the end of the twelfth century.[1] Metamorphoses were to be found everywhere in the natural world—the foodstuff in a seed becomes a tree; food "morphs" into blood and bile. And from Middle Eastern cultures came complex, often spiritual, operations for gradually changing matter that came to comprise alchemy.

Intellectual cross-fertilization was one benign by-product of a horrendous series of Crusades first launched by Pope Urban II in 1095 to remove Moslem control of the Christian shrine of the Holy Sepulchre in Jerusalem.[3] Jerusalem fell to the Crusaders in 1099, and they murdered its Moslem and Jewish inhabitants. Increasing control of the Holy Lands by the Crusaders continued until Zangi, a strong Moslem ruler, recaptured the city of Edessa (in Macedonia). A second Crusade was essentially defeated in 1154 by Zangi's successor Nureddin. By 1187, Nureddin's nephew Saladin had captured Jerusalem and virtually all of the Christian strongholds in the Holy Land. A third Crusade began in 1189 and achieved significant military successes. Although King Richard I (the Lion Heart) failed to reach Jerusalem, he obtained a peace treaty in 1192 with Saladin. However, this treaty quickly crumbled and more Crusades, including the pathetic Children's Crusade of 1212, continued until about 1270 with King Louis of France losing the eighth and final round.

Among the cultural artifacts that the Crusades brought back to Europe were the medicinal practices of Geber, Rhazes, and Avicenna and a cultural belief in the alchemical manifestation of metamorphosis—transmutation. "Geber" is actually a fourteenth-century name attributed to a number of works, some parts of which may well be ascribed to the eighth-century physician and alchemist Jābir ibn Hayyān (ca. 721–815), who was born in present-day Iraq and was educated in Arabia. The concept that all metals were a combination of mercury and sulfur is attributed by some to Jabir.[4] An influential thirteenth-century work, *Summa Perfectionis*, said to be authored by Geber (referred to by chemical historians as pseudo-Geber to avoid confusion) described procedures for characterizing and purifying metals.[5] Figure 6 is from one of the earliest printed books to employ figures derived from copper plates and purports to show Geber (pseudo-Geber or Jabir?) in the laboratory.[6] Avicenna (Latinized version for Abu Ali-al Hussin ibn

FIGURE 6. ■ An image of the eighth-century physician and alchemist Jābir ibn Hayyān ("Geber"), who was born in Arabia and educated in Iraq. This portrait is from Thevet's 1584 *Vies Des Hommes Illustres*. Numerous sixteenth- and seventeenth-century writings in alchemy and medicine were falsely attributed to Geber. To minimize confusion, modern historians credit these to a "pseudo-Geber" (or ψ-Geber).

FIGURE 7. ■ Painting in a fifteenth-century manuscript depicting the eleventh-century Persian physician Avicenna (Abu Ali-al Hussin ibn Abdallah ibn Sina, 980–1037) in an apothecary shop. See color plates. (© Archivo iconografico, S.A./CORBIS).

Abdallah ibn Sina, 980–1037) was a Persian physician of great erudition. His name was also attributed to works of the early Renaissance, and chemical historians refer to the author as pseudo-Avicenna.[5] Figure 7 is from an eleventh-century manuscript said to show Avicenna amid his medicinal preparations.

Metamorphosis, according to Professor Bynum, involves a transformation from one form to another, while maintaining a common characteristic or aspect.[1] For example, Bynum relates[2] the poet Ovid's tale of Jove's punishment of King Lycaon, who was savage with his subjects and also attempted Jove's murder. Although Lycaon is thoroughly and bodily transformed into a wolf[2,7] ("*Lykos*" = "wolf" in Greek):

> He turns into a wolf, and yet retains some traces of his former shape. . . . There is the same grey hair, the same fierce face, the same gleaming eyes, the same picture of beastly savagery.

Bynum notes that "thirst for blood and delight in killing,"[2] what one might call the "essence of wolfishness," were characteristic of both King Lycaon and the wolf.

The notion that metals can be transformed into one another by metamorphosis is quite alien to we moderns. However, it is important to remember that, hundreds of years ago, there was no real concept of an element and metals were commonly found in various states of purity. Alloys such as bronze (copper and tin) and pewter (lead and tin is one formulation) offered a smooth continuity of metallic properties—almost a form of stop-motion metamorphosis or transmutation. The very nature of metals—luster, malleability, and thermal conductivity—argued for a common aspect (or substance)—an "essence of metallicity." And so, we have Johann Joachim Becher, in the mid-seventeenth century, believing that all metals contain quicksilver (mercury)[8] and another important chemist of that period, Johann Kunckel, reporting that he had extracted mercury from all metals.[9] And what magical stuff quicksilver is—the volatile, penetrating, very "essence of metallicity." Mercury dissolves gold and other metals, the nature and appearance of which are dramatically changed upon amalgamation. Heating an amalgam distills the mercury and returns the metal unscathed or perhaps even purer. One can imagine many samples of "pure" metals having some mercury contamination as the result of their history, and so, obtaining a trace of mercury from a combined sample of gold having a sordid history is fairly believable.

Figure 8 is the beautiful frontispiece from the first English edition, published in 1658, of Giambattista della Porta's famous *Natural Magick*.[10] It was first published in Latin in four "books" in 1558, later expanded to twenty books in 1589, and published in numerous editions in Italian, French, and Dutch in addition to the English translation. Porta himself also claims versions in Spanish and Arabic.[11] His 1608 book *De Distillatione* begins with testimonials to him in Hebrew, Greek, Chaldee, Persian, Illyrian, and Armenian as well as a beautiful portrait of the author.[12] Figure 8 similarly honors Porta's ego, seemingly equating him with the four ancient elements, astrological cosmology, and the spirits of Art and Nature that form the basis of "natural magic." In fact, Porta (ca. 1535–1615) had wide-ranging interests in science, particularly physics. He is often credited with designing the *camera obscura* and produced a design for a steam engine. In addition, Porta wrote "some of the best Italian comedies of his age."[12] However, much of Porta's *Natural Magick* is derived from the *Historia Naturalis* of the ancient Roman author Pliny, and he is almost totally credulous about "natural magic."[11]

Here are two brief excerpts from *The Fifth Book of Natural Magick: Which Treateth of Alchymy; Shewing How Metals May Be Altered and Transformed, One into Another*:[13]

CHAP. II.

Of Lead, and How It May Be Converted into Another Metal

> The Ancient Writers that have been conversant in the Natures of Metals, are wont to call Tinne by the name of white Lead; and Lead, by the name of black Tinne: insinuating thereby the affinity of the Natures of these two Metals, that they are very like each to another, and therefore may very easily

FIGURE 8. ■ Title page of the first English edition (1658) of Giambattista della Porta's *Magiae Naturalis* (first published in 4 books in 1558 and expanded to 20 books in 1589). Porta had a considerable knowledge of sixteenth-century chemical operations, has often been credited with invention of the *camera obscura*, and was a renowned playwright. This figure suggests that Porta was the veritable embodiment of order and logic—the antithesis of Chaos. (Courtesy the Roy G. Neville Historical Chemical Library.)

> be one of them transformed into the other. It is no hard matter therefore, as to change Tinne into Lead . . .
>
> To Change Lead into Tinne
>
> It may be effected onely by bare washing of it: for if you bath or wash Lead often times, that is, if you melt it, so that the dull and earthy substance of it be abolished, it will become Tinne very easily: for the same quick-silver whereby the Lead was first made a subtil and pure substance, before it contracted that soil and earthiness which makes it so heavy, doth still remain in the Lead, as Gebrus hath observed; and this is it which causeth that creaking and gnashing sound, which Tinne is wont to yield, and whereby it is especially discerned from Lead: so that when the Lead hath lost its own earthy lumpishness, which is expelled by often melting; and when it is endued with the sound of Tinne, which the quick-silver doth easily work into it, there can be no difference put betwixt them, but that the Lead is become Tinne.

Note some interesting points here. Quicksilver (mercury) is common to both tin and lead. The removal of earthy impurities from lead is a continuum of metamorphosis. Thus, the lead-to-tin transmutation has the essential aspects of a metamorphosis between werewolf and human while conserving a common characteristic—the "essence of metallicity" imbued by quicksilver. Note too, the interesting point that the proof of metal identity is not density, melting point, or chemical reactivity but the sound obtained in its mechanical working!

---

1. C.W. Bynum, *Metamorphosis and Identity*, Zone Books, New York, 2001, pp. 15–36.
2. Bynum, op. cit., pp. 166–176.
3. *The New Encyclopedia Britannica*, Encyclopedia Britannica, Inc. Chicago, 1986, Vol. 16, pp. 880–892.
4. *The New Encyclopedia Britannica*, op. cit., Vol. 6, p. 451.
5. F.L. Holmes and T.H. Levere (eds.), *Instrumentation and Experimentation in the History of Chemistry*, The MIT Press, Cambridge, MA, 2000, pp. 44–49.
6. A. Thevet, *Vrais Portraits et Vies Des Hommes Illustres*, Paris, 1584, p. 73.
7. This is definitely not the two-legged Lon Chaney, Jr. *Wolfman*, B-movie fans.
8. J.R. Partington, *A History of Chemistry*, Vol. 2, Macmillan & Co. Ltd., London, p. 666.
9. Partington, op. cit., p. 362.
10. J. Baptista Porta, *Natural Magick*, Thomas Young and Samuel Speed, London, 1658. I am grateful to The Roy G. Neville Historical Chemical Library (California) for supplying an image of the frontispiece of this book.
11. J. Baptista Porta, *Natural Magick* (reprint edited by Derek J. Price), Basic Books, Inc., New York, 1957.
12. A. Greenberg, *A Chemical History Tour*, John Wiley & Sons, New York, 2000, pp. 37–40.
13. Porta, op. cit., p. 163.

ALBERT THE GREAT AND "ALBERT THE PRETTY GOOD"

Toward the end of the Middle Ages (ca. 500–1450), European thinkers gathered the written lore of the ancients, combined it with knowledge acquired from Moslem cultures during the Crusades, and began to develop methods of inquiry that would begin to define modern science. One of the most important of these figures is Albertus Magnus (ca. 1200–1280).[1] He was born in Swabia (southwestern Germany) and educated at the University of Padua, where he was first exposed to and adopted Dominican beliefs. Ordained as a bishop, Albert was sent to the Dominican convent at the University of Paris some time before 1245. There he read deeply in the Aristotlean and Arabic tracts and began to interpret ancient physics and other sciences and write a summary of human knowledge. He was known as "Albert the Great" even during his own lifetime.[1] Albertus was canonized in 1931 and declared Patron Saint of the Natural Sciences by Papal

Chiloninus Philyninus Lectori.

Qui mirāda cupit populoq; indigna ꝓphano
Noſcere, & in paruo diſcere magna libro.
Nos adeat, gēmas dabimus, gēmeq; colores
Queq; ſit agnati patria concha ſoli
Si cupis, & venas æris, cauſaſq; metalli
Quas patitur propter vulnera tanta parens
Adde quod ex plumbo, ridebit barbar9, aurū
Soluere, natura vertere vera potes
Ars ea Philoſophis nomen Chemia pelaſgis
Accipit, arcanis facta Magna notis.

FIGURE 9. ■ Chemist at a still from the *Liber Mineralium Alberti Magni*, a 1518 text attributed to Albert The Great; note the poem above the figure (courtesy The Roy G. Neville Historical Chemical Library).

decree in 1941. One of Albert's students at the University of Paris was St. Thomas Aquinas.[1]

Numerous books have been falsely attributed to Albert the Great, and only very few seem to be derived from his genuine writings.[2] Figures 9 and 10 are from a 1518 illustrated edition of one of his few authentic works on alchemy and mineralogy.[3,4] Figure 9 depicts an alchemist performing a distillation. Figure 10 is from the last leaf of the book, often missing, and includes a six-line alchemical poem.[4]

Invoking the name Albertus Magnus or even providing tantalizing hints that implied a connection with this revered medieval genius was an effective way to sell books. In Figure 11a we see the title page from a nineteenth-century reprint of a "Marvelous Book of Natural Magic" from a "Petit Albert" or Albert Parvus and first published in 1668.[5,6] (I have adopted "Albert the Pretty Good" to avoid any possible confusion with Albert the Great and because it may be less

De Alchimie phantaſtica fatiga
Exhortatio Virgilij Saltzburgenſis

FIGURE 10. ■ Note the six-line poem at the bottom of this figure from the 1518 *Liber Mineralium Albert Magni*, Oppenheim, 1518. (Courtesy of the Roy G. Neville Historical Chemical Library.)

(a)

LES SECRETS

MERVEILLEUX

DE LA MAGIE NATURELLE

DU

PETIT ALBERT

Tirés de l'ouvrage latin intitulé :

ALBERTI PARVI LUCII

Libellus de mirabilibus naturæ Arcanis

Et d'autres écrivains philosophes

ENRICHIS DE FIGURES MYSTÉRIEUSES, D'ASTROLOGIE, PHYSIONOMIE, ETC., ETC.

Nouvelle édition corrigée et augmentée

A LYON

Chez les Héritiers de BERINGOS fratres

A l'Enseigne d'Agrippa.

M. DC. LXVIII.

(b)

(c)

FIGURE 11. ■ (a) Title page from a nineteenth-century reprint of the 1668 "Marvelous Secrets" of "Little Albert" ["Albert The Pretty Good" (?)] as well as images of (b) Venus (copper) and (c) Jupiter (tin) in their triumphal chariots.

insulting than "Albert the Little"). Although this book is said to be a "well-known collection of magical absurdities and impossibilities,"[6] the fact that it was reprinted for two centuries is certainly nothing to sniff at. How many university or commercial presses can claim such a best seller? Figures 11b, 11c, and 12a–d are depictions of gods and goddesses in triumphal chariots heralding the six ancient metals besides gold: Venus (copper), Jupiter (tin), Saturn (lead), Mercury, Luna (silver), and Mars (iron).

Here is Albert Parvus' recipe for Le Toothpaste:[7]

> Take blood of Dragon and Cinnamon three ounces, calcined alum two ounces; reduce all to a very fine powder, and polish your teeth twice each day.

Sound advice and a good formula—but where to get that first ingredient?

FIGURE 12. ■ Triumphal chariots of (a) Saturn (lead), (b) Mercury (quicksilver or mercury), (c) Luna (silver), and (d) Mars (iron) from "*Petit Albert*" (see Figure 11).

---

1. *The New Encyclopedia Britannica*, Encyclopedia Britannica Inc., Chicago, 1986, Vol. 1, pp. 218–219.
2. J. Ferguson, *Bibliotheca Chemica*, Derek Verschoyle, London, 1954, pp. 15–17.
3. Albertus Magnus, *Liber Mineralium Alberti Magni . . . Sequitur tractatus de lapidum et gemmarum material accidentibus . . . virtutibus ymaginibus, sigillis. De alchimicis speciebus, operationibus et utilitattibus. De metallorum origine et inventione, generatione . . . colore . . . virtute, transmutatione. Ad Emtores Thilonius*, Jacob Koebel, Oppenheim, 1518. I am grateful to The Roy G. Neville Historical Chemical Library for furnishing these images.
4. The Roy G. Neville Historical Chemical Library (California), catalog in preparation. I am grateful to Dr. Neville for helpful discussions.
5. A. Parvus, *Les Secrets Merveilleux de la Magie Naturelle du Petit Albert, Tirés de l'ouvrage latin intitulé: Alberti Parvi Lucii Libellus de mirabilibus naturæ Arcanis Et d'autres écrivains philosophes . . .* , Chez les Héritiers de Beringos fraters, Lyon, 1668 (The copy employed here is a nineteenth-century reprint.)
6. Ferguson, op. cit., p. 17.
7. Parvus, op. cit., p. 154.

A CANTERBURY TALE OF ALCHEMY

Was England's greatest poet a true Adept, or merely adept at rhyming verses? The *Canon's Yeoman's Tale* (or *CYT*) of Geoffrey Chaucer (ca. 1340–1400) implies such a detailed knowledge of alchemical operations[1] that Elias Ashmole[2] included this work in his *Theatrum Chemicum Britannicum*, published in 1652, among those of the other "Famous English Philosophers who have written the Hermetique Mysteries in their owne Ancient Language."[3] Figures 13 and 14 are illustrations from the *Theatrum*. In the first figure, the Master Adept bestows alchemical secrets on the young alchemist: "Receive the gift of God under the

FIGURE 13. ■ "Receive the gift of God under the sacred seal" sayeth the Master Adept to the young alchemist (from Ashmole, *Theatrum Chemicum Britannicum*, 1652, courtesy The Roy G. Neville Historical Chemical Library).

FIGURE 14. ■ Here is a well-funded Renaissance research laboratory. Years later, the young alchemist in Figure 13 can now afford a research scientist on the GOld from Lead Discovery ("GOLD") program funded by the National Treasury. (Ashmole, *Theatrum Chemicum Britannicum*, courtesy The Roy G. Neville Historical Chemical Library.)

sacred seal."[4] The next figure shows an active and well-funded laboratory suggesting that the young adept has indeed heeded good academic counsel. He has become a successful grantsman and is well on his way to tenure and promotion.

No less an expert than John Read suggested that "Chaucer himself had first-hand experience of the joys and sorrows of a 'labourer in the fire.'"[5] To these bits of circumstantial evidence, we now add the apparent "smoking gun"—sixteenth-century manuscripts in the library of Dublin's Trinity College titled *Gal-*

fridus Chauser his worke, describing two alchemical procedures for obtaining the Philosopher's Stone followed by a poem concerning the Elixir.[1]

However, Gareth Dunleavy's careful research suggests that these manuscripts are pseudepigraphons falsely attributed to Chaucer.[1] False attributions to Geber, Albert the Great, Arnold of Villanova and Hermes himself were not uncommon attention-getting devices during the Renaissance. Dunleavy indicates that while Chaucer might have been familiar with general aspects of alchemy, the details in *CYT* closely resemble the writings of Arnold of Villanova.[1] Thus, he is skeptical about Chaucer the alchemist but notes that the manuscript itself may once have belonged to the library of John Dee, astrologer, mathematician, and alchemist to Elizabeth I.[1]

Now, back to *The Canterbury Tales*. The *CYT* Prologue sets the scene. The canon, a clergyman who, in this tale, is also an alchemist, is accompanied by his yeoman or assistant as they encounter a group of travelers on the road. The canon is dismissed by the group's host and the ash-darkened, poverty-stricken, indentured yeoman, who has been badly used by his master, tells a bitter and ironic tale of alchemical chicanery. The canon appears to be part "puffer" (earnest but misguided seeker of The Stone) and part charlatan.

The canon has offered to transmute a Priest's quicksilver into precious silver metal using a mysterious powder of projection. In reality, he has placed an ounce of pure silver into a hole drilled in a lump of coal and sealed the hole with blackened wax. The yeoman describes the canon's bait for the priest:

For here shul ye se by experience,
That this quicksylver I wol mortifye
Right in your syght anon withouten lye,
And make it as good Sylver and as fyne,
As there is any in your purse or myne,

The canon produces his mysterious powder:

I have a poudre that cost me deere,
Shall make all good, for it is cause of all
My connyng, which I you shewe shall

The greedy and gullible priest watches as the canon removes his own crucifix ("crosslet") and sets it in the fire. To this, the priest adds his quicksilver and the canon adds some of the powder. In the yeoman's bitter words:

This Preest at this cursed Chanon's byddyng,
Uppon the fyre anon set this thyng;
And blewe the fyre and besyed him ful faste,
And this Chanon into this crosslet caste
A pouder, I not whereof it was,
Ymade either of Chalke, Erthe, or Glasse
Or somwhat els, was not worth a fly,

And now the Canon's trick:

This false Chanon, the foule fende him fetche;
Out of his bosome toke a bechen cole,

In which ful subtelly was made an hole,
And therein was put of Sylver lymayle,[6]
An unce, and stopped was without fayle,
The hole with waxe to kepe the Lymayle in.

The priest is industriously tending the fire and the canon distracts him by noting that the burning coals need to be rearranged and offering the priest a cloth to wipe his sweaty face. Whereupon the hollowed-out lump of coal is added, the fire is vigorously stirred up, and the canon then joins the priest for a hearty drink. Returning to the fire, the canon finds and recovers metallic silver for the delighted priest.

Now a second demonstration occurs in which the crafty canon actually leaves the priest to perform the transmutation on his own. He provides a hollowed-out stirring stick to the priest that . . . you guessed it . . . is filled with an ounce of silver secured with blackened wax. So now, completely outside the influence of the canon, powder of projection works for the priest himself. The canon works a final demonstration-transmuting copper to silver, leaving the priest in an ecstasy of joyful greed:

This sotted Preest who was gladder than he,
Was never Byrd gladder agenst the day,
Ne Nightyngale agenst the ceason of May,
Was never none, that lyft better to synge,
Ne Lady lustier in Carolyng:

The priest pays the canon forty pounds, a vast sum, for his secret (including powder I suspect). Note Chaucer's distrust of a clergy so widely perceived during the Renaissance as corrupt. There are two clergyman here—one, poor but dishonest; the other, gullible yet incredibly wealthy.

One can just hear the canon as he bids the priest adieu—"A Fulle Moneybacke Guarantee! And if you're ever in Canterbury . . . try to find me."

---

1. G.W. Dunleavy, *Ambix*, Volume XIII, No.1, pp. 2–21 (1965). I thank Professor Dunleavy for helpful discussions.
2. Elias Ashmole (1617–1692) was a gentleman of incredibly wide-ranging interests whose collection formed the basis for the first public museum in England, the Ashmolean Museum of Oxford University. A book published in 1650, titled *Fasiculus Chemicus: Or Chymical Collections . . .* was authored by a James Hasolle [an interesting, perchance smutty(?), anagram of Eljas Ashmole].
3. E. Ashmole, *Theatrum Chemicum Britannicum. Containing Severall Poeticall Pieces of our Famous English Philosophers, who have written the Hermetique Mysteries in their owne Ancient Language. Faithfully Collected into one Volume, with Annotations thereon.*, J. Grismond for Nath:Brooke, at the Angel in Cornhill, 1652. The images were supplied by The Roy G. Neville Historical Chemical Library (California). See also the facsimile reprint with a preface by C.H. Josten, Georg Olms Verlagsbuchhandlung, Hildesheim, 1968.
4. S. Klossowski De Rola, *The Golden Game. Alchemical Engravings of the Seventeenth Century*, Thames and Hudson, London, 1988, pp. 214–221.
5. J. Read, *The Alchemist in Life, Literature and Art*, Thomas Nelson and Sons Ltd., London, 1947, p. 29.
6. Powder or filings—"lymayle" rhymes better with "fayle."

THE SHIP OF FOOLS

In 1494, some 20 years prior to the Protestant Reformation, Sebastian Brant,[1,2] a German poet and humanist, published a long poetic satire titled *The Ship of Fools* (*Das Narrenschiff*). He has been termed "a man of deep religious convictions and of stern morality, even to the point of prudishness."[3] The book imagined a collection of "fools" reflecting mores and excesses that would have tickled the fancies of readers of the day by deflating recognizable character types. The ship, loaded with these fools, was bound for "Narragonia," the Land of Fools. The book's language was accessible, the woodcuts (some *possibly* by Albrecht Durer)[4] handsome and amusing. Six editions appeared during Brant's life (first English in 1509) with numerous additional authorized and pirated editions through 1629.[5] The book was "rediscovered" two centuries later and an edition published in 1839 with others following throughout the nineteenth century and into the early twentieth century.

The stern Brant did not have a very high regard for the sensual pleasures procurable in the streets of Basel. The prologue to his fiftieth poem, "Of Sensual Pleasure," expresses his self-righteous scorn:[6]

FIGURE 15. ■ "Our quite deceptive" alchemist from *The Ship of Fools*. This figure is from the 1506 Basel edition, courtesy The Roy G. Neville Historical Chemical Library.

The stupid oft by lust are felled
And by their wings are firmly held:
For many, this their end hath spelled.

Romantic "night music" was also *verboten*—according to the prologue to Number 62 "Of Serenading at Night":[7]

The man who'd play the amorous wight
And sing a serenade at night
Invites the frost to sting and bite.

So, no sensual pleasures or serenading at night on the streets of Basel! You wouldn't expect Brant to be very open-minded or have a sense of humor about alchemy either, and he doesn't disappoint. Thus, we see in Figure 15[8] alchemists in dunce hats (Oh, the shame of it!) and a snippet from poem 102:[9] "Of Falsity and Deception" (Ha! There's a dead giveaway!):

But let there not forgotten be
Our quite deceptive alchemy:
Pure gold and silver doth it yield
But this in ladles was concealed.

Ah, the Canterbury Canon's old "gold-hidden-in-the-ladle-or-stirrer" trick. I wonder if Brant's was the voice of experience or whether he had only witnessed the bamboozling of wealthy priests and other easy marks.

---

1. *The Catholic Encyclopedia*, Vol. II, 1907, Robert Appleton Co.
2. S. Brant, *The Ship of Fools*, translated into rhyming couplets with introduction and commentary by Edwin H. Zeydel, Dover Publications, New York, 1962 (reprint of 1944 edition).
3. Brant, op. cit., p. 7.
4. Brant, op. cit., p. 20.
5. Brant, op. cit., pp. 21–24.
6. Brant, op. cit., pp. 178–180.
7. Brant, op. cit., pp. 206–208.
8. This figure is from the 1506 Basel edition translated by J. Locher into Latin and reinterpreted by Badius Ascensius, courtesy of The Roy G. Neville Historical Chemical Library (California), catalog in preparation. I am grateful to Dr. Neville for helpful discussions.
9. Brant, op. cit., pp. 327–330.

THE FIRST MODERN ENCYCLOPEDIA

The elegantly simple illustration of an alchemist tending his furnace, with distillation apparatus in the background, depicted in Figure 16 is found in the first edition of the *Margarita Philosophica*, published in 1503.[1,2] It is "the first modern encyclopedia of any importance"[3] and was printed less than fifty years after Johannes Gutenberg printed his first books in 1455. The *Margarita Philosophica* reflects the university curriculum at the end of the fifteenth century. It covers

FIGURE 16. ■ An early-sixteenth-century alchemist from "the first modern encyclopedia of any importance" (Reisch, *Margarita Philosophica*, 1503, courtesy The Roy G. Neville Historical Chemical Library).

grammar, logic, rhetoric, mathematical topics, astronomy, music, childbirth, astrology, and hell.[4] Books 8 and 9 cover chemical topics, including transmutation.[3] The author, Gregorius Reisch, was the Prior of a Carthusian monastery at Freiburg and confessor of Maximilian I,[4] Holy Roman Emperor (1493–1519), who established the dominance in Europe of the Habsburg Family.[5]

1. G. Reisch, *Margarita Philosophica (totius philosophiae rationalis, naturalis et moralis principia dialogice duodecim libris complectens)*, Joannem Schott, Freiburg, 1503. The author is grateful to The Roy G. Neville Historical Chemical Library (California) for supplying a copy of the woodcut in Figure 16 and to Dr. Neville for helpful discussions.
2. J.R. Partington, *A History of Chemistry*, Macmillan & Co., Ltd., London, 1962, Vol. 2, p. 94.
3. D.I. Duveen, *Bibliotheca Alchemica et Chemica*, facsimile reprint, HES Publishers, Utrecht, 1986, p. 501.
4. The Roy G. Neville Historical Chemical Library (California), catalog in preparation. I am grateful to Dr. Neville for helpful discussions.
5. *The New Encyclopedia Britannica*, Encyclopedia Britannica, Inc., Chicago, 1986, Vol. 7, p. 965.

AN ALCHEMICAL BESTIARY

Symbols and metaphors allow us to represent phenomena we do not fully understand and thoughts having no rational translations. Four centuries ago, the wolf represented the "biting" behavior of antimony (or its sulfide) on "base" metals. At a much deeper, subconscious level we may employ sexual imagery to convey perceptions of the male and female nature of things. For millennia, these dualities were projected to explain properties of matter that could be understood only symbolically. It is no wonder that the psychologist Carl Jung wrote extensively on the symbolism of alchemy.[1]

THE WOLF AND THE IMPURE KING

In 1617 Michael Maier wrote a gloriously illustrated book titled *Atalanta Fugiens* for which he composed 50 fugues to accompany 50 illustrations (emblems) of the alchemical process.[2] Alluding to the three principles—sulfur, mercury, and salt—each of Maier's fugues was composed as an epigram in three verses for three voices. (In Brooklyn, this would translate as "three voices for three verses.")

Figure 17, Emblem 24 from *Atalanta Fugiens*, is a chemically astute depiction of a purification of gold.[3] The dead king symbolizes impure gold—say, gold contaminated with copper and other base metals. The wolf, often representing metallic antimony, here represents stibnite or antimony sulfide. Antimony ("not alone") is virtually always found in a combined state—hardly a "lone wolf." The wolf devours the impure king; that is to say, with application of heat, antimony loses its sulfur to copper and other base metals and the freed antimony alloys in a melt with gold. The sulfides of copper and the other base

FIGURE 17. ■ Here is a good question for the Chem I final exam: Write a description of the two simple chemical reactions depicted in this figure. (*Hint:* Think about the purification of gold.) Are you still "stumped"? Then see the accompanying text. (Figure from Maier's 1617 *Atalanta Fugiens*, courtesy The Roy G. Neville Historical Chemical Library.)

metals are dross easily separated from the melted alloy. The alloy is then placed in the fire, where chemically reactive antimony forms an oxide that sublimes off, leaving molten and chemically inert gold (the revivified king). Pierre Laszlo has provided evidence that the land and river in this image (and others in *Atalanta Fugiens*) implies a distinction between the dry way and wet way of chemical operations.[4] I wonder how this image would work as a question on a Chem I final?

LIONS AND DRAGONS AND SNAKES, OH MY!

Paired entities in struggle or passionate embrace (or both) represent the joining of the opposite principles [male–female; sophic ("philosophic") sulfur and sophic mercury] thought to comprise all matter. Figure 18 (Emblem 16 from *Atalanta Fugiens*) depicts the struggle between two lions. The winged creature on the left is the Green Lion, representing the volatile (winged) sophic mercury: the female principle. The male is the Red Lion, symbolizing sophic sulfur, itself a symbol of fixity and combustibility.[3]

BLOOD OF THE DRAGON

In Maier's 1618 book *Viatorium* (Figure 19),[5] he describes how dragon's blood, another symbol for the philosopher's stone, is formed. An elephant engorges itself with water. In ambush lies a dragonlike serpent that attacks and wraps and

FIGURE 18. ▪ The passionate struggle of the winged Green Lion (volatile female principle; "sophic" mercury) and the Red Lion (fixed male principle; "sophic sulfur") (from *Atalanta Fugiens*, courtesy The Roy G. Neville Historical Chemical Library).

FIGURE 19. ■ Title page from Maier's *Viatorium* (second edition, 1651; first edition, 1618) in which the seven ancient metals are represented (from top right clockwise): gold, silver, iron, copper, tin, lead, and mercury. The top center figure is the author Count Michael Maier, whom chemical historian John Read dubbed "a musical alchemist."

tightens its coils about the elephant and drinks its blood (Figure 20). The weakened elephant eventually tumbles onto the serpent and crushes it to a bloody pulp. This dragon's blood, suffused with the matter of the elephant, is effectively a red tincture or philosopher's stone. The sexual imagery of this allegory should be quite obvious to anybody having a pulse.

SALAMANDER AS SPIRIT OF FIRE

The salamander is used to depict the "fiery masculine seed" that survives and is nourished by the fire.[6] The philosopher's stone is frequently likened to a seed that may multiply metals. Sometimes the salamander simply represents fire or the spirit of fire.[7] Emblem 29 from *Atalanta Fugiens* (Figure 21) and its epigram help explain this mystical (and far from obvious to me) connection: "The Salamander cools the flame and goes."[2]

THE ONE AND ONLY FAMOUS, FABULOUS PHOENIX

One, and only one, phoenix[8] can exist in our world. This fabulous bird rises from the ashes of the penultimate phoenix, which self-immolated after 500 years of a lonely, sex-deprived existence. Closely associated with Egyptian mythology, the phoenix appears to have even more ancient oriental origins. Culturally, it is a symbol of rebirth and even life after death. In alchemical imagery, the phoenix is

FIGURE 20. ■ Formula for "Dragon's Blood" (The Philosopher's Stone): (1) elephant gorges on water; (2) engorged elephant ambushed by huge serpent that tightens its coils and drinks the blood of its prey; (3) weakened elephant collapses and completely "smooshes" the serpent; *et voila*!; (4) "Dragon's Blood" fit for alchemical projection. (From Maier's *Viatorium*, second edition, 1651.)

the last of four birds representing the successive color changes during The Great Work (see the top of Figure 22):

1. Crow, black, putrefaction
2. Swan, white, calcination
3. Peacock, yellow or rainbow colors signifying change
4. Phoenix, red, the red tincture or philosopher's stone

Figure 23 shows a nineteenth-century Japanese fan (in black and white rather than the actual color) decorated with a watercolor painting of the phoenix. The figure bears a striking resemblance to a twentieth-century cultural icon—*Rodan*, the subject of a modern Japanese film genre that I will dub *Plastique Monstresque*.[9] Indeed, *Rodan* rises from the ashes of nuclear tests to menace the earth and teach us all a good lesson. It is a phoenix born on the funeral pyre of the atomic bomb.

The symbol of that august scientific body, the American Chemical Society (Figure 24), places a phoenix above Justus Liebig's nineteenth-century *kaliappa-*

FIGURE 21. ■ The salamander (from Maier's *Atalanta Fugiens*, courtesy The Roy G. Neville Historical Chemical Library), which represents the resistance to fire attributed to the Philosopher's Stone. When Yale chemistry professor Benjamin Silliman visited the sweltering (115°F) laboratory of James Woodhouse of Philadelphia in the summer of 1802, he referred to "that salamander's home" (see page 222).

rat.[10] What do we make of *that*? The *kaliapparat* precisely measured carbon dioxide emitted from combustion of organic compounds, leading to accurate formulas and the scientific understanding of the vast "primeval forest"[11] of organic chemistry. In sharp counterpoint, the phoenix represents the culmination of the alchemical operation. Methinks my fellow chemists are hedging their bets—rational chemistry first but magic if it fails.

BEWARE THE AMOROUS BIRDS OF PREY[12]

Copulation is a major theme in alchemical imagery. The two snakes entwined about the rod of Mercury form the caduceus, the symbol of the medical profession (Figure 25).[13] These amorous serpents form three circles representing three cycles of separation and union of male and female principles.[14] Above the two serpents are two amorous birds of prey[15] also closely packed. These birds devour each other as they copulate, representing the process of chemical solution/combination and the loss of individual identities. Common decency prevents the depiction of the chemical wedding between man and woman (sol and luna).[16]

AND SHUN THE FRUMIOUS BASILISK[12]

Transmutation of "base metals" into silver or gold is ultimately achieved through "projection"—an unexplainable process that could occur at a distance—"the red

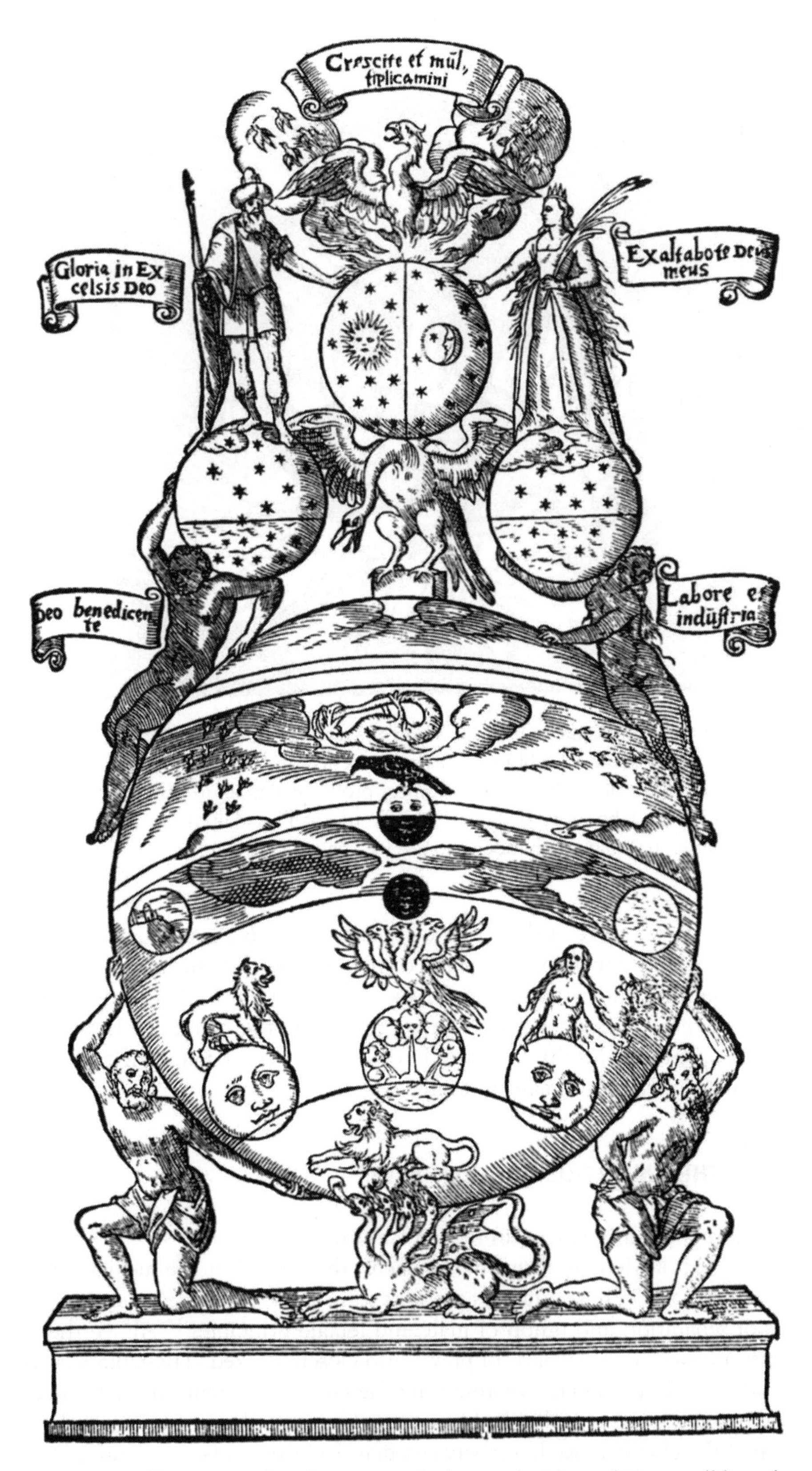

FIGURE 22. ■ Do you see the phoenix perched atop the Vase of Hermes (likened to a kind of Western mandala by alchemical interpreter Adam McLean)? (From Libavius' *Alchymia,* 1606.) The phoenix represents the last of the four color changes occurring during the Great Work. The work begins in darkness and abasement (black crow); the chemical mass whitens during calcination (swan), passes through bright color changes (peacock), and culminates with the rise of the red phoenix.

FIGURE 23. ■ Nineteenth-century Japanese watercolor on rice paper depicting the phoenix. Was the 1950s movie icon *Rodan* a phoenix rising from the ashes of the nuclear bomb?

tincture projected from the heart of the 'red king' (the red stone) onto his subjects, who personify the base metals."[17] The basilisk (or cockatrice) (Figure 26)[18] is a serpent of Roman mythology.[19] Some versions have it hatched by a serpent from an egg laid by a cock. Others view it as "a poisonous mixture of cock and toad."[20] The mere glance of a basilisk (or distant exposure to its emanations) is deadly.[21]

> there is none that perisheth sooner than doth a man by the poison of a Cockatrice, for with his sight he killeth him, because the beams of the Cockatrices eyes, do corrupt the visible spirit of a man, which visible spirit corrupted, all the other spirits coming from the brain and life of the heart, are thereby corrupted, and so the man dyeth.

And it gets even scarier:[21]

> For it killeth, not only by his hissing and by his sight, (as is said of the Gorgons), but also by his touching, both immediately and mediately; that is to say; not only when a man toucheth the body it self, but also by touching a Weapon wherewith the body was slain, or any other beast slain by it; and there is a common fame, that a Horse-man taking a Spear in his hand, which had been thrust through a Cockatrice, did not only draw the poison of it into his own body and so dyed, but also killed his Horse thereby.

FIGURE 24. ■ The symbol of the American Chemical Society, which includes the phoenix as well as Justus Liebig's early-nineteenth-century *kaliapparat*, which revolutionized the analysis of organic compounds. This is a wonderful evocation of the mystical and rational roots of chemistry. (Used with permission from the American Chemical Society.)

FIGURE 25. ■ "Beware the amorous birds of prey . . . ." The amorous birds of prey are another representation of male–female (Sol–Luna; sulfur–mercury) duality. This image is from the 1755 *Medicinisch-Chymisch-und Alchemistisches Oraculum*.

DES VENINS. 105

DV BASILIC ROY DES SERPENS.

CHAPITRE XVIII.

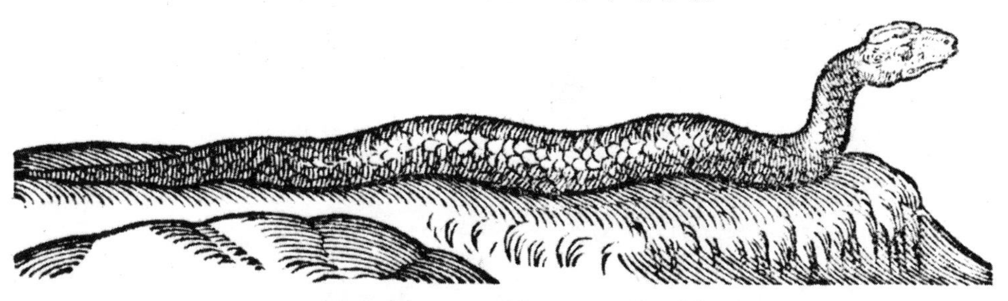

Βασιλίσκος, *Basiliscus*, *Basilic.*

FIGURE 26. ■ "and shun the frumious basilisk." The basilisk (or cockatrice) is a symbol for "projection," the mysterious power of the Philosopher's Stone to transmute metals from a distance. Sometimes represented as a lizard, sometimes a combination of lizard and rooster, and sometimes as a serpent (a spitting cobra?). The mere glance of a basilisk is deadly. (From Grévin, *Deux Livres des Venins*, 1568, 1567.) (Courtesy of The New York Academy of Medicine and B & S Gventer: "Books"—each supplied copies of this image.) Was the Japanese film icon *Godzilla* a basilisk?

Clearly, it is a symbol for alchemical projection at a distance. Come to think of it, our newly named *Plastique Monstresque* film genre[9] affords us a 400-foot-tall basilisk called Godzilla whose breath vaporizes air force jets and army tanks hundreds of meters away.

Actually, it is not hard to imagine cobras being the basis of basilisk mythology. They are among the world's most poisonous snakes; spitting cobras can cause blindness at a distance, and the sheer size of a king cobra (one reported to be 18 feet long[22]), coupled with this serpent's tall, vertical posture and hooded appearance, are the "stuff" of mythology.

THE OUROBOROS (OR KEKULÉ'S DREAM EXPLAINED?)

The ouroboros is a serpent constantly devouring itself as it regenerates. The concept surely flows from the molting of a snake to form a bright new skin. The circle represents unity and continuity. The ouroboros evokes the circular reflux distillation accomplished in a pelican (or even a modern reflux) apparatus. Did Auguste Kekulé dream about the ouroboros in imagining the benzene ring?[23,24] Figure 27, Emblem 14 from *Atalanta Fugiens*,[2] shows a variant on the ouroboros theme. Its Epigram translates as follows:[2]

> The famished Polyps gnawed at their own legs,
> And hunger too, taught men to feast on men.
> The Dragon bites its tail and swallows it,
> Taking for food a great part of itself.
> Subdue it by hunger, prison, iron, until,
> It eats itself, vomits, dies, and is born.

FIGURE 27. ■ The ouroboros—a symbol for completeness, cycle of life, and even the conservation of matter. The ouroboros continuously devours itself as it regenerates. Did August Kekulé actually dream about the ouroboros when he postulated that benzene was a cyclic compound? (From *Atalanta Fugiens*, courtesy The Roy G. Neville Historical Chemical Library.)

Could the ouroboros also be a symbol for the law of the conservation of matter, far predating Lavoisier? Perhaps only Kekulé would have known for sure.

1. N. Schwartz-Salant, *Encountering Jung On Alchemy*, Princeton University Press, Princeton, 1995.
2. M. Maier, *Atalanta Fugiens, hoc est, Emblemata Nova De Secretis Chymica, Accomodata partim oculis & intellectui, figures cupro incises, adjestisque sententiis, Epigrammatis & notis, partim auribus & recreatoni animi plus minus 50 Fugis Musicalibus trium Vocum,* . . . Oppenheimii Ex typographia Hieronymi Galleri, Sumptibus Joh. Theodori Bry, 1617. (I am grateful to The Roy G. Neville Historical Chemical Library, California, for supplying this image.) Second edition, 1618. A more recent translation of this book with English translation and commentary and a cassette audiotape of the fugues were published by Phanes Press, Grand Rapids in 1989.
3. J. Read, *Prelude to Chemistry*, The Macmillan Company, New York, 1937, pp. 236–254.
4. P. Laszlo, Aspects "de la tradition alchimique au XVIIe siecle," F. Greiner (ed.), *Chrysopoeia*, Vol. 4, pp. 278–285 (1998).
5. M. Maier, *Viatorium, hoc est, De Montibus Planetarum septem seu Metallorum..*, Oppenheimii, Ex typographia Hieronymi Galleri, sumptibus Joh. Theodori Bry, 1618. A second edition was published in Rouen in 1651.

6. L. Abraham, *A Dictionary of Alchemical Imagery*, Cambridge University Press, Cambridge, UK, 1998, p. 176.
7. Read, op. cit., pp. 128, 168, 244–245.
8. *The New Encyclopedia Britannica*, Vol. 9, Encyclopedia Britannica, Inc., Chicago, 1986, p. 393.
9. *Plastique Monstresque*: homage to the French who host the Caens Film Festival but also venerate Jerry Lewis, and of course to the Japanese who have given us this film genre.
10. A. Greenberg, *A Chemical History Tour*, John Wiley and Sons, New York, 2000, pp. 196–198.
11. Greenberg, op. cit., pp. 205–207.
12. With apologies to Lewis Carroll and his poem *Jabberwocky*. See also Read, op. cit., pp. 199–200. Apparently, Read actually knew what slithy toves and mimsy borogroves looked like.
13. *Medicinisch-Chymisch-und Alchemistisches Oraculum, darinnen man nicht nur alle Zeichen und Abkürzungen, welche sowohl in den Recepten und Büchern der Aerzte und Apotheker, als auch in den Schriften der Chemisten und Alchemisten verkommen, findet, sondern dem auch ein sehr rares Chymisches Manuscript eines gewissen Reichs***beygefüget*. Ulm und Memmingen, in der Gaumischen Handlung, 1755.
14. Abraham, op. cit., pp. 30–31.
15. Abraham, op. cit., pp. 23–25.
16. However, for a good time call ISBN 3-8228-8653-X. This is the Library of Congress call number for A. Roob, *The Hermetic Museum: Alchemy & Mysticism*, Taschen, Cologne, 1997. Pages 442–455 have lots of great pictures in color and black and white of the *Conjunctio*. This is soberly followed by pictures of the Rebis in which fusion seems to be permanent. See also, Greenberg, op. cit., pp. 39, 54.
17. Abraham, op. cit., pp. 157–158.
18. J. Grévin, *Deux Livres des Venins, Ausquels il est amplement discouru des bestes venimeuses, thériaques, poisons & contrpoisons. Ensemble, Les oeuvres de Nicandre, Medecin & Poete Grec, traduictes en vers François*. Christopher Plantin, Antwerp, 1568, 1567. Both Bruce Gventer and The New York Academy of Medicine Library are gratefully acknowledged for providing images of this figure. I gratefully acknowledge helpful conversations with Miriam Mandelbaum of the Academy Library on images of the basilisk and other "fantastical" creatures.
19. *The New Encyclopedia Britannica*, op. cit., Vol. 3, pp. 420–421.
20. Roob, op. cit., pp. 354, 369.
21. E. Topsell, *The History of Four-Footed Beasts and Serpents, Volume 2. The History of Serpents Taken Principally from the Historiæ Animalium of Conrad Gesner* (reprint of 1658 London edition), Da Capo Press (Plenum), New York, 1967, pp. 677–681.
22. *The New Encyclopedia Britannica*, Encyclopedia Britannica, Inc., Chicago, 1986, Vol. 3, p. 415.
23. Read, op. cit., pp. 108, 117.
24. Greenberg, op. cit., pp. 207–211.

WHO IS ATHANASIUS KIRCHER, AND WHY ARE THEY SAYING THOSE TERRIBLE THINGS ABOUT HIM?

Figures 28–30 are from the 1665 book *Mundus Subterraneus*[1] authored by Athanasius Kircher (1602–1680), a Jesuit priest whose early professorial appointment was in Würzburg and last appointment was at the Jesuits College in Rome.[2] He was a person of both great learning and incredible credulity. It is worthwhile reminding the reader that the towering seventeenth-century scientists Boyle and Newton were both credulous about alchemy. I like to think of the mid-seventeenth-century figure Johann Baptist van Helmont as perhaps half scientist–half pseudoscientist.[3] Kircher appears to be somewhat less of a scientist than van Helmont. He wrote voluminously. However, the science historian John Ferguson says of Kircher:[4]

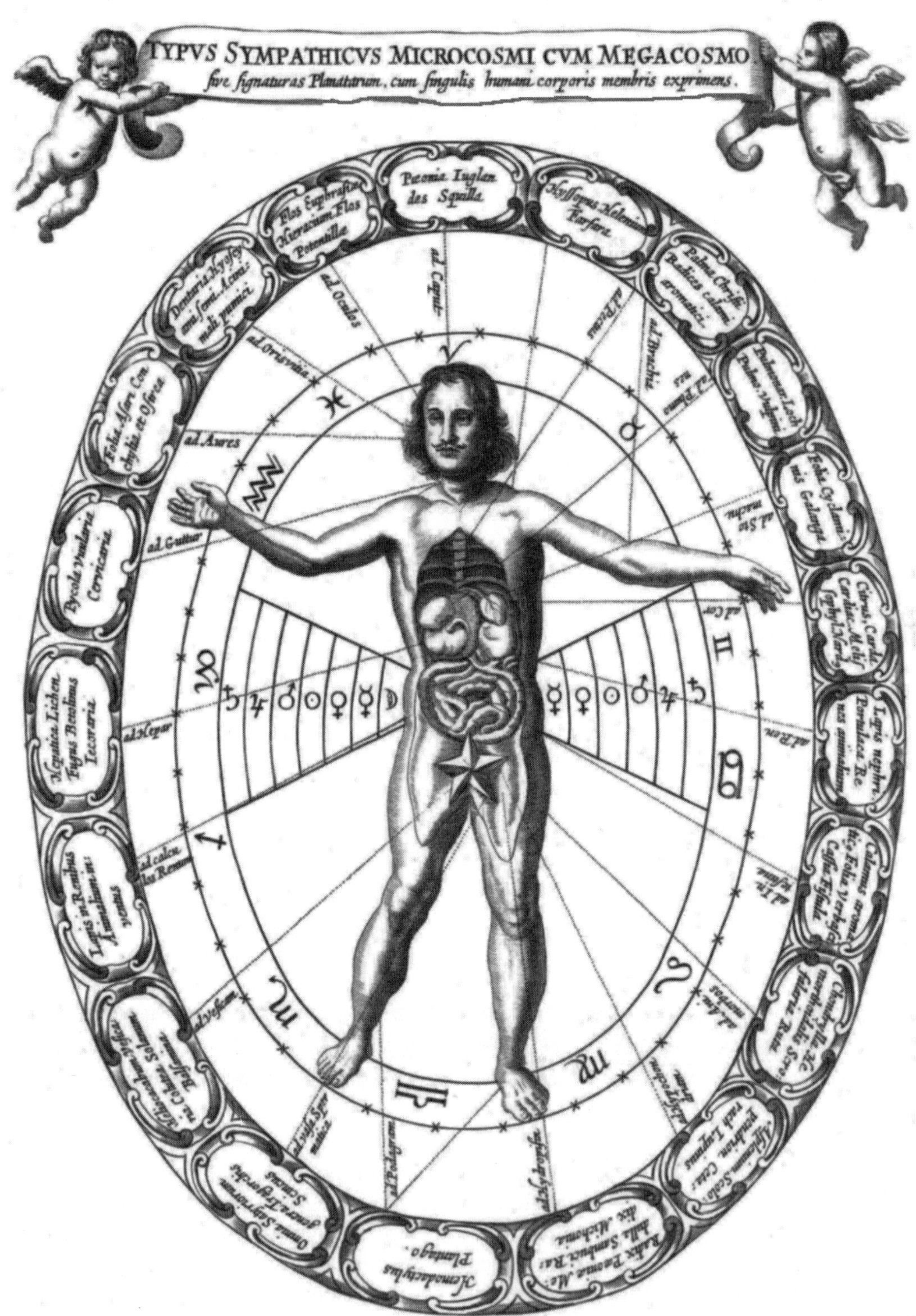

FIGURE 28. ■ Astrological unity of the microcosm and macrocosm in Athanasius Kircher's *Mundus Subterraneus* (1665). The sun is the human heart and the moon the human brain. (Courtesy J.F. Ptak Science Books.)

FIGURE 29. ■ A fabulous spagyrical (pharmaceutical) furnace from Kircher's *Mundus Subterraneus*. Said to be housed at the Jesuits College in Rome, what was its real purpose? (Courtesy J.F. Ptak Science Books.)

FIGURE 30. ■ It seems that the laboratory at the Jesuits College in Rome was equipped for all manner of distillation. Apparatus *E* evokes the wolf in the Romulus-and-Remus legend, and the tall pelican apparatus in M (left column, second from bottom) has the aspect of a stern cleric on a pulpit admonishing his congregants. (From Kircher's *Mundus Subterraneus*, courtesy J.F. Ptak Science Books.)

> Kircher was a man of vast—almost cumbrous—erudition, of equal credulity, superstition, and confidence in his own opinion. His works in number, bulk, and uselessness are not surpassed in the whole field of learning.

The *Mundus* included a vast array of descriptions of mining chemistry, metallurgical chemistry, and spagyrical (pharmaceutical) chemistry as well as chemistry useful to artists and artisans.[2] Most notable, from the historical perspective, is his disbelief in alchemy, which was expounded in the *Mundus*. Perhaps not surprisingly, Boyle "demurred at paying forty shillings for it."[2,5]

Figure 28 is a very astrological depiction of the microcosm–macrocosm description of the human body and its processes representing the larger universe. His relations[6] are as follows: sun = heart; moon = brain; Jupiter = liver; Saturn = spleen; Venus = kidneys; Mercury = lungs; Earth = stomach; veins = rivers; bladder = the sea; The seven major limbs represent the seven ancient metals.

Figure 29 is said to be a "pharmaceutical furnace" at the Jesuits College in Rome although Partington avers that it was "so named to disguise its real function".[2] One wonders whether this apparatus, that looks more like an alien loaded with hatchlings, ever truly existed. I confess that one of the stills in Figure 30 reminds me of the wolf who suckled Romulus and Remus—part of the mythology of ancient Rome.

---

1. A. Kircher, *Mundus Subterraneus, in XII. Libros digestus . . .* Joannem Janssonium & Elizeum Weyerstraten, Amsterdam, 1665. I thank John Ptak, J.F. Ptak Science Books, Washington, DC, for providing these three figures.
2. J.R. Partington, *A History of Chemistry*, Vol. 2, Macmillan & Co. Ltd., London, 1961, pp. 328–333.
3. A. Greenberg, *A Chemical History Tour*, John Wiley & Sons, New York, 2000, pp. 80–83.
4. J. Ferguson, *Bibliotheca Chemica*, Vol. 1, Derek Verschoyle, London, 1954, pp. 466–468.
5. The book is probably worth about $25,000 today.
6. A. Roos, *The Hermetic Museum: Alchemy & Mysticism*, Taschen, Cologne, 1997, p. 565.

UNNATURAL MAGICK—WITCHES AND STIRRING RODS OF BONE

The mythology of witches dates back millennia and exists in virtually all cultures in highly varied guises.[1] Western civilization witchcraft mythology further evolved as the Crusades ended and alchemy, astrology, and other occult beliefs were carried from Moslem lands. In early renaissance Europe a witch was likely to be a person who did not fit the established order in a small village and thus was seen to cause disharmony.[1] A village family might refuse the outcast, sometimes an elderly man or woman, who needed help, say, in gathering firewood—help that one would normally expect in the tight-knit society. One week later, the family's eldest child might be taken ill and lo, a witch would be born. A "witch's revenge" could be even more dire—a naturally occurring "50 year" flood, for example. Somewhere during the fifteenth century, however, witches were recast as consorts of the Devil.[1] Their practices were manifestations of sheer evil. Trials and executions were the remedy. These beliefs traveled to the New World with the Puritans and had their most striking impact in Salem, Massachusetts, where the voodoo tales of a West Indian slave, Tituba, convinced three young girls that they were possessed by the Devil.[2] The resulting hysteria gave rise to the "witch trials" of some 30 women between May through October 1692. Torture was a reasonable method during this "state of emergency" and nineteen convicted "witches" were hanged.

The "Witches' Sabbath,"[3] was a nocturnal revelry filled with illicit and orgiastic behavior. It is the subject of the *Walpurgis Nacht* in Faust and Moussourgsky's opus "Night on Bald Mountain." A small section of the early-seventeenth-century engraving by Jacques De Gheyn, II, "A Witches' Sabbath," is shown in Figure 31. Although modern Halloween versions of witches are typically gaunt elderly women, the healthy, powerful female figure in this engraving does not appear to have needed any help gathering firewood. She is using a large human bone possibly to stir a batch of "Fly Me to the Moon Unguent." It was reputed that witches prepared body oils that gave them the ability to fly (it wasn't the broomstick, after all). And how did De Gheyn's witch offend her village? She might simply have been a better chemist than the local apothecary, who also happened to be the mayor's brother.

It is fascinating to realize that Joseph Glanvill (1636–1680),[4] who played a vital role in defending the new experimental methodologies of science, believed in witches. An ordained minister, he argued that science was not in conflict with religion and that experimental studies of the natural world would effectively further glorify the works of God. He was elected fellow of the Royal Society of Lon-

FIGURE 31. ■ A small section from the early-seventeenth-century engraving "A Witches' Sabbath" by Jacques De Gheyn II. The healthy, powerful witch in the foreground uses a mortar and pestle of bone to grind the ingredients for the body oil that enables her to fly. (Courtesy Krown and Spellman Booksellers, Culver City, CA.)

don in 1664. But he believed that evil spirits haunted the world and defended these views in his works *Philosophical Considerations Touching Witches and Witchcraft* (1666) and *Saducismus triumphatus*, the latter published posthumously.

In contrast, John Webster (1610–1682),[5] another ordained minister and supporter of the Royal Society, was far more practical and less credulous than Glanvill. He criticized Glanvill's views in his book *The Displaying of Supposed Witchcraft* (1677). Webster practiced medicine and chemistry and in his book *Metallographia* (1671) offers the following pungent critique of other authors of works on metallurgy:

> If the Authors that have written of the Mineral Kingdom were to be considered according to their number and multitude , then a man would think that this kind of Learning had already attained its height and Zenith. But if we come to balance them by their substance and weight, we shall find them for the most part but light, and their writings to contain very much Chaff, and but a little Corn.

---

1. *The New Encyclopedia Britannica*, Encyclopedia Britannica, Inc., Chicago, 1986, Vol. 25, pp. 92–97.
2. *The New Encyclopedia Britannica*, op. cit., Vol. 10, p. 351.
3. *The New Encyclopedia Britannica*, op. cit., Vol. 12, p. 715.
4. C.C. Gillespie, *The Dictionary of Scientific Biography*, Charles Scribner & Sons, New York, 1972, Vol. V, pp. 414–417.
5. Gillespie, op. cit., 1976, Vol. XIV, pp. 209–210.
6. J. Webster, *Metallographia*, Walter Kettilby, London, 1671, p. 25.

SECTION II
STILLS, CUPELS, AND WEAPONS

THE MAGIC OF DISTILLATION

Most of us who have been lucky enough to perform distillations know the thrill of winning a clear "spirit" from a dark and dingy solution, capturing a pure oil from a messy residue and even witnessing the collected distillate's abrupt solidification into white crystalline needles. A crude fermentation mixture will, upon distillation, yield an intoxicating "spirit of wine." Small wonder that a synonym for distillation is "rectification"—making things right. Indeed, distillation itself may almost be regarded as a religious act of ascent and descent:[1]

> Ascend with the greatest sagacity from the earth to heaven, and then again descend to the earth, and unite together the powers of things superior and things inferior. Thus you will obtain the glory of the whole world, and obscurity will fly away from you.

The archeological evidence suggests that distillation may have been performed as early as 5000 years ago.[2] We all know that the lid covering a pot of boiling broth will condense water vapor. It is not difficult to imagine fabricating lids with their edges turned up to form a gutter to collect condensed liquids.[2,3] Archeological research in the regions corresponding to ancient Mesopotamia has uncovered evidence of such "apparatus" dating from about 3500 years B.C.[2] Chemical historian Aaron Ihde provided a pictorial "Evolution of the 'Still'" and the major advance was the development, perhaps two thousand years ago, of the alembic or still head that transfers the distilled condensate down a "beak" into a separate receiver.[3] Medications up to the time of Paracelsus were derived from plants and animals as well as distillates from these natural sources.

Figure 32 is the frontispiece of the 1512 edition of the magnificent book by Heironymous Brunschwig, *Liber de Arte Distillandi*.[4] It depicts a double still allowing two separate operations likely to be rectifications of wine. The central unit is a tower containing cold water through which two "snakes" or condenser tubes pass. It is curiously reminiscent of the caduceus symbol wherein male and female snakes are entwined about the rod of Hermes and the four loops symbolize four "copulations." The hot alcohol vapors heat the tower's coolant, and the resulting warm water may be tapped off at the bottom as cool water is added from the top of the tower. The still operator on the left is comparing the temperature gradient between his cucurbit and the first condenser loop.[5] Figures 33–38 are also from Brunschwig's book. Figure 33 depicts the collection of distillate from rose extracts using four "rosenhuts" connected by beaks to receivers. The rosenhut was a form of air-cooled condenser. Figure 34 shows an efficient furnace having thirteen alembics for what appears to be a very profitable commercial operation.

Astrological influences are very much in evidence in Brunschwig's book. Figure 35 instructs that a particular distillation be performed under the influence

FIGURE 32. ■ Title page (hand-colored; see color plates) from *Liber de Arte Distillandi* (Heironymous Brunschwig, 1512), depicting a double-still in which the central tower contains cooling water that is continuously replenished. Come to think of it, this apparatus evokes an image of the caduceus: two serpents entwined about the staff of Hermes where the loops symbolize "couplings." (Courtesy Chemical Heritage Foundation.)

of the Ram, corresponding to the Sun being in Aries (March 21–April 19), and thus, calling for mild heating.[6,7] The distillations in Figure 36 (Twins, Gemini, May 21–June 21) and Figure 37 (Crab, Cancer, June 22–July 22) are carried out under progressively warmer conditions with maximum heat (Figure 38) under the influence of Leo the Lion (July 23–August 22).[6,7]

Books of distillation enjoyed considerable popularity during the sixteenth and seventeenth centuries.[5] Figure 39 is from the 1528 edition of the *Coelum Philosophorum* ("Heaven of the Philosophers") of Philip Ulstad.[8] The two sets of distillation apparatus in Figure 40 are from the *Distillier Buch* of Walter H. Ryff.[9]

FIGURE 33. ■ Distillation employing *rosenhuts* (air-cooled still heads) in Brunschwig's 1512 *Liber de Arte Distillandi* (Courtesy Chemical Heritage Foundation).

FIGURE 34. ■ Thirteen alembics and a furnace (from Brunschwig's 1512 *Liber de Arte Distillandi*, courtesy Chemical Heritage Foundation).

FIGURE 35. ■ Distillation using very mild heat under the influence of Aries (March 21–April 19) from Brunschwig's 1512 *Liber de Arte Distillandi* (courtesy Chemical Heritage Foundation).

FIGURE 36. ■ Distillation using moderate heat under the influence of Gemini (May 21–June 21) from Brunschwig's 1512 *Liber de Arte Distillandi* (courtesy Chemical Heritage Foundation).

FIGURE 37. ■ Distillation using moderately strong heat under the influence of Cancer (June 22–July 22) from Brunschwig's 1512 *Liber de Arte Distillandi* (courtesy Chemical Heritage Foundation).

FIGURE 38. ■ Distillation using conditions of strongest heat under the influence of Leo (July 23–August 22) from Brunschwig's 1512 *Liber de Arte Distillandi* (courtesy Chemical Heritage Foundation).

COELVM PHILOSO
PHORVM SEV DE SECRETIS
naturæ. Liber.

PHILIPPO VLSTADIO PATRICIO
Nierenbergensi. Authore.

FIGURE 39. ■ Title page from Philip Ulstad's "Heaven of the Philosophers," Nuremburg, 1528 (courtesy The Roy G. Neville Historical Chemical Library).

Figure 41 is the frontispiece from the 1576 book by Conrad Gesner, *The newe Iewell of Health . . .* (109-word title!).[10] Although it is interesting to speculate over whether the woman depicted was an alchemist or chemist, it is more likely that she was simply an operator of the still since this was common sixteenth-century practice.

And what of the fruits of these countless distillations? Let us examine a few recipes. From the 1599 Gesner *The practice of the new and olde phisicke*:[11]

> *An oile or ointment sharpening the wit, and increasing memory*
> Take of Stœchas, of Rosemarie flowers, of Buglosse flowers, of Borage flowers, of Camomill flowers, of Maioram, of sage, of baulme, of violet flowers, of red rose leaves, and of bay leaves, of each one ounce and a half, al these put by into a glasse bodie strongly luted, with foure pints either of Malmesie, Rennish wine, or Aqua vita, let these so stand to infuse for five daies, and distilled, add to it of the best Turpentine, one pound and a halfe, of Olibanu, of chosen Myrre, of Mastick, Bolelium, of gum jute, of each two ounces, of Ver-

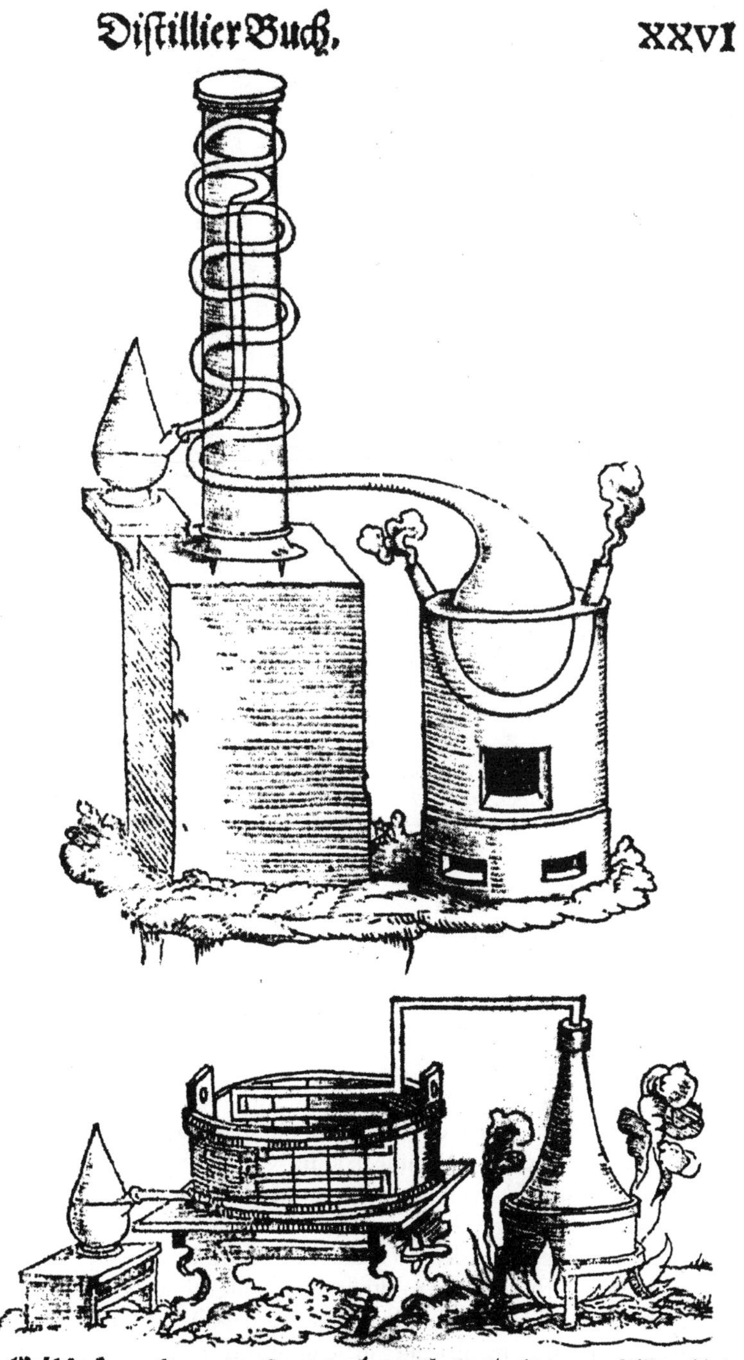

Diſtillier Buch. XXVI

Etliche bereydeen ein ofen wie obgemelt/in zimlicher gröſſe/ alſo dz
ſie auch mit holtz vnder fewern mögen/ſetzen ein irdin helm daruff/ a-
ber doppel/ aller maß als wir in der erſte angezeyget haben/ durch ein

E 2

FIGURE 40. ■ Two stills from Ryff's 1545 *Distillier-Büch* (courtesy Chemical Heritage Foundation).

FIGURE 41. ■ Frontispiece from Conrad Gesner's 1576 *The New Iewell of Health* (courtesy The Roy G. Neville Historical Chemical Library). Women commonly operated stills in the sixteenth century.

> nicis integrae, one ounce, of Mellis anacardi, three ounces, all these brought to powder & infused for five dayes with the foresaid distillation, in a bodie with a head close luted, distill againe, adding to it of Cynamon, of Cloves, of Mace, of Nutmegs, of Cardamomum, of graines of Paradice, of the long and round Pepper, of Ginger, Xyloaloes, and of Cubebæ, of each one ounce, all these finelie brought to powder. To these adde of Muske & Amber gréece, of each two drams, all these mired together distill (after that these added & put into the former distillation have remained five dayes) the fire in the beginning soft, increase after by little and little unto the end of the work. The use of it, is, that the same may be applied in the winter time once in the wœke, but in the sommer time once in a month, the head before being washeth, the temples & hinder part part of the head anoint with it.

Now, if I can only remember this!

From John French's *The Art of Distillation*, published in 1653:[12]

> *How to turn Quick-Silver into a water without mixing any thing with it, and to make thereof a good Purgative and Diaphoretick medicine*

> Take an ounce of Quick-silver, not purified, put it into a bolt head of glass, which you must nip up, set it over a strong fire for the space of two months, and the Quick-silver will be turned into a red sparkling Precipitate. Take this powder, and lay it thin on a Marble in a cellar for the space of two months, and it will be turned into a water, which may be safely taken inwardly, it will work a little upward and downward, but chiefly by sweat.

Hmm—I recognize the calcination of mercury to its red oxide. The rest is a bit of a mystery. And let us take one more from Mr. French, Doctor of Physick:

> *A famous spirit made out of Cranium humanum.*
> Take of Cranium humanum as much as you please, break it into small pieces, which put into a glass Retort well coated, with a large Receiver well luted, then put a strong fire to it by degrees continuing of it till you see no more fumes come forth; and you shal have a yellowish spirit, a red oyl, and a volatile salt.
>
> Take this salt and the yellow spirit and digest them by circulation two or three months in Balneo, and thou shalt have a most excellent spirit.
>
> It helps the falling sickness, gout, dropsie, infirm stomache, and indeed strengthens all weak parts, and openeth all obstructions, and is a kinde of Panacea.

No limit apparently on the availability of *Cranium humanum*. I wonder who the supplier is. I'm a bit surprised that *Cranium humanum* was not employed in Gesner's *Memory and Wit Ointment*.

Harkening back to Gesner, let us try one additional medication made by bruising, rather than distilling, the cantharides beetle (*Spanish fly*):[14]

> Their virtue consists in burning the body, causing a crust, or . . . to corrode, cause exulceration, and provoke heat; and for that reason are used mingled with medicines that are to heat the Lepry, Tettars, and Cancerous sores.

But, despite its toxicity, the unique qualities of Spanish fly "are good for such as want erection, and do promote venery very much." And here is a more specific observation by Gesner:[14]

> When Anno 1579, I staid at Basil, a certain married man (it was that brazen bearded Apothecary that dwelt in the Apothecaries shop) he fearing that his stopple was too weak to drive forth his wifes chastity the first night, consulted one of the chief Physicians, who was most famous, that he might have some stifte prevalent Medicament, whereby he might the sooner dispatch his journey.

Propriety forces us to end here, but the outcome was both painful and fruitless.

---

1. R.G.W. Anderson, in F.L. Holmes and T.H. Levere (eds.), *Instruments and Experimentation in the History of Chemistry*, The MIT Press, Cambridge, MA, 2000. pp. 7–8.
2. Anderson, op. cit., pp. 5–34.
3. A.J. Ihde, *The Development of Modern Chemistry*, Harper & Row, New York, 1964, pp. 13–18.
4. H. Brunschwick, *Lieber de arte distillandi de composites. Das buch waren kunst zu distillieren die composita und simplicia und ds Buch thesaurus pauperum, ein Schatz der armen genant Micarium . . . ,*

Strassburg, 1512. I am grateful to Ms. Elizabeth Swan, Chemical Heritage Foundation for supplying an image of this hand-colored plate.

5. J.R. Partington, *A History of Chemistry*, Macmillan and Co. Ltd., London, 1961, Vol. 2, pp. 82–89.
6. A. Roob, *The Hermetic Museum: Alchemy & Mysticism*, Taschen, Cologne, 1997, p. 146.
7. *The New Encyclopedia Britannica*, Encyclopedia Britannica, Inc., Chicago, 1986, Vol. 12, p. 926.
8. P. Ulstadt, *Coelum Philosophorum seu de Secretis naturae. Liber*, Ioannis Grienynger, Strassburg, 1528. I am grateful to The Roy G. Neville Historical Chemical Library (California) for providing the image from this book.
9. W.F. Ryff, *New gross Distillier-Būch, wolgegrūndter kūnstlicher Distillation . . .*, Bei Christian Egenolffs Erben, Frankfort, 1545. I am grateful to Ms. Elizabeth Swan, Chemical Heritage Foundation, for supplying an image of this page.
10. C. Gesner, *The newe Iewell of Health, wherein is contained the most excellent Secretes of Phisicke and Philosophie, divided into fower Bookes. In the which are the best approved remedies for the diseases as well as inwarde as outwarde, of all the partes of mans bodie: treating very amplye of all Dystillations of Waters, of Oyles, Balmes, Quintessences, with the extraction of artificiall Saltes, the use and preparation of Antimonie, and Potable Gold. Gathered out of the best and most approved Authors, by that excellent Doctor Gesnerus. Also the Pictures, and maner to make the Vessels, Furnaces, and other Instruments thereunto belonging. Faithfully corrected and published in Englishe, by George Baker, Chirurgian*, Henrie Denham, London, 1576. The Roy G. Neville Historical Chemical Library.
11. C. Gesner, *The practice of the new and old phisicke, wherein is contained the most excellent Secrets of Phisicke and Philosophie, divided into foure Bookes, In the which are the best approved remedies for the diseases as well inward as outward, of al the parts of mans body: treating very amplie of al distillations of waters, of oyles, balmes, Quintessences, with the extraction of artificiall saltes, the use and preparation of Antimony, and potable Gold Gathered out of the best & most approved Authors, by that excellent Doctor Gesnerus. Also the pictures and maner to make the Vessels, Furnaces, and other Instrumentsd thereunto. Newly corrected and published in English, by George Baker, one of the Queenes Maiesties chiefe Chirurgians in ordinary*, printed by Peter Shaw, London, 1599, p. 240 (i.e., p. 140).
12. J. French, *The Art of Distillation or, A Treatise of the Choicest Spagiricall Preparations Performed by way of Distillation. Together with the Description of the Chiefest Furnaces & Vessels Used by Ancient and Moderne Chymists, Also a Discourse of Divers Spagiricall Experiments and Curiosities: And the Anatomy of Gold and Silver, with the Chiefest Preparations and Curiosities thereof; together with their Vertues. All which are contained in VI. Bookes; Composed by John French Dr. of Physick*, E. Cotes, London, 1653, pp. 73–74.
13. French, op. cit., p. 91.
14. T. Muffet, *The History of Four-Footed Beasts and Serpents and Insects*, Vol. 3, *The Theatre of Insects* (reprint of 1658 London edition), Da Capo Press (Plenum), New York, 1967, pp. 1003–1005.

PRACTICAL METALLICK CHYMISTRY

The development of chemistry rests upon an ancient tripod. One leg of the tripod is formed out of the spiritual, mystical, and conceptual roots of chemistry, which began with the two contraries and four elements, evolved into the *tria prima* (mercury, sulfur, salt) out of which arose Becher's *terra pinguis* or "fatty earth," which, in turn, became Stahl's phlogiston. A second leg is comprised of the practical iatrochemical experience, techniques and apparatus derived from extractions and distillations using animal parts and plants that provided medications.

This animal and plant chemistry eventually became our modern organic chemistry and biochemistry. The third leg is the metallurgical chemistry derived from mining and the ancient metallic arts. Aside from techniques learned and apparatus developed, it was this chemistry—the chemistry of metals and minerals that first truly connected experiment and chemical theory. It ultimately evolved into inorganic chemistry.

In his 1671 book *Metallographia,*[1] John Webster uses biblical quotation to trace metallurgical chemistry back to Moses who, in turn, references Tubal-Cain (*Genesis* 4:22), "the eighth of Mankinde from Adam,"[2] a biblical worker of iron and brass. Here is how Tubal-Cain *might* have discovered metallurgical chemistry:[3]

> While through a Forest Tubal with his yew
> And ready Quiver did a Bore pursue,
> A burning Mountain from his fiery vain,
> An Iron River rolls along the Plain.
> The witty Huntsman musing, thither hies,
> And of the wonder deeply can devise.
> And first perceiving that this scalding mettle
> Becoming cold, in any shape would settle,
> And grow so hard, that with his sharpened side,
> The firmest substance it would soon divide.

Georg Bauer (1494–1555), Latinized as Georgius Agricola (the German *bauer* means "farmer"), studied medicine, probably obtained the M.D. degree in Italy and in 1526 returned to Germany, where he settled in a mining district in Bohemia.[4] Agricola served as a physician to the miners and developed an interest in mining and metallurgical chemistry. Although he wrote a Latin grammar, religious works and a medical work on the plague, his truly lasting works concern metallurgy. Figure 42 shows the title page of Agricola's first book on metallurgy, *Georgii Agricolae Medici Bermannvs, Sive De Re Metallica,* published in Basel in 1530 by Froben.[5,6] The "Bermannus" was the first book on the science of mineralogy published in Europe and is of great rarity.[7] The first truly comprehensive book on metallurgical chemistry was the *De La Pirotechnia* of Vannuccio Biringuccio (Venice, 1540), and we will return to it soon.

Agricola died a quarter of a century after publication of the 1530 "Bermannus," and during the next year his most famous book, *De Re Metallica Libris XII,*[8] was published in Basel. Although significant sections were adapted from Biringuccio's work as well as other contemporary treatises, Agricola's work summarized a lifetime of experience, observation and learning. His work began with methods of surveying mountains and veins of ore and planning mine shafts. Figure 43, from *De Re Metallica,* illustrates the use of a carefully constructed hemicircle (protractor) and its use for surveying and planning a mine.[8,9] One can only assume that the mining company had progressive managers who encouraged their surveyors to be unencumbered in both their thinking and manner of dress. Figure 44 depicts a horse-driven apparatus for pumping water out of mines. The subterranean chamber was carefully reinforced by timbers to prevent its collapse and the death of the miners. The hollow plunger had a tightly sealed leather bag at the bottom that pushed air out in the downstroke and drew in drainage water in the upstroke.

GEORGII
AGRICOLAE MEDICI
BERMANNVS, SIVE
DE RE METALLICA

Basileæ, in ædibus Frobenianis
Anno M. D. XXX.

FIGURE 42. ■ Title page from Agricola's first book on metallurgy—the 1530 *Bermannus*. The extensive mining and mineralogy book collection of President Herbert Hoover and Lou Henry Hoover lacked this exceedingly rare book. (Courtesy The Roy G. Neville Historical Chemical Library.)

Figure 45 shows a laboratory containing stills for the synthesis and purification of "*aqua valens*" or "powerful water."[8,9] *Aqua valens* was a term used by Agricola for powerful acidic agents, including both *aqua vita* (nitric acid) and *aqua regia* (hydrochloric acid–nitric acid, 3 : 1). (The etymology of "*valens*" is related to the modern term "valence," which refers to "combining power," for instance, of an atom with one hydrogen atom, two hydrogen atoms, etc.). In Figure 45, a typical distillation of *aqua valens* is represented. It includes an *ampulla* (or cucurbit, *K*) containing a mixture of niter or saltpeter, vitriol, and water along with some alum (aluminum sulfate–potassium sulfate), joining it to an *operculum* (or alembic, *H*). The *operculum* is heated by charcoals (stored in earthenware, *F*) in furnace A, red fumes are observed and liquid nitric acid is collected dropwise. Tiny quantities of silver are typically added to the distilled acid in order to precipitate small quantities of chloride that have co-distilled as a result of sea-salt impurities in the starting materials.

The purified nitric acid is used to "part" gold from silver and other base metals because gold is unreactive under these conditions. First, lead is added and the impure alloy heated in a cupel until the least reactive metals, gold and silver, form a melt while the "baser" metals have oxidized and merged with the bone cupel. The silver-gold alloy is then mixed with the nitric acid—silver dissolves, while gold sinks to the bottom, is filtered, and is then washed.

LIBER QUINTUS. 101

Libella stativa A. *Ejus ligula* B. *Libella & ligula* C.

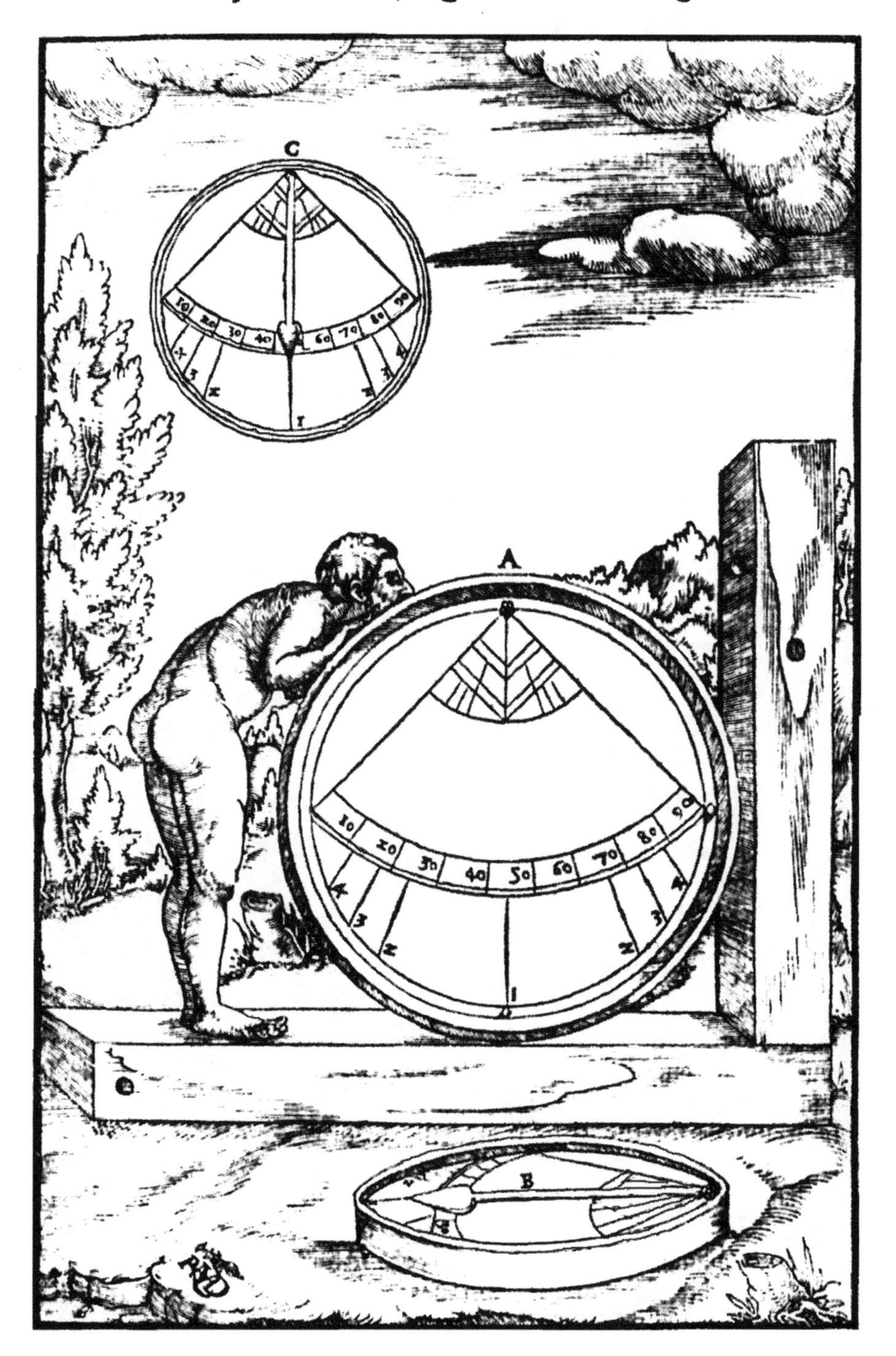

FIGURE 43. ■ Surveying the coordinates of a mine (on a sweltering summer day?) from Agricola's most famous book, the 1556 *De Re Metallica Libris XII* (courtesy Chemical Heritage Foundation).

FIGURE 44. ■ A horse-driven apparatus for pumping water out of mines (from Agricola's 1556 *De Re Metallica*, courtesy Chemical Heritage Foundation). Agricola believed in mine goblins whose emanations (carbon monoxide?) were deadly to miners.

FIGURE 45. ■ Distillation of *aqua valens* ("powerful water"), a general term used by Agricola to describe powerful acids such as *aqua fortis* (nitric acid) and *aqua regia* (hydrochloric acid/nitric acid, 3 : 1), depicted in Agricola's 1556 *De Re Metallica*. (Courtesy Chemical Heritage Foundation.)

To this useful scientific knowledge, we must add Agricola's belief in mine "goblins" whose exhalations were deadly to miners.[4] A book concerning subterranean animals published by Agricola in 1549 includes a description of salamanders that survive fire[4] (perhaps taking the salamander allegory, Figure 21, a bit too seriously). Nevertheless, the value of Agricola's famous book was well stated by Webster—never a shy critic, in 1671, over a century after *De Re Metallica* was published:[10]

> As for the beating, grinding, sifting, and washing of Ores in general, from their earthy filthiness and superfluitities, Georgius Agricola hath written very largely and learnedly, more than any other Author that I know of. And I could with that some person that hath ability and leisure, would translate it into English; for it might be very serviceable to our common Miners, that in that particular have little to direct them, but what they learn from one another.

Webster's wish was finally answered some 241 years later by two people of very considerable ability and very little leisure: Herbert Hoover, the future president of the United States of America, and his wife Lou Henry Hoover, the first female geology graduate of Stanford University (see the next essay).

Vannucio Biringuccio (1480–1539) is much less well known than Agricola.[4] However, his *Pirotechnia* (Venice, 1540)[11] was the first comprehensive book on mining and metallurgy and was sumptuously illustrated.[4] Amazingly, its first English translation appeared over four centuries later in 1942![12] Biringuccio was very much involved in the political affairs of his day, had military knowledge and skill and was Director of the Pope's (!) Arsenal[4] (much as Lavoisier over two centuries later would direct the Arsenal of Louis XVI). He did not believe in transmutation and was one of the earliest to note the increase in weight of lead upon its calcination:[13]

> The calcination of lead in a reverberatory furnace seems to me to be such a fine and important thing that I cannot pass by it in silence. For it is found in effect that the body of the metal increases in weight to 8 or perhaps 10 per hundred more than it was before it was calcined. This is a remarkable thing when we consider that the nature of fire is to consume everything with a diminution of substance, and for this reason the quantity of weight ought to decrease, yet actually it is found to increase.

We know now that oxidation of lead to form lead oxide (PbO) should involve a weight increase relative to the metal of 7.7%.

Figures 46 and 47 are from the 1540 *Pirotechnia*. They depict five different types of large cupeling furnaces. Typically, we think of cupels as small molded cups made, for example, from calcined crushed bone ground into a paste with beer, molded, dried, and baked.[13] Crude silver ore may be heated to high temperature in these cupels. More reactive metals oxidize and their calxes are physically absorbed into the cupels leaving molten silver to be cooled and form the purified solid. The huge cupels in Figures 46 and 47 were made from wood ashes, crushed brick, limestone, and egg white and were employed to purify large quantities of silver.[14] The figure at the left (verso page) of Figure 46 shows a worker forming the hearth of a large cupeling furnace.[14] The upper and lower figures to the right (recto page) of Figure 46 are large cupeling furnaces with a brick dome and an iron hood, respectively. The upper figure on the verso page in Figure 47 is a cupeling hearth covered with clay plates, and the lower figure depicts a cover of wooden logs over a cupeling hearth.[14]

Agricola's *De Re Metallica* and Biringuccio's *Pirotechnia* are both recognized as "Heralds of Science—two hundred epochal books and pamphlets in the Dibner Library, Smithsonian Institution."[15] One additional mining and metallurgy book has also been included on this rarified list—*Beschreibung: Allerfürnemisten Mineralischen Ertzt Unnd Berckwercks Arten* ("Treatise Describing the Foremost Kinds of Metallic Ores and Minerals," 1574, Prague) by Lazarus Ercker. Figure 48 is from the title page of the second (1580) edition of this beautiful folio book.[16,17] It depicts a full array of operations in a sixteenth-century mineral assayer's laboratory. The sumptuous book was published in eight Frankfort editions beginning in 1574 with the final one appearing in 1736.[17,18] The wonderful wood blocks used to print the illustrations in 1574 were preserved and

LIBRO TERZO

di ſopra v'ho detto gia nella Alemagna viddi affinare a vn fornello che haueua in ſcambio di capello vna volta murata. & atorno vi ſtauano ghettando a lauorare a ſei fenestrette ſei maeſtri, & queſto talcenetaccio haueua tre gran mantici con canne & doppie canne lunghe & groſſe, & alla bocchetta dell'uſcita del vento ognuna haueua di ferro vna uentula, quale s'apriua quando venuta il vento, & quando non caſcando ſi riſerraua. & queſte ventole teneuano che poter coprendere ſerrauano inſieme il corpo dentro de mantici che nel tirare aſſe non v'entraſſero carboni acceſi che li bruciaſſero, & ancho perche tali impedimenti alle bocche facciſer batter il lor vento piu nel mezzo del bagno, & di piu erano anchora di modo adattati, che mandar ſi poteuano in qua & in la, & far che'l vento arriuaſſe, doue piu li pareua a propoſito.

ERA fatto di muro ſotto doue poſauano li mantici, & doue entrauano le canne era uno aperto a modo d'una fineſtra alto un braccio a circha, larga uno & mezzo, & a ognuna ui era congegnato in due anelli di ferro un ruzzolo grande, ſopra alquale ſi metteua la ponta d'un mezzo traue dabete o daltro legno groſſo, longo un quatro o cinque braccia, & ſpingendolo quanto era largo il diametro del ceneraccio, facilmente il mandauano dentro, & queſte erano le legna che adoperauano, che ueramente mi parſe coſa bella, & conſiderando anchora conobbi che tal uia non potena ſeruire bene, ſe non all'opere grandi & continuate come in que lochi ſi faceuano, la doue ogni ſettimana due uolte o almeno una non era che non ſe adoperaſſe, & che non riduceſſero a fino. 150. & 200. marche d'argento per uolta, & coſi ſi lauoraua in affinare a gli edificii dell'Imperatore in Spruch.

DEL FAR LI CENERA 58

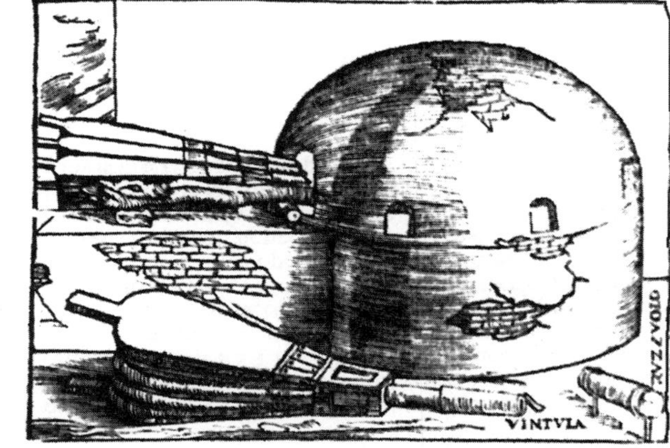

QVELL'ALTRO modo che s'adopera per coprire il ceneraccio, il cappel di ferro mi piace aſſai piu, Perche molto piu ſi puo riſtregnere il fuoco & tenere il bagno caldo, & con eſſo ſi puo affinare il poco, & l'aſſai come al maeſtro piace.

ET COME u'ho detto auanti ſi copreno, anchora quando s'affinano li ceneracci con certe piaſtre di terra cotta groſſe tre dita, & larghe mezzo braccio, & longhe quanto il ceneraccio: & queſte mi piacciono molto piu che alcuni de gli altri modi ch'io habbi ueduto adoperare, perche s'accoſtano meglio per tenerla calda ſecondo che la ua mancando.

H

FIGURE 46. ■ Large cupeling furnaces depicted in Vannucio Biringuccio's 1540 *Pirotechnia*. Cupels are cups made typically from crushed bone ground into a paste with beer, then molded, dried, and baked. Calxes of base metals (e.g., iron oxides) are absorbed into the cupel while molten gold is not and is thus readily separable. (Courtesy The Roy G. Neville Historical Chemical Library.)

LIBRO TERZO

IL SIMILE ſi fa anchora con li ceppi di quercia,ma non coſi bene,ne con tanta facilita.

ET PERCHE molte ſon le conſiderationi & l'auertentie che a coſdur perfetta l'opera biſogna hauere,& chi non ha uedute p eſperientia, o che prima molto bene nõ ne ſia ſtato auertito , difficilmẽte ſi guarda dalli inconuenienti. P E R O ſappiate ſe in qllo argento o piombo che affinate,fara ſtagno,durarete gran fatiga a condurlo, & la via (quando queſto interueniſſe a purgarlo) e qſta,che ſe gli ſtrenga il fuoco adoſſo, & ſcaldi bene il bagno,& come ſi uede che ſia ben caldo, vi ſi gitta ſopra della carbonige trita,& coſi ſoffiando con li mantici ſi fa il bagno ben gonfiare,& dipoi con vn caſtagniolo gentilmente ſcoprendolo ſe gli va leuando da doſſo la carbonige,cõ laquale tirãdola fuore ne vien con ſeco anchor lo ſtagno,ilquale prima tutto creſpo ſi ſta nel bagno, & non ſi diſtende in quella ſottigliezza che fa il piombo . E T ANCHO ſe aueniſſe che'l ceneraccio p troppa caldezza faceſſe li bollori

DEL FAR LI CENERA 59

habbiate a mente di far allargare li ceppi,ouer fermare li mantici tanto,che ſi temperi . E T A N C H O ſe aueniſſe che'l bagno fuſſe molto ramigno come ſon le ritratte delle minere , o di ghette, o di loppe, auertite nel principio a ſopraſedere il gettare per fino à tanto che'l ceneraccio pigli certo neruo di ghetta , pche le materie ramigne gli fa teneri,per ilche ſono al ghettar pericoloſi,& pero auertirete di far ch'el taglio nel ceneraccio ſia ſottile & un poco appendino , & battete ſpeſſo la ponta del voſtro ferro accio non s'ingroſſi . A P P R E S S O di voi habbiate ſempre vn caſtagniuolo o due, & coſi ancho di quelli che nella ponta habbino legata con vn poco di fil di ferro vna pezzeta di panno bagnato per poter dare in ſul taglio & fermare quando vedeſſe che del bagno s'auiaſſe per volere vſcire fuore piu ghetta che quella che vorreſte, ouero per bagnare alle volte qualche luoco per li ceneracci fatti teneri dal piombo , ouer per inhumidire doue voleſte tagliare che fuſſe duro per farlo piu facile,Ricordateui anchora di fare il ceneraccio ſimile alle materie,cioe ſe le ſon dolci,dolce,& ſe le ſon dure,duro,& a ogni ceneraccio che farete,ricordateui di fregare ſpeſſo la verga alli ceppi,& di far caſcare di quella carbonigia acceſa ſopra il bagno , & maſſime quando non fuſſe alle ſponde ghetta che ſubito ve la vedrete apparire, & coſi ſe va ſeguitando tanto che l'arriuate al termine di fino quãto il ceneraccio per il ſuo ordinario puo.

M A V O L E N D O L O anchora vn poco piu sforzare , apparecchiate quando ſete a l'ultimo vn ceppo o due che nõ ſien ſtati in fuoco,& ſien ben ſecchi , & li mettete ſopra al ceneraccio aponto che coprino bene l'argento , & di nuouo li ridate una quantita di piombo ſecondo che uolete , & fate riuenire l'argento , liquali come gli uedrete inſieme uniti, & uoi con un caſtagnolo ſottile deſtramente gli rimenate & gli unite inſieme; & di poi pian piano menando li mantici sfumando il piombo,laſſarete l'argento ben chiarire , & dipoi fatto queſto, & che uedete che ghe finito , leuate li ceppi & cauatene il uoſtro argento & lo fondete & nettate dal ceneraccio come auanti u'ho detto. M I V I R E S T A a dire come nel leuar del ceneraccio adoperato , auertiate che non ſi meſcoli di quella cenere di ceppi che ſpeſſo reſta ſopra al ceneraccio con quella che ui meteſte per ſotto ricotta & ben diſpeſta a rifare la compoſition del ceneraccio, perche la guaſtarebbe,& ſieui a mente per un de ricordi generale che mai con ferro freddo o con carboni che non ſien prima acceſi o con legna,o coſe molli,non toccare il uoſtro bagño,perche ui creſciarebbe fatiga a condurlo al ſuo fine,& in luoco d'utile ui darebbe forſe danno,et pero in ogni parte uſate la diligentia et prudentia uoſtra.

H iii

FIGURE 47. ■ Large cupeling hearths from Biringuccio's 1540 *Pirotechnia* (courtesy The Roy G. Neville Historical Chemical Library).

used through 162 years in all eight editions.[17,18] A Dutch edition was published in 1745.

It is interesting that the first two of the three great mining, assaying, and metallurgy books of the sixteenth century were finally translated into English roughly four centuries after their original publication: Biringuccio's *Pirotechnia* (Venice, 1540; Chicago, 1942); Agricola's *De Re Metallica* (Basel, 1556; London, 1912). However, Ercker's *Beschreibung* (Frankfurt, 1574) was translated by Sir John Pettus[19] about one century (London, 1683) after the Frankfurt original. Anneliese Grünhaldt Sisco and Cyril Stanley Smith conjecture that the reason for the earlier English translation of Ercker's book in the seventeenth century might have been that it was the most recent, and thus current, of the three great texts. Pettus (1613–1690) played a significant military role in England's Civil War and, at one point, was held captive by Oliver Cromwell for 14 months. Following the restoration, Pettus served as Restoration Deputy to the Vice Admiral and was seriously wounded in the leg during a naval battle with the Dutch.[19]

FIGURE 48. ■ A sixteenth-century mineral assayer's laboratory from the second (1580) edition of Lazarus Ercker's treatise on mining and metallurgy. The woodblock used to print the first edition (1574), and this (second) edition was preserved and employed for over 160 years through the final 1736 edition. (Courtesy The Roy G. Neville Historical Chemical Library.)

Lazarus Erskerus

aliàs

Erckern.

BOOK I.

CHAP. I.

Of Silver Oars.

Sculpture I.

Deciphered.

The *Aſſayer* 1. the *Scales* 2. the Caſes for *Weights* 3. *Glaſſes* for Aqua Regis, Aqua Fortis Aqua Vitrioli, Aqua Argentea or Quickſilver, *&c.* 4.

FIGURE 49. ■ Depiction of an assayer in John Pettus' 1683 translation and extension of Ercker's treatise on metallurgy. Pettus' book title begins "*Fleta Minor,*" referring to the Fleet Prison, in which he was an inmate while writing this book. Pettus was imprisoned "through the accusations of an un-scrupulous woman" who was, incidentally, his wife. (Courtesy Chemical Heritage Foundation.)

An interesting aspect of Pettus' translation of Ercker's book (*Fleta Minor, or, the Laws of Art and Nature, In Knowing, Judging, Assaying, Fining, Refining and Inlarging the Bodies of confin'd Metals*), is the replacement of Ercker's sixteenth-century woodcuts with late-seventeenth-century costumed English figures engraved in copper plates. Figure 49 depicts a contemporary English assayer. Figures 49–53 include partial explanations with the plates. Figure 50 describes the making and molding of cupels made from crushed bone paste. Figure 51 is a scene in a gold-assaying laboratory. The cone-shaped vessel on the right is a parting flask, for assaying gold, seated on its stand. The wooden piece hanging to the right of the assayer in the rear of this figure has a slit through which to view the furnace while protecting the eyes. The person in the foreground is testing the density of "auriferous" silver in water. The *aqua fort* referred to in Figure 52 is nitric acid. Figure 53 depicts the smelting of bismuth in open air. It is fun to compare Figures 50, 52 and 53 with their sixteenth-century German counterparts.[18]

Now, how the *Copel-Case* and the Copel is to be ordered and performed the following *Sculpture* will shew.

FIGURE 50. ■ Manufacture of cupels (using a paste made from crushed bone and beer) from Pettus' 1683 *Fleta Minor* (courtesy Chemical Heritage Foundation).

Of Gold Oars. 153

Sculpture XIX.

Deciphered

1. *How the* Aſſayer *ſtands before the* Aſſay-Oven *to prove* Metals.
2. *The Iron on which the Proof is to be caſt.*
3. *A wooden Inſtrument to ſee through into the fire to prevent hurt to the Eyes.*

R r 4. *A*

FIGURE 51. ■ Assaying gold ore in Pettus' 1683 *Fleta Minor* (courtesy Chemical Heritage Foundation).

The curious title of Pettus' book, *Fleta Minor,* derives from the final years of his life that were spent in the "Fleta" or Fleet Prison wherein he wrote this work. Pettus informs his readers that "it seems a strange disposition of Providence that a man who had done so much for his King and Country should be suffered through the accusations of an unscrupulous woman, and that woman his own wife, to spend the closing years of an active and useful life a Prisoner in the Fleet."[19]

Of Gold Oars: 173

Sculpture XXII.

CHAP. XXIX

CHAP. XXIX.

To distil Aqua fort. *in* Retorts *with other* Advantages.

DISTILLING *Aqua fort.* in *Retorts* is no old *Invention*, and no long Labour, but a ſhort way; if *Retorts* may be had which are made of one piece, and will hold *Aqua fort.* and *Oyl*; then lute ſuch over with good and ſound *Clay*, let it be well dry, put in it the *Ingredients* or *ſtuff*, which ſhall be *calcin'd* and mingled with *Calx viva*, and lay the *Retort* in an *Oven* made on purpoſe(whoſe Deſcription ſhall follow hereafter) and fill a *Receiver* with water before it, then make a fire in the *Oven* (and ſpeedi- Yy ly

Section 1.

FIGURE 52. ■ Distillation of *aqua fortis* (nitric acid) in Pettus' 1683 *Fleta Minor*. Compare this figure with the same plate in the 1736 edition (i.e., the original 1574 edition; see A. Greenberg, *A Chemical History Tour*, Wiley, New York, 2000, p. 16), and you will note that the costumes have been updated by a century while the apparatus remains unchanged—fashion outstripping technology. (Courtesy Chemical Heritage Foundation.)

Of Lead Oars. 307

CHAP. X.

Sculpture XXXVII.

Deciphered.

1. *The little* Iron Pans *for Spelter or Wismet* Oar.
2. *The* fire *of* vvood *for them.*
3. Melted Spelter *that is to be made clean in the* iron Pan, *and the workman that tends it.*
4. He *that draws the* Oar *out of the* Mine.

CHAP.

FIGURE 53. ■ Smelting of bismuth ore in the open wind from Pettus' 1683 *Fleta Minor* (courtesy Chemical Heritage Foundation).

---

1. J. Webster, *Micrographia: Or, An History of Metals*, Walter Kettilby, London, 1671.
2. J. Read, *Humour and Humanism in Chemistry*, G. Bell and Sons Ltd., London, 1947, pp. 3–4.
3. Webster, op. cit., p. 3.
4. J.R. Partington, *A History of Chemistry*, Macmillan and Co. Ltd., London, 1961, Vol. 2, pp. 32–66.
5. G. Agricola, *Georgii Agricolae Medici Bermannus, sive De Re Metallica*, Frobenianus, Basel, 1530. I thank The Roy G. Neville Historical Chemical Library (California) for supplying the image of the title page for this book.
6. Johann Froben (Johannes Frobenius, ca. 1460–1527) was a famous Basel printer-publisher

whose techniques revolutionized printing. Among his gifted illustrators were Hans Holbein, and after 1513 he was the sole publisher of the great Dutch humanist-philosopher Desidarius Erasmus (*The New Encyclopedia Britannica*, Encyclopedia Britannica, Inc., Chicago, 1986, Vol. 5, p. 16.).
7. I thank The Roy G. Neville Historical Chemical Library (California) for supplying this image, and I am grateful to Dr. Neville for helpful discussions.
8. G. Agricola, *De Re Metallica Libri XII, Quibis Officia, Instrumenta, Machinae, Ac Omnia Denique Ad Metallicam Spectantia*, Basel, 1556. I am grateful to Ms. Elizabeth Swan, Chemical Heritage Foundation, for providing these images.
9. H.C. Hoover and L.H. Hoover (transl.), *Georgius Agricola De Re Metallica* (translated from the first Latin edition of 1556), *The Mining Magazine*, London, 1912 (reprinted by Dover Publications, Inc., New York, 1950), see pp. 439–447.
10. Webster, op. cit., p. 155.
11. V. Biringuccio, *De La Pirotechnia. Libri X.*, Venice, 1540. I am grateful to Ms. Elizabeth Swan, Chemical Heritage Foundation, for supplying images from this book.
12. C.S. Smith and M.T. Gnudi, *The Pirotechnia of Vannoccio Biringuccio Translated from the Italian with an Introduction and Notes by Cyril Stanley Smith and Martha Teach Gnudi*, The American Institute of Mining and Metallurgical Engineers, New York, 1942 (see also the 1959 reprint published by Basic Books, New York).
13. Smith, op. cit., p. 58.
14. Smith, op. cit., pp. 161–169.
15. *Heralds of Science*, revised edition, Burndy Library and Smithsonian Institution, Norwalk and Washington, DC, 1980. It has been duly noted that, of the *Great Books of the Western World*, published by Encyclopedia Britannica in 1952, only one work (of a collection of 130 authors and 517 works) is a treatise on chemistry (Lavoisier's *Traité élémentaire de Chimie*, Paris, 1789, first English translation, 1790) [R. Wedin, *Chemistry* (published by the American Chemical Society), Spring 2001, pp. 17–20]. Wedin surveyed a small, selected list of chemists and librarians to obtain his list of "The Great Books of Chemistry." The six books on "The Gold Shelf" included Lavoisier's *Traité*, Boyle's *The Sceptical Chymist*, Jane Marcet's *Conversations on Chemistry* (a useful and influential textbook that drew the young Michael Faraday into chemistry), Dalton's *A New System of Chemical Philosophy*, Mendeleev's *Osnovy Khimii*, and Pauling's *The Nature of the Chemical Bond*. "The Silver Shelf" comprised six additional books including Agricola's *De Re Metallica*. "The Bronze Shelf" included 12 books. Of the total of 24 books, thirteen were American publications, and two of these were published by the American Chemical Society itself. Hmmm.
16. L. Ercker, *Beschreibung Allefürnemisten Mineralischen Ertzt vnnd Bergwercks arten . . .* , Johannem Schmidt in verlegung Sigmundt Feyrabends, Frankfurt, 1580. I am grateful to The Roy G. Neville Historical Chemical Library (California) for supplying the image of the title page.
17. A.G. Sisco and C.S. Smith (transl.), *Lazarus Ercker's Treatise on Ores and Assaying* (translated by Anneliese Grünhaldt Sisco and Cyril Stanley Smith from the German edition of 1580), The University of Chicago Press, Chicago, 1951.
18. A. Greenberg, *A Chemical History Tour*, John Wiley and Sons, New York, 2000, pp. 12–22.
19. Sisco and Smith, op. cit., pp. 340–342.

A PROMISING PRESIDENT

The first English translation of Agricola's 1556 *De Re Metallica*[1] (Figure 54) was published in 1912 by Herbert C. Hoover (1874–1964),[2] the future president of the United States, and his wife, Lou Henry Hoover. It is hard to imagine a more promising future president. Born in rural Iowa to Quaker parents of extremely modest means, Herbert Hoover was orphaned by the age of nine. Although shy, he developed a very early sense of independence, rejected the choice of Quaker

GEORGIUS AGRICOLA

DE RE METALLICA

TRANSLATED FROM THE FIRST LATIN EDITION OF 1556

with

Biographical Introduction, Annotations and Appendices upon
the Development of Mining Methods, Metallurgical
Processes, Geology, Mineralogy & Mining Law
from the earliest times to the 16th Century

BY

HERBERT CLARK HOOVER

A. B. Stanford University, Member American Institute of Mining Engineers,
Mining and Metallurgical Society of America, Société des Ingénieurs
Civils de France, American Institute of Civil Engineers,
Fellow Royal Geographical Society, etc., etc.

AND

LOU HENRY HOOVER

B. Stanford University, Member American Association for the
Advancement of Science, The National Geographical Society,
Royal Scottish Geographical Society, etc., etc.

Published for the Translators by
THE MINING MAGAZINE
SALISBURY HOUSE, LONDON, E.C.
1912

FIGURE 54. ▪ Title page from the first English translation of Agricola's 1556 *De Re Metallica*, written and tested for scientific accuracy by engineer Herbert Hoover, the future President of the United States, and his wife Lou Henry Hoover, the first female geology graduate of Stanford University (courtesy Chemical Heritage Foundation).

colleges suggested to him by his relatives, and chose to attend a brand new college, Stanford University. Hoover majored in geology and met Lou Henry, the only female geology major at Stanford. They married in 1899 and remained happily united until her death in 1944. She was a smart, independent, and forceful woman, raised as a tomboy and adept at riding horses, and later became a powerful champion of women's suffrage.[2] Not long after graduating from Stanford in

1895, Herbert Hoover began a career in mining engineering and management that soon would make him wealthy and possibly the world's most famous engineer. He spent most of his time overseas during the following decades and was in China during the Boxer Rebellion (1900), where he directed relief for foreigners.[2]

The Hoovers amassed a huge and famous mining book collection and, during the course of writing their translation, performed occasional experiments to test Agricola's veracity.[3] The challenge of the Hoovers' translation is not "merely" mastery of Latin but also a profound understanding of the engineering and chemistry, which allowed them to incorporate hundreds of now-defunct terms and concepts and make sense out of them for the modern reader. Not many years after this intellectual triumph, with the outbreak of World War I Herbert Hoover was appointed head of the Allied relief operation. After the American entry into the war in 1917, he was appointed national food administrator. Hoover's efforts at increasing food production, conserving foodstocks and relieving famine in Europe were so successful that the term "hooverize"[4] entered the vocabulary as an expression symbolizing the acts of being productive, economical, and generous with foodstocks. Even more generally, it became a term for efficiency, effectiveness, and compassion.

How ironic, then, that Herbert Clark Hoover, thirty-first president of the United States (1928–1932), is now principally remembered for his failure to ease the hardships of the Great Depression. Very strict ethical values inculcated in early childhood and reinforced by his own very early independence (and subsequent success) made widespread federal aid, particularly to the urban unemployed, anathema to him.[2] He was widely regarded as distant from the suffering populace.[2] And so, sadly enough, "hooverville,"[4] a shanty town populated by the unemployed poor, is a word both more recent and more widely remembered than "hooverize."

1. H.C. Hoover and L.H. Hoover, *Georgius Agricola De Re Metallic* (translated from the first Latin edition of 1556), *The Mining Magazine*, London, 1912.
2. J.H. Wilson, *Herbert Hoover—Forgotten Progressive*, Little, Brown and Co., Boston, 1975.
3. Wilson, op. cit., pp. 22–23.
4. *Oxford English Dictionary*, second ed., Vol. VII, Clarendon Press, Oxford, 1989, p. 374.

THESE ARE A FEW OF OUR NASTIEST THINGS

It is generally agreed that gunpowder (black powder) was invented in China over a thousand years ago.[1] It is a mixture consisting of about 75% saltpetre (potassium nitrate) with the remaining 25% containing comparable quantities of charcoal and sulfur. Saltpetre was readily obtained from old dung heaps; charcoal readily made by heating vegetables or wood under oxygen-poor conditions; sulfur was found in crystalline deposits and could also be obtained by heating many metal ores. William Brock has speculated, rather ironically, I think, that the Chinese accidentally discovered gunpowder through seeking an elixir of life by a combination of the "Yin-rich saltpetre and Yang-rich sulphur."[1] There is further irony in that gunpowder held secret keys to understanding the origins of fire and the very respiration that supports life. However, these would remain hidden for almost a millennium. Early hints would be provided by Boyle, Hooke, and Mayow in the mid-seventeenth century and the riddle solved by Lavoisier over a century later.

Gunpowder was introduced quite early into Western warfare. Figure 55 is from the first Stainer edition of the ancient book on the technology of warfare authored by Flavius Vegetius Renatus.[2] This excessively rare edition, published in 1529, contains the first printed text on making gunpowder, along with directions for purifying its components.[3] Figure 56 is from a 1598 work on artillery and fireworks by Alessandro Capo Bianco,[4] Captain of Bombadiers at Crema in the Veneto. The figure depicts a sixteenth-century mill for grinding components of gunpowder.[3] In Book Ten of his *Pirotechnia* (1540),[4] Biringuccio provides detailed directions for making gunpowder. Saltpetre is derived from the "manurous" soils of barns and the floors and walls of caves (rich in bat guano), which contain calcium nitrate as a decomposition product. If the "manurous" soil, once dried, is tasted and found to be "sufficiently biting," it is suitable for use.[5] The soil is added to boiling water, and wood ashes (rich in "pearl ash" or potassium carbonate) are stirred in. The hot solution is then filtered and allowed to cool; the resulting crystalline potassium nitrate is filtered and recrystallized once more with water and a bit of nitric acid.[5] Charcoal is made preferably from willow twigs by heating over fire in a large sealed earthen pot. The components of gunpowder must be moistened before being ground together, to avoid ignition, and Biringuccio recommends slow addition of finely ground sulfur to a paste of moistened charcoal and saltpetre.[5]

Biringuccio begins his chapter on gunpowder thus:[5]

> A great and incomparable speculation is whether the discovery of compounding the powder used for guns came to its first inventor from the demons or by chance.

At many points in Book Ten, Biringuccio laments the irony that learned and decent men discover and invent explosives that maim and kill. He then dutifully describes their fabrication in full detail. For example, Book Ten, Chapter Eight is titled "The Method of Preparing Fire Pots and of Making Balls of Incendiary Composition to Be Thrown by Hand." Biringuccio begins this chapter (in the 1559 edition):[6]

> There have always been in this world men of such keen intelligence that with their discourse they have been capable of infinite and various inventions that are as beneficial as they are simultaneously harmful to the human body.

He then describes pots made of dried clay filled with course gunpowder, pitch, and sulfur, and sealed with congealed pig fat mixed with powder (see Figure 57).[7] Prior to use, a small hole is bored into the fatty seal and either a fuse or black powder placed inside. The fuse or powder is lit, the pot tossed or launched with a sling and this penetrating, sticky mass will adhere to and burn its target.

FIGURE 55. ■ Figure from the first Stainer edition (Augsburg, 1529) of the ancient work by Flavius Vegetius Renatus on the technology of warfare (courtesy The Roy G. Neville Historical Chemical Library).

Other early explosives and incendiary weapons included "Greek Fire" dating from the Hellenistic period. Chemical historian John Hudson describes "Greek Fire" as a liquid that caught fire on contact with water and speculates that calcium phosphide (from heating bones, lime and urine together), added to crude petroleum might have constituted the active ingredients.[8] Leonardo Da Vinci (1452–1519) described "Greek Fire" as consisting of charcoal, sulfur, pitch, saltpeter, spirit of wine, frankincense, and camphor boiled together and applied over Ethiopian wool.[9]

Fulminating gold (*aurum fulminans*) was first described at the beginning of the seventeenth century.[10] Gold was dissolved in an *aqua regia* derived from ammonium chloride and nitric acid. Addition of potassium carbonate led to a precipitate that, when dry, exploded readily with only the mildest application of heat. Johann Rudolph Glauber first described fulminating powder (*pulvis fulmi-*

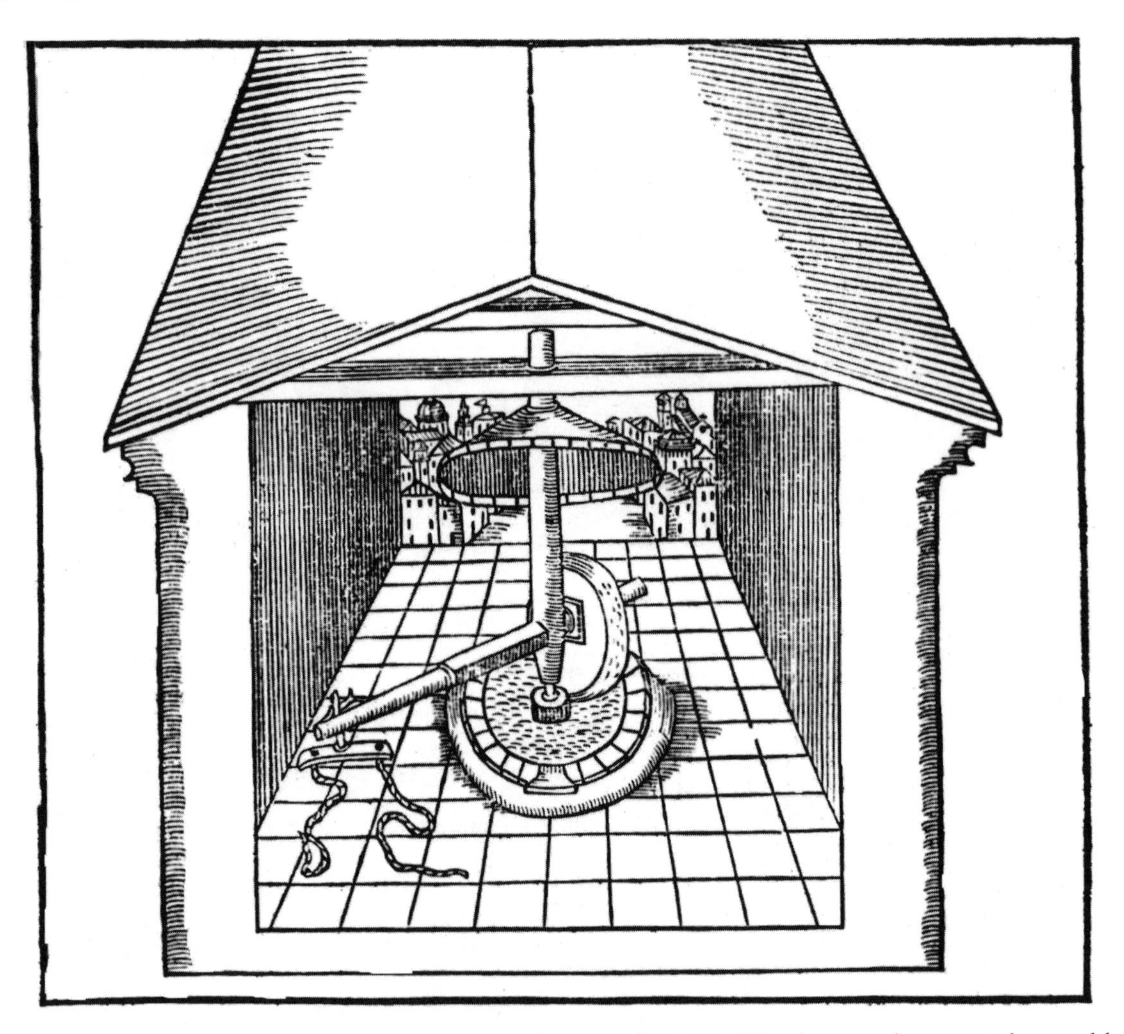

FIGURE 56. ■ Mill for grinding the components of gunpowder (ca. 75% saltpetre; the remainder roughly equal parts of charcoal and sulfur) depicted in Bianco's 1598 work on artillery and fireworks (courtesy The Roy G. Neville Historical Chemical Library).

nans) in 1648.[10] It is a mixture of potassium nitrate, potassium carbonate, and sulfur that violently explodes upon mild heating. Tenney Davis has described various similar mixtures discovered over the course of two centuries.[10] In the late seventeenth century Johann Kunckel made mercury fulminate by dissolving mercury in *aqua fortis* (nitric acid), adding spirit of wine and gently warming the mixture in horse dung.[11] The next day, the concoction exploded violently.

The nineteenth century would witness the development of nitrostarch, nitrocotton, nitroglycerin, trinitrotoluene (TNT), and pentaerythritol tetranitrate (PETN) and end with the discovery of RDX (cyclotrimethylenetrinitramine).[12] Contemporary studies of synthetic azides and picrates would also add to the armamentarium of war technology. From a modern perspective these developments appear to have grimly foreshadowed World War I. In 1867, Alfred Nobel (1833–1896), a Swedish chemist and industrialist, immobilized nitroglycerin onto diatomaceous earth, making it much safer to use, and thus made the first of many successful formulations of dynamite.[12] Just as hope often accompanies

FIGURE 57. ■ Nasty munitions made from clay pots filled with coarse gunpowder, pitch, and sulfur, and sealed with congealed pig fat mixed with powder depicted in Biringuccio's *Pirotechnia*. These are manufactured to be lighted and launched with malice using a sling. (Courtesy The Roy G. Neville Historical Chemical Library.)

tragedy, Nobel willed most of his vast fortune to establish a series of Nobel Prizes—one of which is a prize to further the cause of world peace.

---

1. W.H. Brock, *The Norton History of Chemistry*, W.W. Norton & Co., New York, 1993, p. 6. Brock notes that in Taoism, "Yang" is the male, hot principle, "Yin" is the female, cool principle. In Western alchemical beliefs, sulfur is the male principle (Sol) and mercury the female principle (Luna).
2. F. Vegetius Renatus, *Vier Bũcher der Ritterschaft . . . Mit einem zũsatz von Bũchsen geschoss, Pulver, Fewrwerck, Auff ain newes gemeeret unnd gebessert, Gedruckt durch Heinrich Stainer*, Augsburg, 1529.The author is grateful to The Roy G. Neville Historical Chemical Library for supplying a copy of the woodcut in Figure 55.
3. The Roy G. Neville Historical Chemical Library; catalogue in preparation. I am grateful to Dr. Neville for helpful discussions.
4. Alessandro Capo Bianco, *Corona e Palma Militare di Arteglieria. Nella quale si tratta dell' Inventione di essa, e dell' operare nella fattioni da Terra, e Mare, fuochi artificiati da Giucco, e Guerra; & d'un Nuovo Instrumento per misurare di stanze. Con una giunta della fortificatione Moderna, e delli errori scoperti nelle fortezze antiche, tutto a proposito per detto essercitio dell' Artiglieria, con dissegni apparenti, & assai intendenti. Nova composta, e data in luce. Dallo strenuo Capitano Alessandro Capo Bianco . . .* Appresso Gio. Antonio Rampazetto, Venice, 1598.
5. C.S. Smith and M.T. Gnudi, *The Pirotechnia of Vannoccio Biringuccio* (English transl.), The American Institute of Mining and Metallurgical Engineers, New York, 1942, pp. 409–416. This is the first English translation of Biringuccio's *De La Pirotechnia* published in Venice in 1540.
6. Smith and Gnudi, op. cit., pp. 434–435.
7. The author thanks The Roy G. Neville Historical Chemical Library (California) for supplying this image from the 1540 edition of *De La Pirotechnia*.
8. J. Hudson, *The History of Chemistry*, The Macmillan Press Ltd, Hampshire and London, 1992, p. 22.
9. J.R. Partington, *A History of Chemistry*, Macmillan and Co. Ltd., London, 1961, Vol. 2, p. 6.
10. T.L. Davis, in *Chymia*, T.L. Davis (ed.), Vol. 2, University of Pennsylvania Press, Philadelphia, 1949, pp. 99–110.
11. Partington, op. cit., p. 377.
12. D.M. Considine (ed.), *Van Nostrand's Scientific Encyclopedia*, seventh edition, Van Nostrand Reinhold, New York, 1989, pp. 1104–1105.

SECTION III
MEDICINES, PURGES, AND OINTMENTS

THE JOY OF SEXTODECIMO

It is self-indulgent, inappropriate, and simply rude to "kvell" or joyously brag about a book purchase in a "solemn" text such as this one. However, our text admits to being idiosyncratic, and self-indulgent idiocy is not a long stretch. Among the greatest joys of book collecting is "the hunt." A devoted collector will constantly be alert to quarry and prepared to pounce and feed at any opportune moment. Figure 58 displays the title page for the exceedingly rare true first English edition of the important seventeenth century text by Nicolas Le Fèvre (translated as Nicasius le Febure in the English editions). The original French edition was published in Paris in 1660, and the usual expert sources speak only of a first English edition of 1664 and a second of 1670.[1–3] While the British Museum[4] lists a copy of the 1662 edition[5] (which really constitutes the first half of the later editions), it appears to be almost unknown. My copy was purchased on a well-known World Wide Web auction site, and I stayed up three hours past my normal bedtime to win it. The English editions are in quarto (4to or 4°) format meaning each original sheet for printing has been folded twice to produce four leaves. Octavo (8vo or 8°), the most common modern book format, requires three folds and provides eight leaves, while the sextodecimo format (16mo or sixtodecimo for polite company) has sixteen leaves per original sheet.[6]

Le Fèvre had presented numerous well-regarded public lectures on chemistry in mid-seventeenth-century France and was appointed demonstrator in chemistry at the *Jardin du Roi* in 1650.[3] The mid-seventeenth century was a period of crisis throughout Europe. Failures of crops, famine, desperate poverty, frequent wars, and divided and shifting loyalties between nobles and kings, played out against a background of pervasive conflict between Catholics and Protestants.

The Reformation played out violently in sixteenth-century France and was punctuated by the Religious Wars during its latter half. A founding father of Protestant thought, Jean (John) Calvin, was born in France (1509) converted to Protestantism, and in 1534 emigrated to Geneva, where he established a model church and wrote numerous influential tracts. The Catholic king Henry II (1547–1559), ruled particularly harshly and this, in turn, influenced the ascendancy of less compromising Protestants, the French Calvinists or Huguenots. Ironically, the regency of Catherine de Médicis, the Queen Mother of Charles IX, tried to take a moderate approach, prompting a violent response from powerful Catholics and a counterresponse from the Huguenots. France was in danger of coming apart during the latter half of the sixteenth century, and a strong King, Henry IV, signed the Edict of Nantes in 1598, guaranteeing religious freedom to the Huguenots in designated parts of France and giving them the right to build fortresses (just in case).

A
Compendious Body
OF
CHYMISTRY,
Which will ſerve
As a *Guide* and *Introduction* both for underſtanding
the AUTHORS which have treated of
The *Theory* of this *SCIENCE* in general;
And for making the way Plain and Eaſie to perform,
according to Art and Method, all Operations, which
teach the *Practiſe* of this ART, upon
Animals, Vegetables, and Minerals,
without loſing any of
The ESSENTIAL VERTUES contained in them.

By *N. le FEBURE* Apothecary in
Ordinary, and Chymical Diſtiller to the King of
France, and at preſent to his Majeſty of *Great-Britain*.

LONDON,
Printed for *Tho. Davies* and *Theo. Sadler*, and is to be ſold
at the ſign of the Bible over againſt the little *North-door* of
St. *Pauls-Church*, 1662.

FIGURE 58. ▪ Title page from the exceedingly rare 1662 English edition of Le Fèvre's famous text that is generally said to be first published in 1664. Such finds are cherished by rare-book collectors who will bore their unfortunate families, relatives, and friends to death with blow-by-blow accounts of successful book hunts.

During the first half of the seventeenth century threats to stability were constant during the early reign Louis XIII (1610–1643). Early in this reign, one of the great figures in French history, Cardinal Richelieu, came to the notice of the ruling house and by 1624 had become the King's principal minister. The powerful Richelieu, whose political acumen became legendary, dedicated himself to the consolidation of regal and religious authority. He died in 1642, and Louis XIII died in 1643. Louis XIV ("The Sun King") was not yet five years old when he assumed the throne and within a brief period another all-powerful Catholic leader, Cardinal Mazarin, became the ultimate authority in France. A series of rebellions started in 1648 and were crushed by 1653. In an environment of increasing intolerance, Le Fèvre moved to London in 1660. Others, including the surgeon Moyses Charas, also emigrated during this period. Louis XIV "viewed himself as God's representative on earth and considered all disobedience and rebellion to be sinful."[7] He declared himself absolute Monarch in 1661 and in 1685 revoked the Edict of Nantes, causing great dislocation and misfortune. During his lifetime Louis established the grandeur of France perhaps best symbolized by the palace he built at Versailles—its cost was estimated to equal that of a modern municipal airport.[7] His extravagant style and arrogance probably foreshadowed the fall of the French monarchy. The king died in 1715, and his body was carried in procession to the jeers of the populace.[7]

Le Fèvre entered England in 1660 at the beginning of the Restoration of the Monarchy. Religious fervor on the part of Protestants had overthrown the monarchy in the person of Charles I in 1649. This was the culmination of religious conflict that started with the split by King Henry VIII of the Church of England from Rome in 1534. The power of Protestantism advanced under King Edward VI. However, during the reign of Queen Mary (1553–1558), Catholics assumed power, and many Protestants were killed and others fled, some to Geneva, where they were influenced by Calvin. The accession of Elizabeth I in 1558 again placed Protestants in power, but her moderate treatment disappointed more radical sects, some of whom came to be called "Puritans." The Puritans sought to "purify" Protestantism from the last traces of Catholicism, and they placed the rulers of England under increasing pressure. Some of these Puritan groups emigrated to America, establishing communities in Virginia and New England. The pressure mounted during the reign of Charles I (1625–1649) and culminated in a military coup that overthrew the monarchy and turned power over to the military leader Oliver Cromwell. The Great Persecution occurred toward the end of this decade-long period and moderate Puritans finally helped to restore the monarchy and Charles II ascended the throne.

Starting around 1645, scholars from London and Oxford and other colleges began to meet in what came to be called the "Invisible College." This loose organization evolved into the Royal Society of London for the Promotion of Natural Knowledge, founded in 1660 and chartered by Charles II in 1662. It is not clear how enthusiastic the king was about his Royal Society, but it kept the Puritans who dominated the universities and their faculties occupied. In 1663 Le Fèvre became one of the initial members of the Royal Society.[3] England had become a beacon for learned men.

While on religious topics, it is amusing to read Le Fèvre's introduction in which he explores the antiquity of chemical knowledge:[8]

> and so we have upon record, that Moses took the Golden Calf, an Idol of the Israelites, did calcine it, and being by him reduced to powder, caused those Idolators to drink it, in a reproach and punishment of their sin. But no body, how little soever initiated in the mysteries of this Art, can be ignorant, that Gold is not to be reduced to powder by Calcination, unlesse it be performed either by immersion in Regal Waters, Amamulgation with Mercury or Projection; all of which three Operations are only obvious to those which are fully acquainted both with the Theorical and Practical part of Chimistry.

So, did Moses really calcine the Golden Calf? In the thirteenth century BCE it appears beyond reasonable doubt that *aqua regia* ("regal waters") was unknown. Alchemical projection would not be in vogue for at least another millennium or so. Amalgamation was a chemical possibility, but we assume that Moses wanted to punish his people, not poison them. The possibility that the calf was really made of marble (limestone) has been suggested.[9] Thus, a drink of the powdered calf would have been an excellent treatment for upset stomachs following prolonged hedonistic partying (see "There Is Truth in Chalk," p. 142).

1. J. Ferguson, *Bibliotheca Chemica*, Vol. II, Derek Verschoyle, London, 1954 (reprint of original edition of 1906), pp. 17–18.
2. D. Duveen, *Bibliotheca Alchemica et Chemica*, H&S Publishers, Utrecht, 1987 (reprint of original 1949 edition), pp. 345–346.
3. J. Read, *Humour and Humanism in Chemistry*, G. Bell and Sons Ltd., 1947, pp. 101–114.
4. British Museum Dept. of Printed Books. *General catalogue of printed books*. London Trustees, 1959–1966.
5. N. le Febure, *A Compendious Body of Chymistry Which will serve as a Guide and Introduction both for understanding the Authors which have treated of the Theory of this Science in general; And for making the way Plain and Easie to perform, according to Art and Method, all Operations, which teach the Practice of this ART, upon Animals, Vegetables, and Minerals, without losing any of the Essential Vertues contained in them*, Thos. Davies and Theo. Sadler, London, 1662.
6. J. Carter, *ABC for Book Collectors*, fifth edition, revised, Alfred A. Knopf, New York, 1987, pp. 100–101.
7. *The New Encyclopedia Brittanica*, Vol. 7, Encyclopedia Brittanica, Inc., Chicago, 1986, pp. 500–501.
8. le Febure, op. cit., p. 2.
9. *The New Encyclopedia Brittanica*, Vol. 24, Encyclopedia Brittanica, Inc., Chicago, 1986, p. 374.

THE COMPLEAT APOTHECARY

Le Fèvre's book *A Compendious Body of Chymistry* (see previous essay) was noteworthy for its clarity concerning the construction of apparatus and execution of chemical operations. In Figure 59a we see a "superdeluxe" philosopher's furnace or athanor with all of its accessories. Clearly, Charles II royally supported his "Royal Professor in Chymistry" and "Apothecary-in-Ordinary" to the royal household.[1] (One can only imagine the negotiations for "start-up monies" and moving expenses to bring this young professor from Paris. Was immediate tenure part of the package? Was a faculty committee involved or was tenure granted by Royal Decree?)

(a) (b)

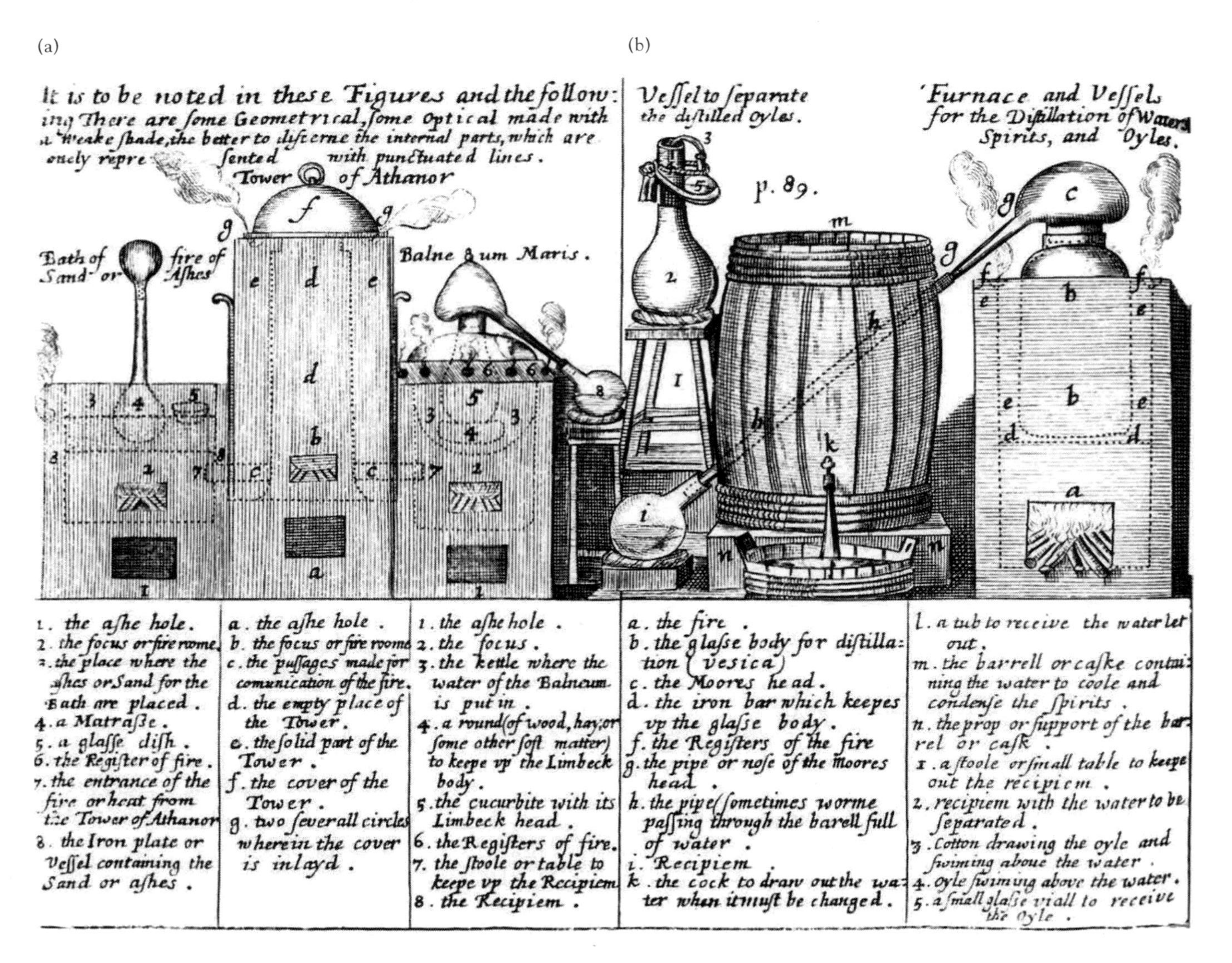

FIGURE 59. ■ Multipurpose stills in Le Fèvre's A *Compendious Body of Chymistry* (1662).

The heat from the athanor was communicated as needed to the *balneum maris* (*balneum marie* or *bain marie*—hot-water bath) accessory on the right and also to the sand bath accessory on the left in Figure 59a. Both the *balneum maris* and sand bath accessories had their own furnaces for specialized operations. The athanor itself was commonly used for operations involving a sealed vessel or philosopher's egg.

Figure 59b depicts an apparatus for distilling alcohol and other volatile spirits. The long, straight "worm" *h* descends from the "Moores head" *c* through a barrel filled with cold water in order to condense the distillate. When water was added, as a "menstruum," to crushed herbs, flowers, or animal parts, an oily substance, sometimes steam-distilled to form an upper layer over the water collected in receiver *i*. The oil was collected through capillary action by dipping cotton into the oil layer and having it drain into the small glass vial 5.

The lamp furnace (Figure 60a), "used by the most curious Artists for many Chymical Operations," was made of clay and designed to carefully control more modest degrees of heat. Control was performed through the screwdrive attached to lamp *b* as well as by the number of wicks burned simultaneously in the lamp. The most fascinating aspect of this figure is instrument n, the "Weatherglass, Thermometer, or Engin to judge of the quality or degrees of heat." Thermometry was in its infancy—the nature of heat was not understood. Boyle explained that an air current was cooler than standing air because it "drove away the 'warm streams of the body' that normally shielded the skin from the ambient cold" and also apparently penetrated the pores of the skin more than calm air.[2] Le Fèvre's thermometer contained some water in the lower (righthand) bulb, a strip of dyed water in the lower loop, and a hole in the upper bulb. The lower bulb would be inserted in the part to be sensed, and the heat of the water in the bulb would be transmitted to the air that would move the column of dyed water. The purpose was to improve the reproducibility of chemical operations. Some 60 years later, Herman Boerhaave would make the thermometer a standard part of chemical operations.[2,3] The sublimatory furnace (Figure 60b) had a number of condensing vessels cooled by surrounding air. The least volatile sublimates would be largely collected in f and the most volatile in *k*.

The wind furnace (Figure 61a) was used for mineral and metallic fusions and vitrifications, and was particularly useful for obtaining the pure form ("regulus") of a metal. The crucible d was made of iron or clay. The alembic (limbeck) in Figure 61b was cooled in a cold-water bath rather than air as in Figure 59a. Figure 62a shows a shelf of glassware, including my personal favorite the double-pelican (5) suggestively symbolized by Porta in 1608 as a man and woman mutually circulating bodily fluids.[4] Item 3 in Figure 62a was termed an "infernell glass" or "hell" since nothing introduced escaped. That is also the function of a philosopher's egg. Indeed, item 9 is LeFèvre's own *Ovum in Ova* ("egg within an egg")—an apparatus that also shares the function of 3. We hesitate to call this Le Fèvre's "private little hell."

The apparatus in Figure 62b has a built-in efficiency of two collection vessels. This double alembic or distilling head can be made of iron if vegetables are distilled or steam-distilled. However, the distillation of oil of vitriol and other acidic substances (Figure 63)[5] requires tin or tin-lined vessels. Distillation of mercury, Le Fèvre notes, can never employ metallic vessels since amalgamation will occur. The diarist Samuel Pepys visited Le Fèvre's laboratory on January 15,

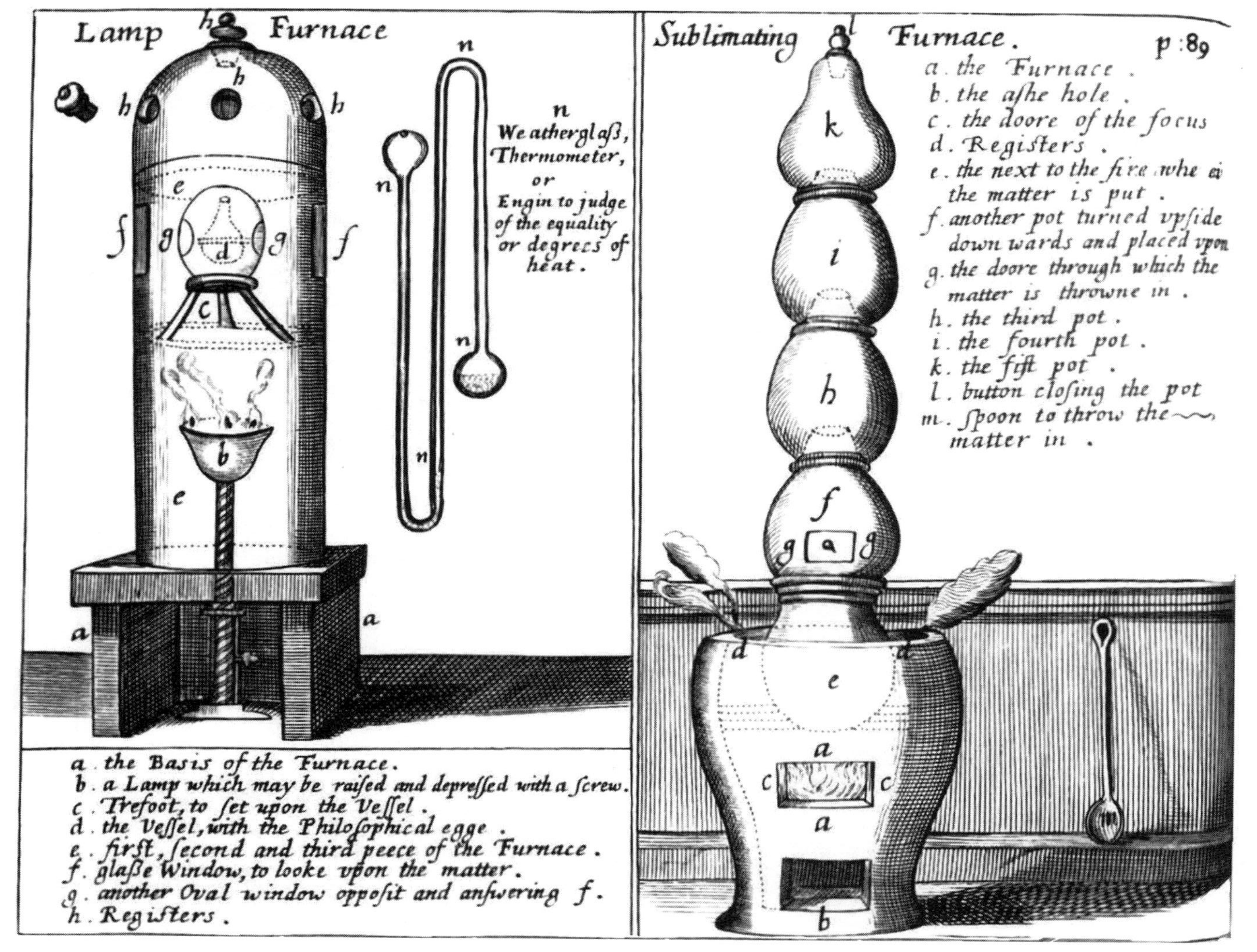

FIGURE 60. ▪ The lamp furnace on the left employed an early thermometer (n) to render operations more reproducible. The sublimates obtained in the vessel on the right were condensed through air cooling. (From Le Fèvre's *A Compendious Body of Chymistry*.)

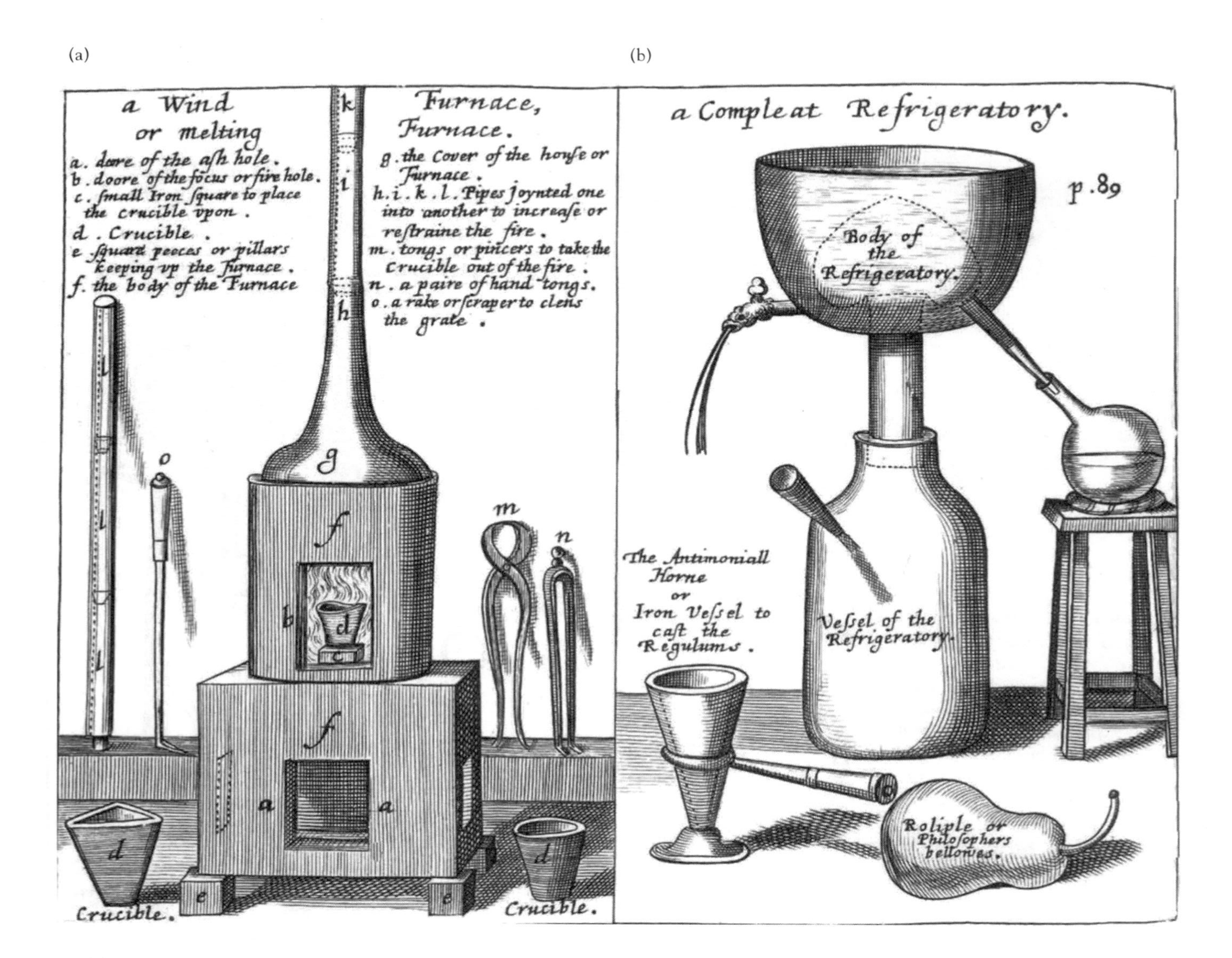

FIGURE 61. ▪ The wind furnace on the left was used to obtain the regulus (purest form) of various metals; the alembic on the right used cold water rather than air to condense distilled "spirits" (from Le Fèvre's *A Compendious Body of Chymistry*).

(a) (b)

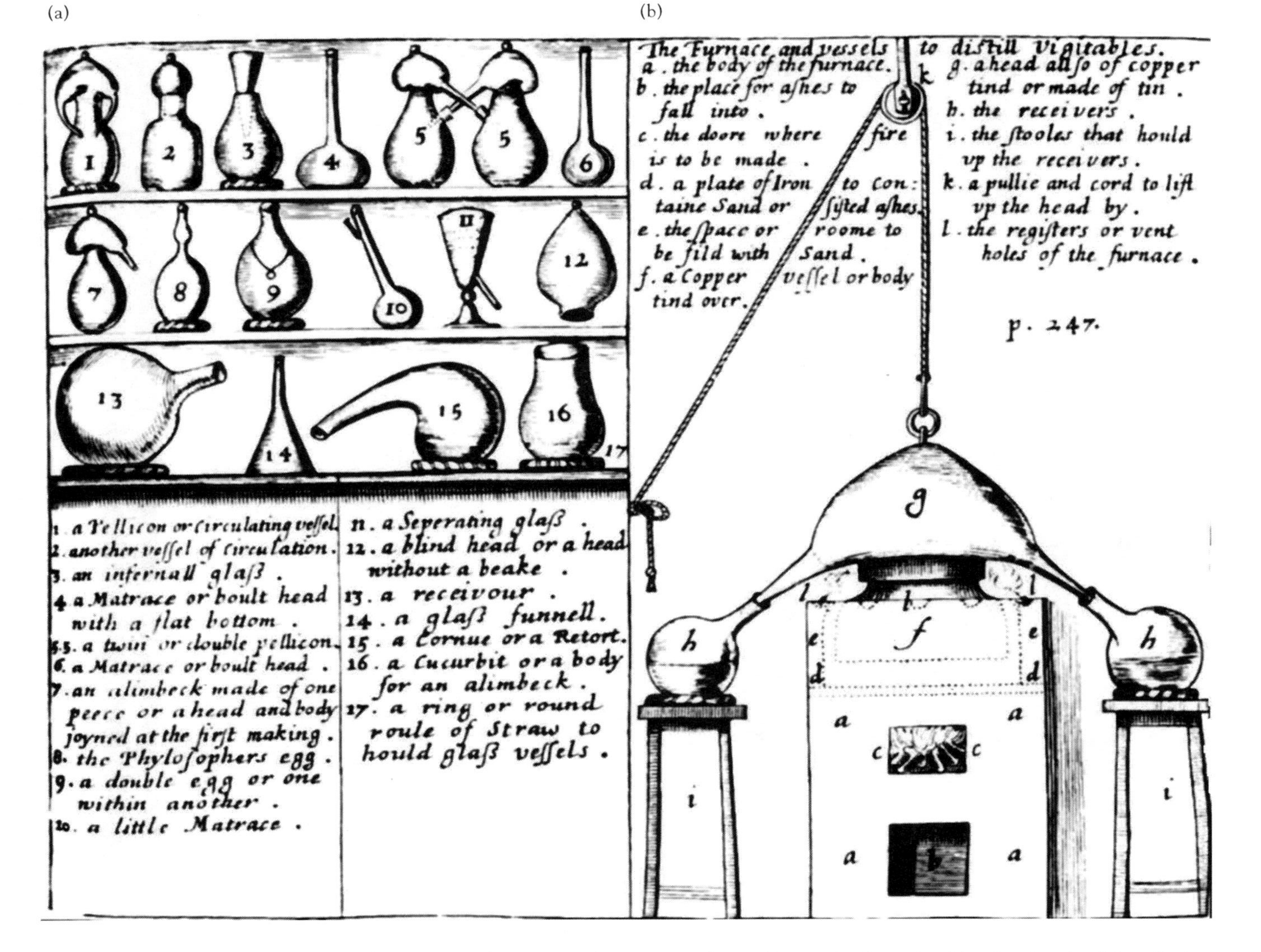

FIGURE 62. ■ A nice shelf of mid-seventeenth-century glassware and an efficient-looking double alembic (still head) that allows the entire apparatus to be lifted off of the furnace when such control is desired (from Le Fèvre's *A Compendious Body of Chymistry*).

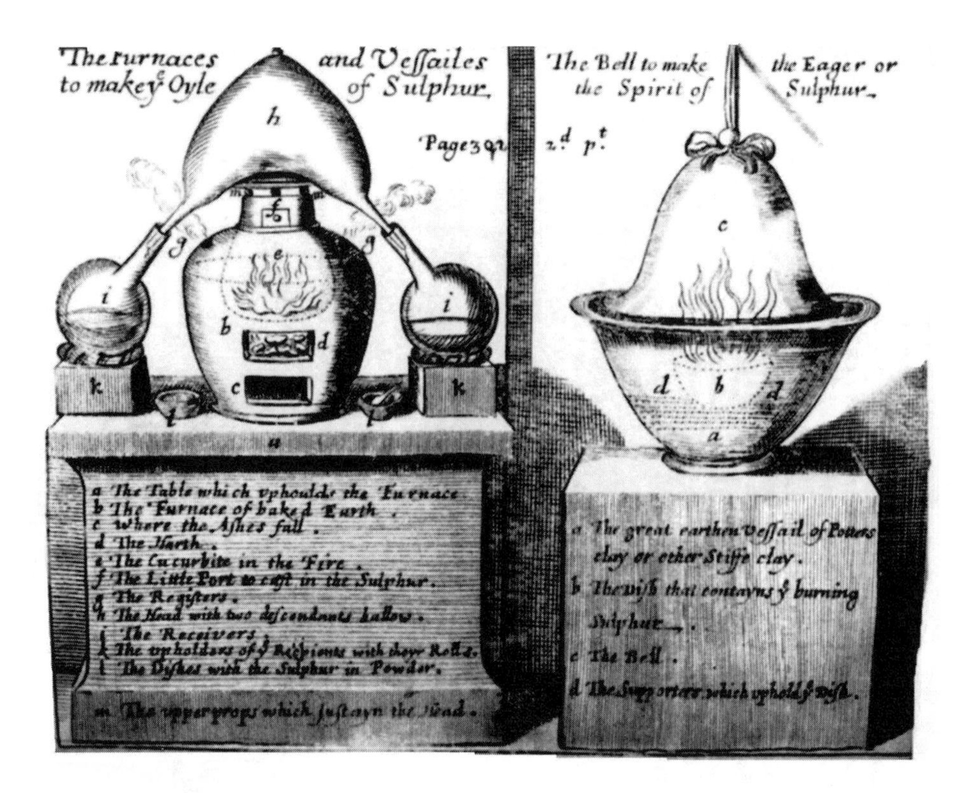

FIGURE 63. ■ Distillation of oil of vitriol (sulfuric acid) in tin-lined vessels (from Le Fèvre's 1670 edition, *A Compleat Body of Chymistry*).

1669 "and there saw a great many chemical glasses and things, but understood none of them."[1]

---

1. J. Read, *Humour and Humanism in Chemistry*, G. Bell and Sons Ltd., London, 1947, pp. 101–114.
2. J. Golinski, In *Instruments and Experimentation in the History of Chemistry*, F.L. Holmes and T.H. Levere (eds.), The MIT Press, Cambridge, MA, 2000, pp. 185–210.
3. A. Greenberg, *A Chemical History Tour*, John Wiley & Sons, New York, 2000, p 117.
4. Greenberg, op. cit., p. 39.
5. Figures 60–62 are common to all three of the English editions of Le Fèvre's book (1662, 1664, and 1670) while Figure 63 is not in the 1662 edition that is in reality the first half only of the latter two editions.

"RARE EFFECTS OF MAGICAL AND CELESTIAL FIRE"

Antimony is readily released from its ore stibnite (Sb_2S_3) by heating with iron and has been known for centuries. It may also be roasted and the oxide, thus formed, heated with charcoal to obtain the metal.[1] But it has also been "a puzzlement" for centuries. While it is a silvery-white, brittle metal, antimony has a number of allotropes (differing arrangements of atoms as in diamond and graphite—carbon allotropes) that differ in properties. Indeed, rapid condensation of antimony vapor produces a soft yellow, nonmetallic solid that changes spontaneously to the metal in sunlight. Moreover the oxides (calxes) of antimony also exhibit interesting properties. Antimony burns with a bright blue flame, producing a vapor that rapidly condenses into a white powder (Sb_2O_3). This oxide dissolves in both acids and bases—remember that oxides of nonmetals such as phosphorus and sulfur are acidic while oxides of metals are typically basic.[2] Metallic oxides such as "rust" are not volatile while Sb_2O_3 is. Do you remember the wolf in Maier's *Atalanta Fugiens* (see Figure 17)? It represents stibnite (sometimes antimony itself). The final purification of the king (gold) in the fire occurs because the antimony alloyed with gold burns to form its volatile oxide, which sublimes. Obviously, such a metal *must* have powerful medicinal properties. Indeed, tartar emetic (salt of potassium tartrate and antimony) is a potent purgative (as antimony itself is), although these substances also exhibit toxicity. Basil Valentine's *Triumphal Chariot of Antimony* celebrated the medicinal value of antimony.[3]

Nicolas Le Fèvre performed careful quantitative work and discovered that calcinations involving sunlight amazingly increased the mass of antimony:[4]

> But those that are ignorant of the noble Works and rare Effects of Magical and Celestial Fire, drawn from the Rayes of the Sun, by the help of a Refracting or Burning-Glass, shall scarce believe that which we have to say, and are to demonstrate upon this Subject.

The "burning glass" is designed to be three to four feet in diameter (although that in Figure 64 is much smaller), made with two concave pieces of glass, filled with water and sealed with fish glue.[4] Le Fèvre was aware that calx of antimony could also be made using niter (e.g., by adding the metal to nitric acid and heating).[4] In this, he was anticipating Mayow's discovery a decade later (see later essay). However, the incompleteness of the reaction, which forms mixtures in any case, and losses during recovery were vastly inferior to the results achieved with divine sunlight. Crude burning of 12 grains of antimony is found to produce white smoke (said by Le Fèvre to be nothing more than "sublimated" antimony) and calx weighing only 6–7 grains. The calx so produced still has the emetic (nauseative) qualities of antimony powder, albeit somewhat reduced. On the other hand, absorption of "magic and celestial fire" produces 15 grains of calx from 12 grains of antimony.[4] Indeed, we know that conversion of 12 grains of antimony should produce about 14.5 grains of Sb_2O_3.

> but that which is yet more to be admired, and less conceivable, is, that these xv grains of white Powder, are neither vomitive nor purging, but contrariwise Diaphoretical and Cordial; which doth cast into admiration, not without reason, the most curious and intelligent searchers of Nature, and the wisest Physicians.

Diaphoretical substances induce perspiration—a more gentle purge than vomiting.

FIGURE 64. ▪ Calcination of antimony using "celestial fire" (from Le Fèvre's 1670, *A Compleat Body of Chymistry*). Le Fèvre was among the earliest chemists to discover that the calx was heavier than the corresponding pure metal.

Le Fèvre was not the first to discover that calxes were heavier than the metals. Biringuccio noted it in 1540 (see p. 60) and Jean Rey had made this discovery almost a century later. Like Boyle, who also made this discovery, Le Fèvre assumed some kind of incorporation of fire (for Boyle, it was "igneous particles") to augment the weight of the metal.[5]

1. F.A. Cotton and G. Wilkinson, *Advanced Inorganic Chemistry*, fifth edition, John Wiley and Sons, New York, 1988, pp. 387–388, 401.
2. T.L. Brown, H.E, LeMay, Jr., and B.E. Bursten, *Chemistry—the Central Science*, seventh edition, Prentice-Hall, Upper Saddle River, NJ, 1997, pp. 841–843, 847–848.
3. A. Greenberg, *A Chemical History Tour*, John Wiley and Sons, New York, 2000, pp. 94–95.
4. N. le Febure, *A Compleat Body of Chymistry: Wherein is contained whatsoever is necessary for the attaining of the Curious Knowledge of this Art; Comprehending in General the whole Practice thereof; and Teaching the most exact preparation of Animals, Vegetables and Minerals, so as to preserve their Essential Vertues. Laid open in two Books, and Dedicated to the Use of all APOTHECARIES, &c.*, O. Pulleyn Junior, London, 1670, pp. 212–217.
5. Greenberg, op. cit., pp. 109–111.

SECRETS OF A LADY ALCHEMIST

Very little is known of the life of Marie Meudrac,[1,2] but it appears that she was the first woman to author a printed book on chemistry. *La Chymie Charitable et Facile, en Faveur des Dames*, was first published in 1656 followed by editions of 1674 and 1687.[3] The title page and frontispiece from the third (French) edition are shown in Figures 65 and 66.[4] There were also at least six German editions and one Italian edition.[3,5]

The Rayner-Canhams note that her work was based on the three alchemical principles, sulfur, mercury and salt, but presented clear discussions of useful chemical operations.[1] The book consists of six parts:[1,2]

Part 1—Principles and operations

Part 2—"Simples"—methods of preparation and treatment

Part 3—Animals

Part 4—Metals

Part 5—Making compound medicines

Part 6—Preserving and increasing the beauty of ladies

Meudrac's doubts about publishing her work were summarized in her preface:[1,2]

> When I began this little treatise, it was solely for my own satisfaction and for the purpose of retaining the knowledge I have acquired through long work and oft-repeated experiments. I cannot conceal that upon seeing it complet-

LA
CHYMIE
CHARITABLE
ET FACILE,
EN FAVEUR DES DAMES.
TROISIÉME EDITION.
Rêveüe & augmentée de plusieurs Pré-parations nouvelles & curieuses.

A PARIS,
Chez LAURENT D'HOURY, ruë
Saint Jacques, devant la Fontaine
S. Severin, au Saint Esprit.

M. DC. LXXXVII.
Avec Privilege & Approbation.

FIGURE 65. ▪ The third edition (1687) of the first chemistry book published by a woman, Marie Meurdrac. The first edition was published in 1656. This wonderful pocket-size book was dedicated to Madame La Comtesse De Guiche and includes a sonnet to Mademoiselle Meurdrac's book by a Mademoiselle D.I.

> ed better than I had dared hope, I was tempted to publish it: but if I had reasons for bringing it to light, I also had reasons for keeping it hidden and for not exposing it to general criticism.

Apparently Ms. Meudrac also kept chemistry apparatus mysteriously hidden behind her exquisite curtain, if the book's frontispiece (Figure 66) is taken as a clue. But further in her preface[1,2]

> On the other hand, I flattered myself that I am not the first lady to have something published; that minds have no sex and that if the minds of women were cultivated like those of men, and that if as much time and energy were used to instruct the minds of the former, they would equal those of the latter.

FIGURE 66. ■ The enchanting frontispiece from Marie Meudrac's 1687 chemistry book (see Figure 65) promising to disclose hitherto hidden secrets of chemistry.

This wonderfully assertive nonapology for daring to author a book precedes by about 140 years a similar nonapology by the brilliant Elizabeth Fulhame[6] introducing her own *Essay on Combustion* in 1794.[7]

1. M. Rayner-Canham and G. Rayner-Canham, *Women in Chemistry—Their Changing Roles from Alchemical Times to the Mid-Twentieth Century*, American Chemical Society and Chemical Heritage Foundation, Washington, DC and Philadelphia, 1998, pp. 9–10.
2. L.O. Bishop and W.S. DeLoach, *Journal of Chemical Education*, Vol. 47, pp. 448–449 (1970).
3. D.I. Duveen, *Bibliochemica Alchemica et Chemica*, facsimile edition, HES Publishers, Utrecht, 1986, p. 401.
4. I am grateful to Ms. Elizabeth Swan, Chemical Heritage Foundation, for supplying these images.
5. J. Ferguson, *Bibliotheca Chemica*, facsimile edition, Vol. 2, Derek Verschoyle, London, 1954, pp. 92–93.
6. Rayner-Canham, op. cit., pp. 28–31.
7. A. Greenberg, *A Chemical History Tour*, John Wiley & Sons, New York, 2000, pp. 156–159.

"PRAY AND WORK"

Any chemist who has ever "punted" understands the poetic dictum of Saint Benedict of Montecassino:[1,2] "Ora et Labora" ("Pray and Work"). A series of well-planned, rational experiments may fail to yield an expected product, while a well-placed "punt"[3] ("going for broke"—with a one-step, less rational . . . vacuum sublimation, for example) sometimes works. The chemist depicted in Figure 67 may be trying just such a "punt" and pays homage to God whose all-encom-

FIGURE 67. ▪ "Work and pray"—helpful advice for chemists who are not atheists or agnostics (from the 1755 *Medicinisch-Chymisch-und Alchemistisches Oraculum*).

Aqua pluuialis. — Regen-Wasser.

Aqua regis. — Goldscheid-Wasser.

Aqua vitae. — Aquavit, Lebens-Wasser.

Arena. — Sand.

Argentum, luna. — Silber.

Argen-

FIGURE 68. ■ Chemical hieroglyphics from the 1755 *Oraculum* (see Figure 67).

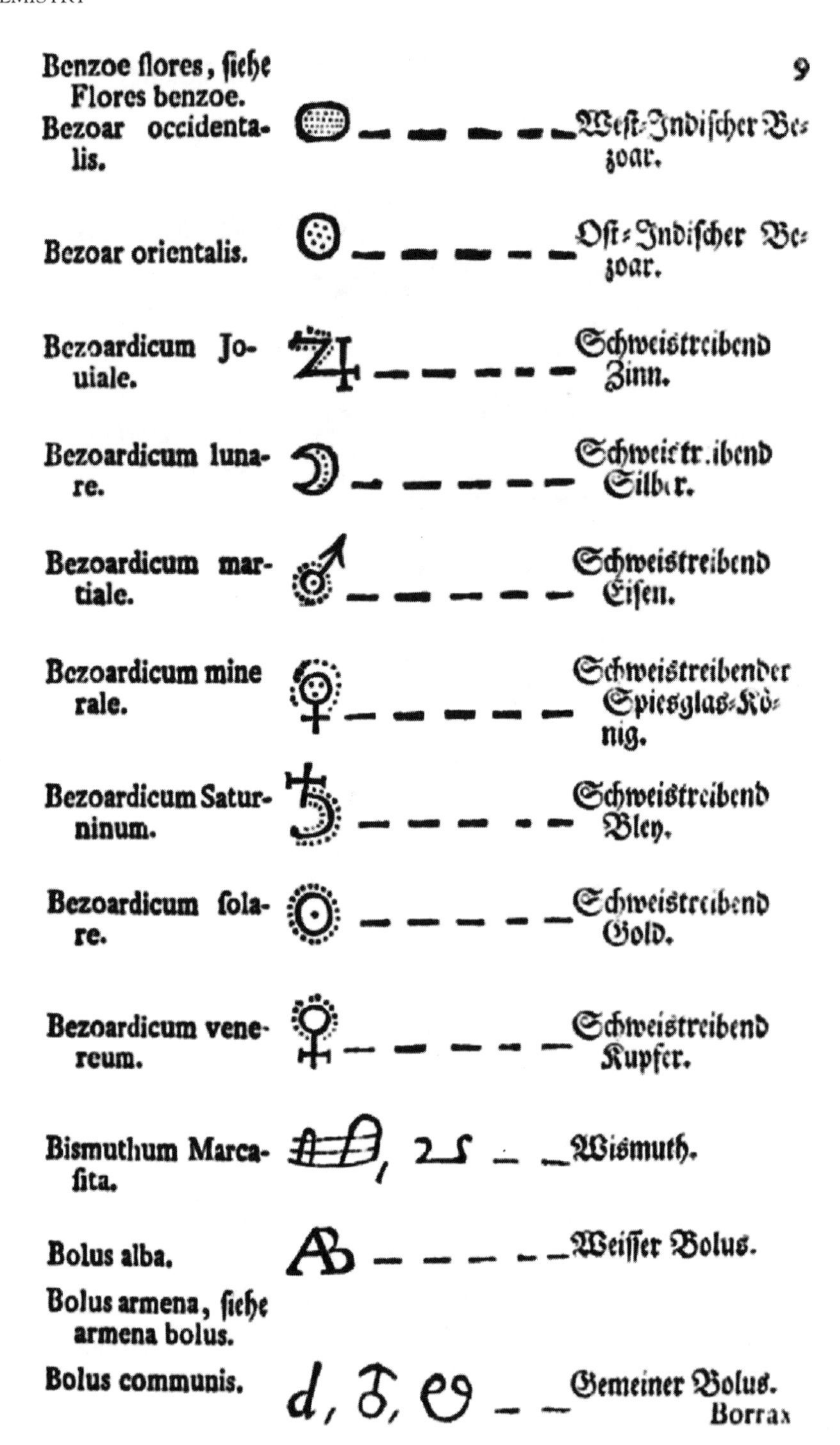

9

Benzoe flores, ſiehe Flores benzoe.	
Bezoar occidentalis.	Weſt-Indiſcher Bezoar.
Bezoar orientalis.	Oſt-Indiſcher Bezoar.
Bezoardicum Jouiale.	Schweistreibend Zinn.
Bezoardicum lunare.	Schweistreibend Silber.
Bezoardicum martiale.	Schweistreibend Eiſen.
Bezoardicum minerale.	Schweistreibender Spiesglas-König.
Bezoardicum Saturninum.	Schweistreibend Bley.
Bezoardicum ſolare.	Schweistreibend Gold.
Bezoardicum venereum.	Schweistreibend Kupfer.
Bismuthum Marcaſita.	Wismuth.
Bolus alba.	Weiſſer Bolus.
Bolus armena, ſiehe armena bolus.	
Bolus communis.	Gemeiner Bolus.

Borrax

FIGURE 69. ▪ Additional chemical hieroglyphics from the 1755 *Oraculum* (see Figure 67). "Bezoar" may be defined as "hard masses deposited around foreign masses found in the walls of stomachs or intestines" of animals, especially ruminants.

passing wisdom is captured in his reaction vessel—a philosopher's egg nestled into an athanor (philosophical furnace). He even captures sunlight, as "philosophical fire," through a magnifying lens. The words passing between the sun and the chemist translate as "Without me you can do nothing" (possibly paraphrasing the word of the Lord).[1] And from the hand of God straight to the flask, we learn "The beginning of wisdom is the fear of God."[1] This delightful full-page woodcut comes from a German pamphlet published in 1755.[4]

This pamphlet, "Medical, Chemical and Alchemical Oraculum" (an "oraculum" is a "divine announcement"), consists of two parts. The second part is said to be a previously unpublished fourteenth-century manuscript.[5] The first part includes a 33-page table of chemical symbols (hieroglyphics?[6]), accompanied by Latin and German definitions (see Figures 68 and 69). For example, there are 9 symbols collected for *aqua regis* (*aqua regia*), the 3:1 solution of hydrochloric acid and nitric acid that "dissolves" gold (actually it oxidizes or "calcines" gold to $AuCl_4^-$: no oxygen is involved). Most of these symbols are representations of water (either a downward-pointing triangle or waves) appended to an "R." *Aqua vita* (nitric acid) has at least 20 symbols perhaps reflecting differences in chemical properties (it is an acid as well as an oxidizing agent), origins and uses. Figure 68 depicts 17 symbols for silver while the pamphlet also includes 34 discrete symbols for gold, 35 for arsenic, 40 for "quicksilver" (mercury), and no less than 54 for various preparations of the fabulous antimony!

Figure 69 depicts symbols for the "bezoardic forms" of the seven ancient metals. "Bezoar" may be defined as "a hard mass deposited around a foreign substance, found in the stomach or intestines of some animals and formerly thought to be a remedy for poisoning."[7] Apparently, ruminants were particularly prized because of the complicated digestions in their chambered stomachs. I suspect that, were I to be poisoned, I would be willing to spare the life of a goat, forgo a dose of bezoardic gold—as valuable as it is, and take my chances with a good, old-fashioned purge.

---

1. I thank Professors Heinz D. Roth and Pierre Laszlo for help in the interpretation of Figure 67.
2. *The New Encyclopedia Britannica*, Vol. 2, Encyclopedia Britannica, Inc, Chicago, 1986, p. 97.
3. It is painful to confess, but over 20 years ago I found that rational, well-precedented syntheses failed to produce an exciting and previously unknown lactam. In desperation, I tried a thermal dehydration of the precursor amino acid under vacuum. Crystals of apparent product were obtained, and they were . . . the starting material only. Drat! Almost simultaneously, and completely independently, a research group at another university also found that the precedented reactions failed and these scientists also "punted" with a thermal dehydration under vacuum. They employed slightly different conditions than mine and were rewarded with a tiny yield of the desired compound (see H.K. Hall, Jr. and A. El-Shekeil, *Chemical Reviews*, Vol. 83, p. 549, 1983, which reviews this synthesis and their other outstanding work in this field). Perhaps I should have employed "philosophical" apparatus rather than simple glassware.
4. *Medicinisch-Chymisch-und Alchemistisches Oraculum, darinnen man nicht nur alle Zeichen und Abkürzungen, welche sowohl in den Recepten und Büchern der Aerzte und Apotheker, als auch in den Schriften der Chemisten und Alchemisten vorkommen, findet, sondern dem auch ein sehr rares Chymisches Manuscript eines gewissen Reichs*** beygefüget*, Ulm und Memmingen, in der Gaumischen Handlung, 1755.
5. *Bibliotheca Alchemica Et Chemica*, H&S Publishers, Utrecht, 1986, p. 440. This is a reprint of the

book published in 1949 by E. Weil, London and supplemented by Catalogue 62, H.P. Kraus, originally printed in 1953 by H.P. Kraus, New York.

6. A. Roob, *The Hermetic Museum: Alchemy and Mysticism*, Taschen, Cologne, 1997, p. 600.
7. *Webster's New World Dictionary of the American Language—College Edition*, The World Publishing Co., Cleveland and New York, 1964, p 143.

A GOOD OLD-FASHIONED PURGE

Paracelsus revolutionized Renaissance medicine through his use of synthetic metallic drugs. For example, he employed calomel (Hg_2Cl_2)[1] as a purgative, and it is easy to imagine unburdening the *archeus*[2] by ridding the body of "ill humours" as well as intestinal worms. The *London Pharmacopoeia*,[3] first authorized by the Royal College of Physicians in London and published in 1618, listed a synthesis of calomel in which mercury was dissolved in *aqua fortis* (nitric acid), sea salt added, and the precipitate collected and washed with water. Violent purging of the bowels was one means for cleansing the body. Another was through emetics, substances that induced repeated vomiting. Again, it is easy to imagine how medically helpful this would be for clearing the stomach of tainted food or poison. Perhaps the best known was tartar emetic, antimony tartrate. A preparation of this substance provided by Hadrian Mynsicht involved boiling cream of tartar (potassium hydrogen tartrate obtained from wine dregs) with antimony carbonate and allowing crystallization in a cool place.[4] The use of antimony for medicinal purposes stirred up great controversy in the mid-seventeenth century since antimony compounds were known to be quite toxic.[5] However, the "Chariot of Antimony" did indeed triumph, and antimony compounds were used as purgatives, emetics, and sudorifics ("sweating" agents).[5] Books throughout the sixteenth and seventeenth centuries also referred to *Aurum Potabile* ("potable gold").[6] This was considered to be the universal medicine, tincture of gold, suitable for treating (and improving) animals, plants, and *minerals*. Logically speaking, it may have been a dilute solution of gold dissolved in *aqua regia* (hydrochloric acid–nitric acid, 3 : 1).

Figures 70 and 71 are from Annibal Barlet's 1657 *Le Vray et Méthodique Cours de la Physique*.[7] Little is left to the imagination concerning the efficacy of antimony compounds as violent emetics (as well as effective purgatives). Figure 70 also suggests mercury salts as all-purpose purges. The remaining six panels in Figures 70 and 71 depict the metallurgy of iron (Mars), copper (Venus), lead (Saturn), tin (Jupiter), silver (*Lune* or the moon), and gold (*Soleil* or the sun). Barlet apparently taught chemistry but made no contribution to the field.[8] Although the book was said by contemporaries to be of little value, its illustrations of laboratories treating animal extracts (Figure 72), plant extracts (Figure 73), mineral chemistry (Figure 74) and metallurgical chemistry (Figure 75) provide some insights into seventeenth-century chemical operations.

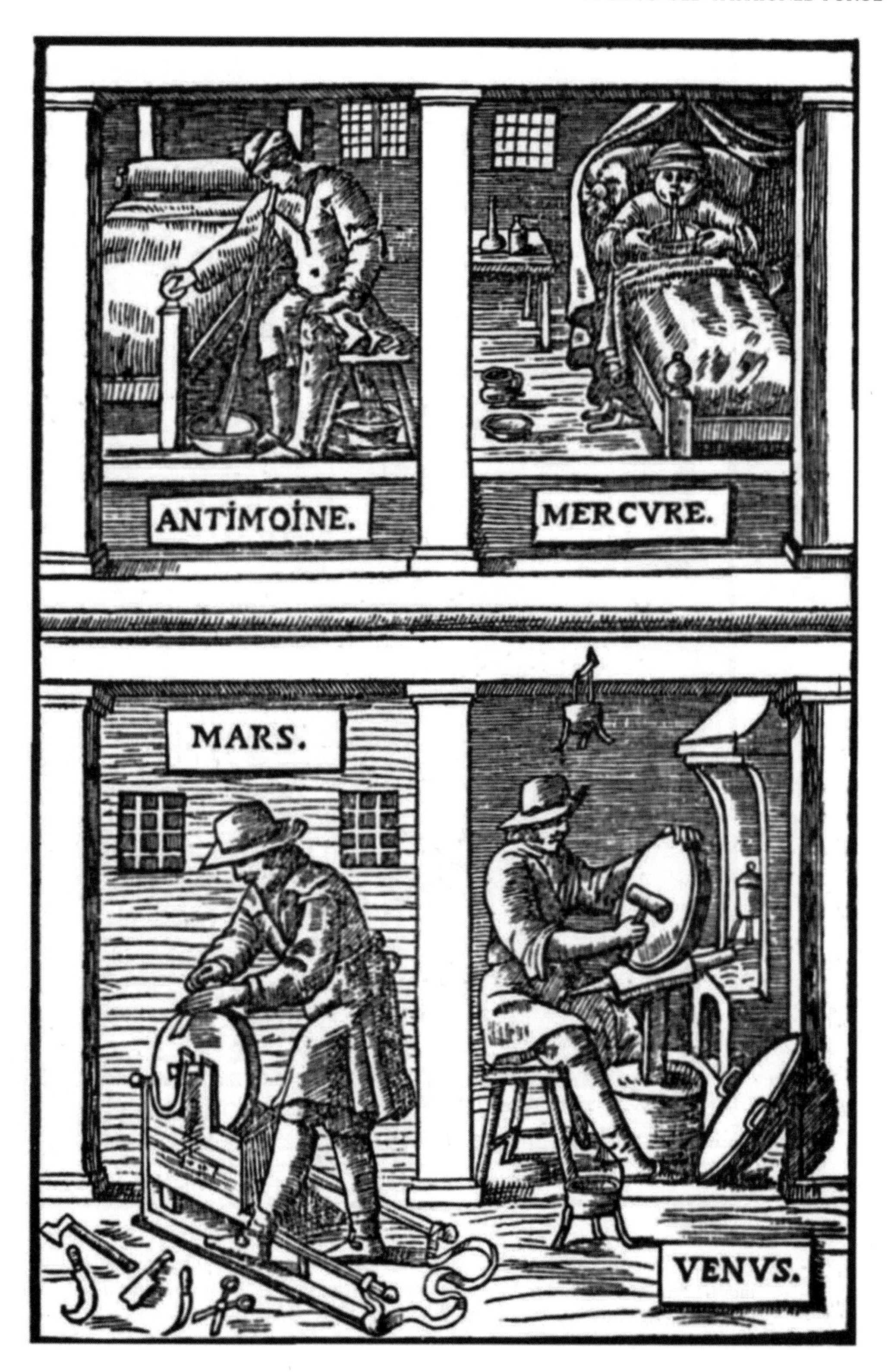

FIGURE 70. ■ Illustration of uses for the metals antimony, mercury, iron ("Mars"), and copper ("Venus"). Clearly, both antimony and mercury compounds are effective emetics, but a closer look at this figure suggests that some antimony compounds are effective laxatives (purgatives) as well. (From Barlet's 1657 *Le Vray Méthodique Cours de la Physique*, courtesy The Roy G. Neville Historical Chemical Library.)

FIGURE 71. ▪ Applications for lead ("Saturne"), tin ("Jupiter"), silver ("Lune"—moon), and gold ("Soleil"—sun). (From Barlet's 1657 *Le Vray Méthodique Cours de la Physique*, courtesy The Roy G. Neville Historical Chemical Library.)

FIGURE 72. ■ A seventeenth-century laboratory for processing animal products (from Barlet's 1657 *Le Vray Méthodique Cours de la Physique*, courtesy The Roy G. Neville Historical Chemical Library).

FIGURE 73. ▪ A seventeenth-century laboratory for processing plant products (from Barlet's 1657 *Le Vray Méthodique Cours de la Physique*, courtesy The Roy G. Neville Historical Chemical Library).

FIGURE 74. ■ A seventeenth-century laboratory for processing minerals (from Barlet's 1657 *Le Vray Méthodique Cours de la Physique*, courtesy The Roy G. Neville Historical Chemical Library).

FIGURE 75. ■ A seventeenth-century laboratory for processing metals (from Barlet's 1657 *Le Vray Méthodique Cours de la Physique*, courtesy The Roy G. Neville Historical Chemical Library).

1. J.R. Partington, *A History of Chemistry*, Macmillan & Co. Ltd., London, Vol. 2, 1961, p. 145.
2. A. Greenberg, *A Chemical History Tour*, John Wiley and Sons, New York, 2000, pp. 98–99.
3. Partington, op. cit., p. 165.
4. Partington, op. cit., pp. 178–179.
5. Greenberg, op. cit., pp. 94–95.
6. J.R. Glauber, *The Works of the Highly Experienced and Famous Chymist, John Rudolph Glauber: Containing, Great Variety of Choice Secrets in Medicine and Alchymy in the Working of Metallick Mines, and the Separation of Metals: Also, Various Cheap and Easie Ways of making Salt-petre, and Improving of Barren-Land, and the Fruits of the Earth. Together with many other things very profitable for all the Lovers of Art and Industry*, London, printed by Thomas Milbourn, 1689, pp. 206–220.
7. (A.) Barlet, *Le Vray et Méthodique Cours de la Physique résolutive, uulgairement dite Chymie. Réprésenté par Figures générales & particulières. Pour connoistre la Théotechnie Ergocosmique. C'est à dire, l'Art de Diev, en l'ouvrage de l'univers. Seconde Édition. Avec l'indice des Matières de ce Volume, & quelques Additions*. Paris, Chez N. Charles, 1657. I am grateful to The Roy G. Neville Historical Chemical Library (California) for supplying these images.
8. J. Ferguson, *Bibliotheca Chemica*, Vol. 1, reprint edition, Derek Verschoyle, London, 1954, pp. 72–73.

SECTION IV
AN EMERGING SCIENCE

THE ANCIENT WAR OF THE KNIGHTS

A mysterious fable of a battle pitting Gold and Mercury, armed as knights, against the Philosopher's Stone first appeared in print (in German) in 1604, although manuscripts probably existed earlier.[1,2] Various versions were published throughout the seventeenth century in German and French. The definitive text (Figure 76)[3] was published in English in 1723 and compare both German and French sources. Nonetheless, the author's identity remains a mystery: was it Alexandre Toussaint De Limojon De Saint Disdier (nice name, but apparently not the true author[1]), or Johann Thölde, the original publisher, who might also have been the legendary Basil Valentine?

While the amusing aspect of this book is its use of allegory, the most fascinating aspect is the manner in which solid scientific reasoning, based upon experimentation, is employed to discredit some fundamental tenets of alchemical lore. While the "magistery" of alchemy wins the day ("The Hermetical Triumph"), early stirrings of the Scientific Revolution are quite audible.

In this fable, Gold is an arrogant and aggressive knight while Mercury, his subordinate knight, dutifully supports Gold. But let us hear Gold in his own bombastic voice as he confronts the Philosopher's Stone:[4]

> 'Tis God himself who has given me the Honour, the Reputation, and the glittering Brightness, which renders me so estimable, it is for that Reason that I am so searched for by every one. One of my greatest perfections is to be a Metal unchangeable in the Fire, and out of the Fire: So all the World loves me, and runs after me; but you, you are only a Fugitive, and a Cheat, that abuses all Men: This is seen in that, that you fly away and escape out of the Hands of those who work with you.

And here is the start of the Stone's measured, yet powerful response:[4]

> 'Tis true, my dear Gold, 'tis God who has given you the Honour, the Durability, and the Beauty, which makes you precious; 'tis for that Reason that you are obliged to return (eternal) Thanks (to the divine Bounty,) and not to despise others as you do; for I can tell you, that you are not that Gold, of which the Writings of the Philosophers make mention; but that Gold is hidden in my Bosom.

The Stone's point is that the substances that embrace in the legendary "chymical wedding" are not the two hopelessly naive metals who confront him but rather Philosophers (or "Sophic") Gold and "Sophic" Mercury each having a much more complex origin. And then, Gold makes the fundamental alchemical argument that the "wedding" (*conjunctio* to be more precise) between Gold and Mer-

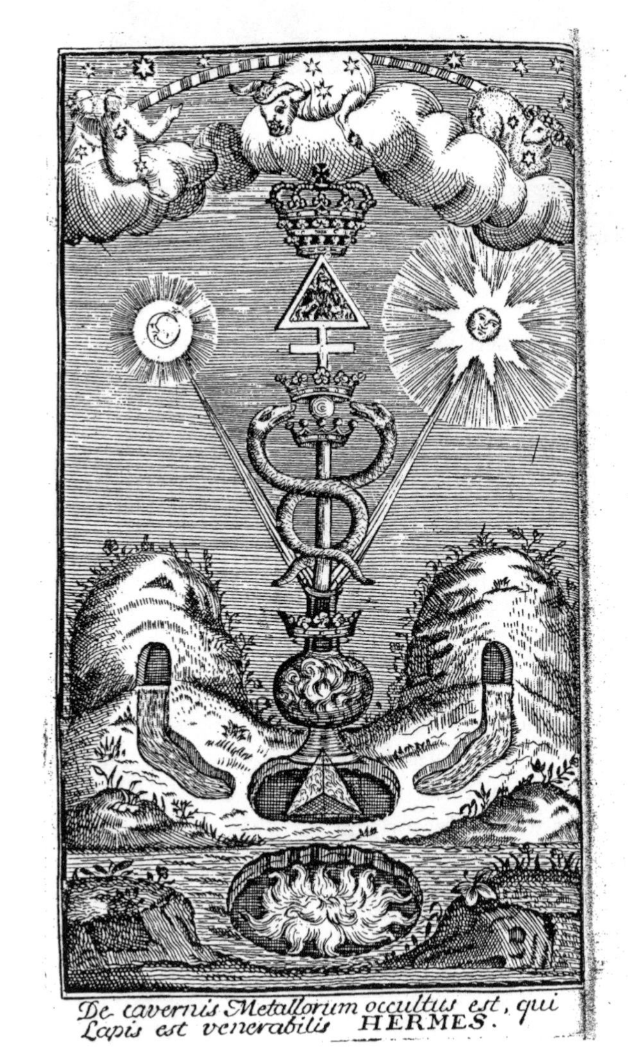

THE
HERMETICAL
TRIUMPH:
OR,
THE VICTORIOUS
Philoſophical Stone.

A TREATISE more compleat and more intelligible than any has been yet, concerning
The HERMETICAL MAGISTERY.
Tranſlated from the FRENCH.

To which is added,
The Ancient War of the KNIGHTS.
Tranſlated from the GERMAN Original.

AS ALSO,
Some ANNOTATIONS upon the moſt material Points, where the two Tranſlations differ.
Done from a GERMAN Edition.

LONDON,
Printed; and Sold by P. HANET, at the Sign of the *Black-Spread-Eagle*, near *Somerſet-Houſe* in the *Strand*. 1723.

FIGURE 76. ▪ Frontispiece and title page for "The Hermetical Triumph" which includes the fable "The Ancient War of the Knights." This allegorical tale tells of combat between Gold and Mercury, girded for battle against the Philosopher's Stone.

cury is necessary in order to multiply gold, indeed, it is Nature's universal way for procreation:[5]

> I am not ignorant, that the Philosophers speak after this manner; yet this may be apply'd to my Brother Mercury, who is as yet imperfect; but if one join both of us together, he then receives from me the Perfection (which he wants). For he is of the Feminine Sex, and I am of the Masculine Sex; which makes the Philosophers say, that the Art is one quite homogeneal Thing. You see an Example here in the (the Procreation of) Men, for there can no Child be Born without (the Copulation of) Male and Female; that is to say, without the Conjunction of the one with the other. We have the like Example thereof in Animals, and in all living Beings.

Let us leave the sexual complexities of this statement to the psychologists and read the Stone's scientifically respectable response:[5]

'Tis true, your Brother Mercury is imperfect, and by consequence he is not Mercury of the wise. So though you should be join'd together, and one should keep you thus in the Fire during the Course of many Years, to endeavor to unite you perfectly to one another, there will always happen (the same Thing, namely,) that as soon as Mercury feels the Action, of Fire, it separates itself from you, it is sublimed, it flies away, and leaves you alone below. That if one dissolve you in Aqua-fortis, if one reduce you into one only (Mass), if one melt you, if one distill you, if one coagulate you, you will never produce any Thing but a Powder, and a red Precipitate: That if one make a Projection of this Powder on an imperfect Metal, it tinges it not; but one finds as much Gold as one put therein at the beginning, and your Brother Mercury quits you and flies away.

In other words, the union of the metals Gold and Mercury has changed nothing in any profound chemical way. Heat gold amalgam and pure mercury is distilled leaving pure gold behind. Alternatively, if the newlyweds were to bathe together in a (heart-shaped?) tub containing *aqua fortis* (nitric acid), mercury would separate as a red calx just as it does in the absence of gold. Enraged by his superior logic, Gold and Mercury violently attack the Philosophers Stone and are consumed, leaving no trace.

The Gold and Silver of this fable represent false alchemists: they suffer equally from ignorance and *hubris*. In contrast, the Philosopher's Stone is the True Adept—a natural philosopher pursuing truth and seeking the wisdom of God. He is the proto-scientist whose experimentation and reasoning will one day lead to a true chemical science. Or, does he somehow presage the birth of Robert Boyle, the "Sceptical Chymist" of our next essay?

---

1. J. Ferguson, *Bilbiotheca Chemica*, Vol. II, Derek Verschoyle, London, 1954, pp. 486–487.
2. L.I. Duveen, *Bilbiotheca Alchemica Et Chemica*, HES Publishers, Utrecht, 1986, p. 361.
3. Limojon De Saint Disdier, Alexandre Toussaint de, *The Hermetical Triumph;, or, The Victorious Stone. A Treatise more compleat and more intelligible than any has been yet, concerning The Hermetical Magistery. Translated from the French. To which is added, The Ancient War of the Knights. Translated from the German original. As also, some Annotations upon the most material Points, where the two Translations differ. Done from a German Edition.* P. Hanet, London, 1723.
4. Limojon De Saint Disdier, op. cit., pp. 4–5.
5. Limojon De Saint Disdier, op. cit., pp. 13–14.

SKEPTICAL ABOUT "VULGAR CHYMICAL OPINIONS"

Robert Boyle (1627–1691) was born in Ireland to a wealthy family, educated at Eton, received further education on the Continent, and returned to England in 1645.[1] He began his scientific studies during the following decade and in 1656 moved to Oxford, where he secured the assistance of Robert Hooke. Hooke built a vacuum pump for Boyle, who used it for numerous studies, including study of the relationship between volume and pressure of gas that now bears his name.[2] Boyle is generally considered to be the Father of Chemistry due in part to his gas law and other physical studies but also because of his classic book, *The Sceptical*

Chymist, which included the first serious attempts to define chemical elements and atomistic concepts with experimental justification.

When The Honorable Robert Boyle published *The Sceptical Chymist* in 1661 (Figures 77 and 78),[3,4] two untested theories of matter dominated the protoscience of chemistry. The earliest of these "vulgar" (i.e., "common") "chymical opinions" was based upon the four elements (earth, fire, air, water) routinely attributed to Aristotle. Aristotelians were often referred to as "peripatetics" because of their master's style of teaching that included walking around. The other prevailing vulgar opinion, dating from the time of Paracelsus (1493–1541), was

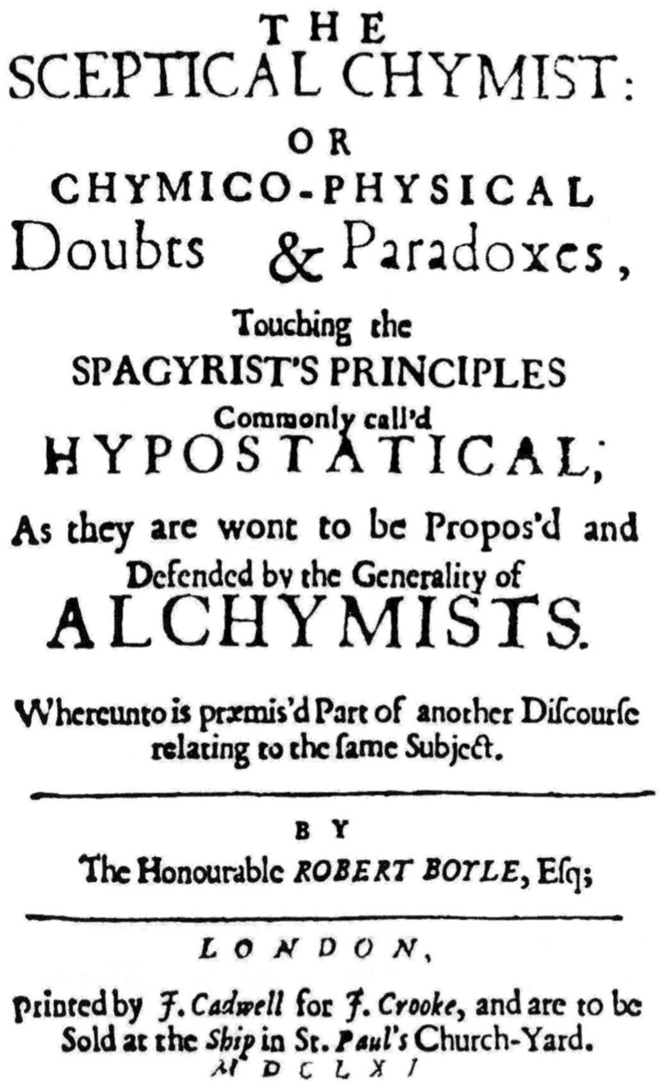

THE
SCEPTICAL CHYMIST:
OR
CHYMICO-PHYSICAL
Doubts & Paradoxes,
Touching the
SPAGYRIST'S PRINCIPLES
Commonly call'd
HYPOSTATICAL;
As they are wont to be Propos'd and
Defended by the Generality of
ALCHYMISTS.
Whereunto is præmis'd Part of another Difcourfe
relating to the fame Subject.

BY
The Honourable *ROBERT BOYLE*, Efq;

LONDON,
Printed by *J. Cadwell* for *J. Crooke*, and are to be
Sold at the *Ship* in St. *Paul's* Church-Yard.
MDCLXI

FIGURE 77. ▪ The polite version of the title page of Robert Boyle's 1661 classic *The Sceptical Chymist*. This book is written in the form of a discussion among fictional characters, including Themistius, representing the "Peripateticks," defenders of the four ancient elements, Philoponus, who defends the three Paracelsian principles and Carneades, the voice of reason (i.e., Boyle), who, of course, gets all the best lines. (Courtesy The Roy G. Neville Historical Chemical Library.)

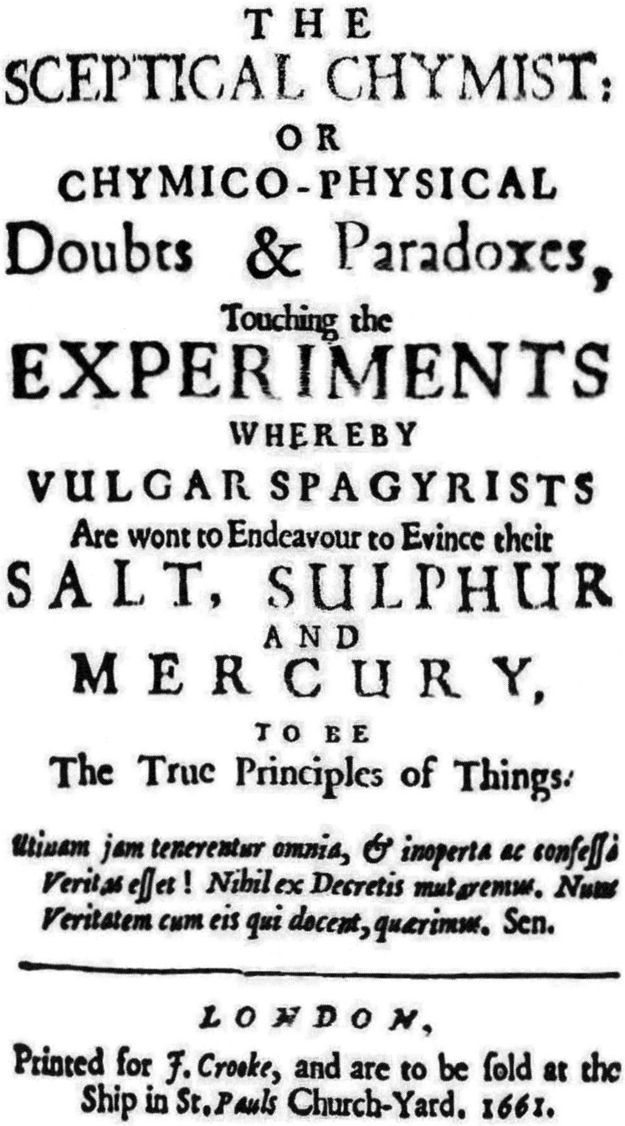

THE
SCEPTICAL CHYMIST:
OR
CHYMICO-PHYSICAL
Doubts & Paradoxes,
Touching the
EXPERIMENTS
WHEREBY
VULGAR SPAGYRISTS
Are wont to Endeavour to Evince their
SALT, SULPHUR
AND
MERCURY,
TO BE
The True Principles of Things.

Utinam jam tenerentur omnia, & inoperta ac confeſſâ Veritas eſſet! Nihil ex Decretis mutaremus. Nunc Veritatem cum eis qui docent, quærimus. Sen.

LONDON,
Printed for *J. Crooke*, and are to be ſold at the Ship in St. *Pauls* Church-Yard. 1661.

FIGURE 78. ▪ The less polite version of the title page also included in Boyle's 1661 *The Sceptical Chymist* (see Figure 77) (courtesy The Roy G. Neville Historical Chemical Library).

based on the *tria prima* (sulfur, mercury, and salt). Boyle referred to the adherents of this theory as "chymists" (we would refer to them as "alchemists" and "iatrochemists"). There was no great honor in being a "chymist," although a "sceptical chymist" was, at least, capable of salvation. In addition to "chymists" and "Peripateticks," there were "hermetick philosophers" who believed that "fire ought to be esteemed the genuine and universal instrument of analyzing mixt bodies"; that is to say, the role of fire is chemical decomposition. And although Boyle believed in a corpuscular theory of matter, akin to the ancient atomic theory of Leucippus, he faulted the Greek philosophers for performing no experimentation:[5]

> And therefore we sent to invite the bold and acute Leucippus to lend us some light by his atomical paradox, upon which we expected such pregnant hints, that 'twas not without a great deal of trouble that we had lately word brought us that he was not to be found;

While Boyle's writing style makes for slow reading, the brief selection above illustrates some of the humor employed in *The Sceptical Chymist*. Moreover, he used an entertaining technique that would probably not work in today's scientific journals and monographs. Boyle set up imagined conversations between himself and convenient "straw men" whose arguments he could readily demolish. Although Boyle the narrator plays a passive role as he accompanies "the inquisitive Eleutherius" on a visit to "his friend Carneades," the latter is really Boyle's voice. Carneades is seated at a little round table in a garden with Themistius, who argues for the "Peripateticks," and Philoponus, who defends the Paracelsian view.

Here is Themistius' "proof" that green wood "disbands" into the four elements upon combustion:[6]

> The fire discovers itself in the flame by its own light; the smoke by ascending to the top of the chimney, and thereby vanishing into the air, like a river losing itself in the sea, sufficiently manifests to what element it belongs and gladly returnes. The water in its own form boiling and hissing at the ends of the burning wood betrays itself to more than one of our senses; and the ashes by their weight, their firiness, and their dryness, put it past doubt that they belong to the element of earth.

Boyle (oops, Carneades) responds that there is confusion here. First, it appears that the "element" fire must be applied to free the "element" fire from green wood. A second point, however, is the assumption that application of fire merely releases the four elements unchanged. Here is Carneades' excellent scientific counterargument:[7]

> When, for instance, a refiner mingles gold and lead, and exposing this mixture upon a cuppel to the violence of the fire, thereby separates it into pure and refulgent gold and lead (which driven off together with the dross of the gold, is thence called *lythargyrium auri*), can any man doubt that sees these two so differing substances separated from the mass, that they were existent in it before it was committed to the fire.

The point is that metallic lead and gold may be melted together to form an alloy. However, prolonged heating in air converts lead to its calx—the yellowish-red powder litharge, a pigment that we recognize today as lead oxide. The molten gold is chemically unchanged, and any impurities (dross) as well as the newly formed litharge will be absorbed into the cupel leaving pure gold. Litharge was clearly never present in the original alloy but was "released" by fire. Indeed, this is almost identical with the argument of the Philospher's Stone in the previous essay.

And here are some other problems with the four elements.[8] It appears to be impossible, even with the aid of fire or other agents, to draw earth, air, fire, or water out of gold. Indeed, if anything might appear to be a true element, it is gold. On the other hand, when blood is "analyzed" by fire, it yields "five distinct substances": phlegm, spirit, oil, salt, and earth.[8]

In answering Philopones, the Paracelsian Spagyrical Chymist, Carneades questions whether the nature of fire always requires that it produce "analysis" of substances or, at least, consistent "analysis."[9] Thus, combustion of "guajacum" wood produces soot and ash, while its distillation (by fire) in a retort yields "oil, spirit, vinegar, water and charcoal." Boyle (oops, Carneades) further argues that if the resulting charcoal is removed from the retort and burned openly, it becomes ash. Similarly, open combustion of camphor produces soot, which may be captured and examined.[9] This soot retains none of the properties of camphor. However, if camphor in a closed glass vessel is exposed gently to fire, a smoke rises that condenses as a white solid, retaining the characteristic penetrating camphor odor. We recognize, in this second case, that camphor has sublimed, unchanged by the "analytical knife" of fire.

In *The Sceptical Chymist*, Boyle offers four propositions[10] that define his views of the organization of matter. He is attempting to fundamentally describe both the physical organization of matter (originating as minute particles) and chemical organization (as elements that cannot be further simplified chemically):

Proposition I. It seems not absurd to conceive that at the first production of mixt bodies, the universal matter whereof they among other parts of the universe consisted, was actually divided into little particles of several sizes and shapes variously moved.

Proposition II. Neither is it possible that of these minute particles divers of the smallest and neighboring ones were here and there associated into minute masses or clusters, and did by their coalitions constitute great store of such little primary concretions or masses as were not easily dissipable into such particles as composed them.

Proposition III. I shall not peremptorily deny, that from most of such mixt bodies as partake either of animal or vegetable nature, there may by the help of the fire be actually obtained a determinate number (whether three, four, or five, or fewer or more) of substances, worthy of differing denominations.

Proposition IV. It may likewise be granted, that those distinct substances, which concretes generally either afford or are made up of, may without very much inconvenience be called elements or principles of them.

The first two propositions deal with the physical structure of matter.[11] There are two levels of organization—minutest particles, Boyle's "corpuscles," which may associate into "coalitions" of "minute masses or clusters." Microscopes, invented around the start of the seventeenth century, provided direct evidence of "the extream littleness of even the scarce sensible parts of concretes."[10] Boyle's associate Hooke published his masterpiece, *Micrographia* (see the next essay), just four years after *The Sceptical Chymist*. Boyle noted further that quicksilver could be distilled, dissolved in acids and filtered, and converted to amalgams that could be finely ground, but all finely divided forms could eventually be recovered as the shiny, metallic liquid. One of Boyle's most wonderful works is his *Effluviums* essay (1673),[12] in which he imagines the smallest physically measurable "minute masses" (or "effluvia") of matter. For example; $1\frac{1}{4}$ grains of gold could be beaten into six $3\frac{1}{4}$-inch squares. Boyle's finest ruler (100 divisions per inch) could, in principle, produce 6 × (3.25 × 100)^2 or 2,535,000 gold squares, each of which would weigh 0.000000032 gram.[12]

The third and fourth propositions deal with Boyle's chemical concepts of the elements. Boyle believed in transmutation. Indeed, in an anonymous essay of 1678, Boyle, in the voice of a certain Aristander, recounts witnessing a "retrotransmutation" (gold degraded alchemically to a lesser metal) by a Pyrophilus.[13] When the other witness, Simplicius, asks in effect "What's the point in degrading gold?", the sage Boyle (oops, Aristander) replies, in effect, "If you know how to transmute in one direction, you can transmute in the other as well." Perhaps this essay should have been titled "The Credulous Chymist." In any case, two things are abundantly clear—Boyle's concept of atoms and elements differed profoundly from the modern concepts because of his belief in transmutation, and his definition of elements suggested no scientific tests. In contrast, Lavoisier's definition of elements ("simples") over a century later was testable—a substance was an element if it could not be further "simplified" chemically:[14]

FIGURE 79. ▪ The title page for the 1668 continental edition of Boyle's *The Sceptical Chymist*. What is going on here?! Boyle has demolished the "contraries," the four elements, and the three principles, and here we see this "mumbo-jumbo" adorning the title page of this translation. One can only imagine publisher Arnold Leers' desire to sell a serious book using tabloid techniques. And one might also imagine Boyle's pained response upon receiving his gratis copy: "We are not amused," he might say, anticipating Queen Victoria by two centuries.

> Thus, as chemistry advances towards perfection by dividing and subdividing, it is impossible to say where it is to end; and these things we at present suppose simple may soon be found quite otherwise. All we dare venture to affirm of any substance is, that it must be considered as simple in the present state of our knowledge, and so far as chemical analysis has hitherto been able to show.

Finally, I cannot resist the temptation to show the frontispiece (Figure 79) from the 1668 Latin translation of *The Sceptical Chymist* published in Rotterdam.[15] What was the publisher Arnold Leers thinking about?! The figures are the classical Sol–Luna (sulfur–mercury), Amorous Birds of Prey, et cetera that honor the dualities that Boyle demolished. Since it seems that Boyle did not have a sense of humor about things scientific,[13] we can probably assume that Leers never consulted Boyle. *What* was Leers thinking about? Profits, no doubt. And one wonders whether the Right Honorable Robert "boyled" when he received his complimentary copy.

---

1. J.R. Partington, *A History of Chemistry*, Macmillan and Co. Ltd., London, Vol. 2, pp. 486–549.
2. A. Greenberg, *A Chemical History Tour*, John Wiley and Sons, New York, 2000, pp. 87–92.
3. R. Boyle, *The Sceptical Chymist: Or Chymico-Physical Doubts & Paradoxes, Teaching the Spagyrist's Principles Commonly call'd Hypostatical, As they are wont to be Propos'd and Defended by the Generality of Alchymists. Whereunto is præmis'd Part of another Discourse relating to the same subject*, F. Caldwell for F. Crooke, London, 1661. I am grateful to The Roy G. Neville Historical Chemical Library (California) for supplying these two images. Although the title page cited above (Figure 77) is commonly quoted, the original first title page appears to be that shown in Figure 78 (Dr. Neville, personal correspondence). It is both anonymous and a bit nasty ("Vulgar Spagyrists"), and perhaps Boyle (or the publisher) had some second thoughts.
4. E. Rhys (ed.), *The Sceptical Chymist by The Hon. Robert Boyle*, J.M. Dent & Sons, London; E.P. Dutton & Co., New York, 1944.
5. Rhys, op. cit., p. 13.
6. Rhys, op. cit., p. 21.
7. Rhys, op. cit., p. 24.
8. Rhys, op. cit., p. 27.
9. Rhys, op. cit., pp. 36–37.
10. Rhys, op. cit., pp. 30–34.
11. Brock, op. cit., pp. 54–70.
12. Greenberg, op. cit., pp. 109–111.
13. Greenberg, op. cit., pp. 92–94.
14. A. Lavoisier, *The Elements of Chemistry in a New Systematic Order Containing All the Modern Discoveries* (Robert Kerr, translator), William Creech, Edinburgh, 1790, p. 177.
15. R. Boyle, *Chymista Scepticus Vel Dubia Et Paradoxa Chymico-Physica circa Spagyricorum Principia*, Apud Arnoldum Leers, Rotterdam, 1668. This is the second Latin edition, the first edition was published in 1662 (Partington, op. cit.).

ENHANCING FRAIL HUMAN SENSES

We chemists seem to have almost surrendered Robert Hooke (1635–1703) to the physicists and biologists, if introductory textbooks are any indication. From middle school onward everybody learns that Hooke coined the term "cell" to describe the microscopic structure of cork. Those who take physics learn that springs, coiled or not, obey Hooke's law. We do know that not long after Otto von Guericke invented the vacuum pump (1654),[1] Hooke, assisting Robert Boyle, constructed the "Boylean" vacuum pump.[1,2] However, Hooke would have termed himself a "natural philosopher," and his incredible scope of activity would have amply justified it. Trained at Oxford, he was appointed curator of experiments to the Royal Society, and was elected FRS in 1663 and professor of geometry of Gresham College in 1665.[3,4] Hooke was said to have "had poor health and slept badly," was something of a hypochondriac, and "For a few years before his death he is said never to have gone to bed or taken off his clothes."[5] This is easy to understand since "The dispersion of his effort seems to have been due at least in part to the varying interests of the Royal Society, which set Hooke to perform a variety of experiments without giving him time to finish any of them. The Society also asked him to repeat the same experiment over and over again, refusing to see the correct interpretation Hooke put upon it."[5]

Hooke's major published work was his 1665 folio *Micrographia*,[6] one of the most beautiful books in the history of science. It is overwhelmingly a book of microscopy, although the final two essays describe telescopic studies of the stars and the moon. Hooke's later sketches of Mars were employed in the nineteenth century to determine the planet's period of rotation.[3]

From the distant mirror of the seventeenth century, Hooke[7] assures us that we can "recover some degree of those former perfections" (lost upon Adam and Eve's expulsion from Eden) if

> The next care to be taken, in respect of the Senses, is a supplying of their infirmities with Instruments, and as it were, the adding of artificial Organs to the natural;

And while Hooke's microscopic tour-de-force is state-of-the-art in 1665, he avers:[7]

> 'Tis not unlikely, but that there may be yet invented several other helps for the eye, as much exceeding those already found, as those do the bare eye, such as by which we may perhaps be able to discover living Creatures in the Moon, or other Planets, the figures of the compounding Particles of matter, and the particular Schematisms and Textures of Bodies.

An ambitious agenda, indeed. But let us select a few micromorsels from *Micrographia*.

Observation XIII[8] explores the microscopic appearances of crystalline materials and offers the profound hypothesis that these regular, three-dimensional structures can be explained by (hexagonal) closest packing of spheres (Figure 80). Observation XIV[9] (*Of Several kindes of frozen* figures) depicts crystals of ice having different origins (Figure 81). Crystals observed on the surface of frozen urine are sometimes quite huge (especially those "observ'd in Ditches which have been full of foul water"). They have near sixfold symmetry (*Fig. i*). (*Note:* Italic figure numbers cited in text refer to the original figures shown collected in these composite illustrations.) What is the nature of the urine crystals? In Hooke's words, "Tasting several cleer pieces of this *Ice*, I could not find any *Urinous* taste in them, but those few I tasted, seem'd as *insipid* as water."[9]

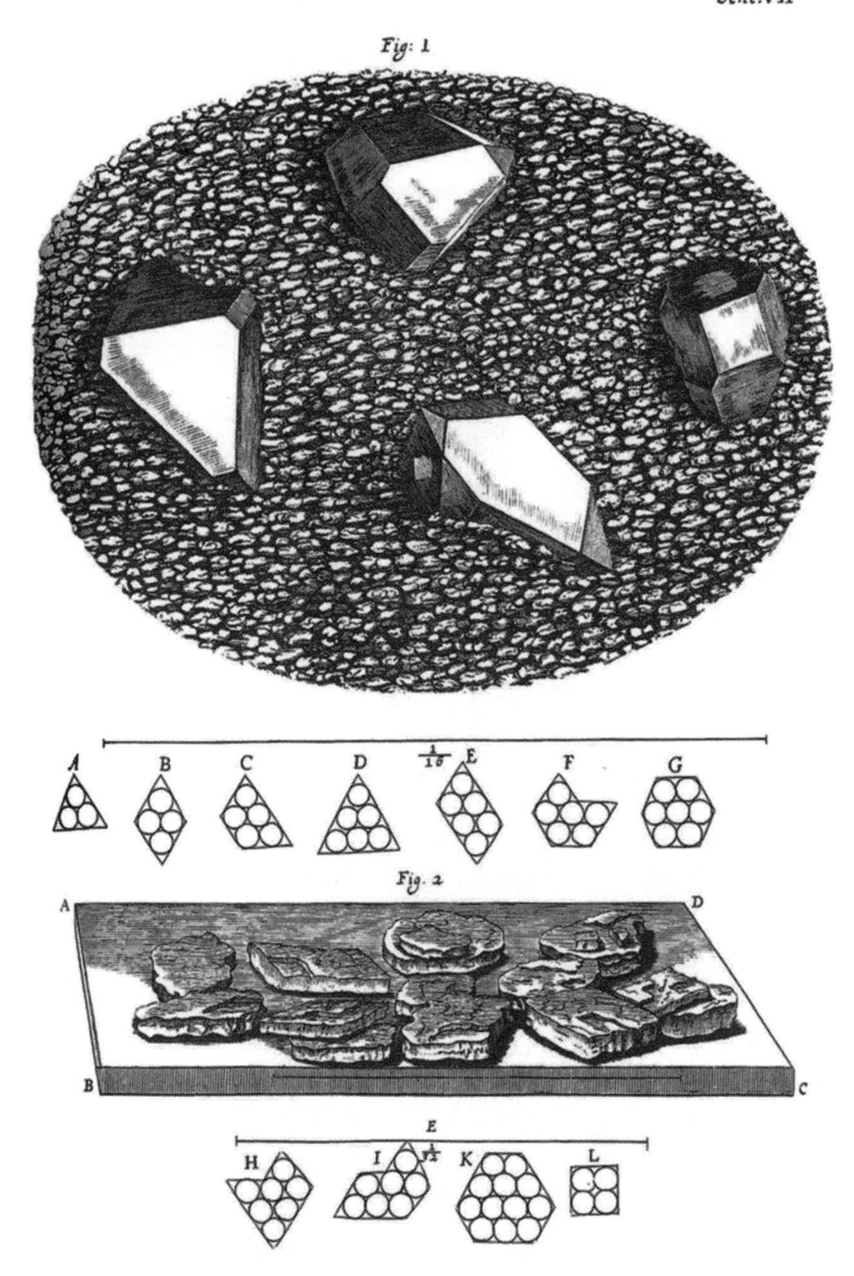

FIGURE 80. ■ Crystals from Robert Hooke's 1665 *Micrographia*. Hooke explained crystalline structures on the basis of close packing of spheres, an insightful anticipation of Dalton's explanation 140 years later.

Figure 2 (in Figure 81) depicts snow flakes caught on a black cloth, the six arms in each flake identical but different from those of other flakes. *Figure 3* is an enlargement of a single snowflake. *Figure 4* and *Figure 5* are ice crystals removed with a knife from the surface of a glass vessel filled with water and chilled. Although hexagonal symmetry is not obvious at first, the angles made with the "central stem" in *Fig. 4* are 60° and 120°. *Figure 6* depicts the water surface just

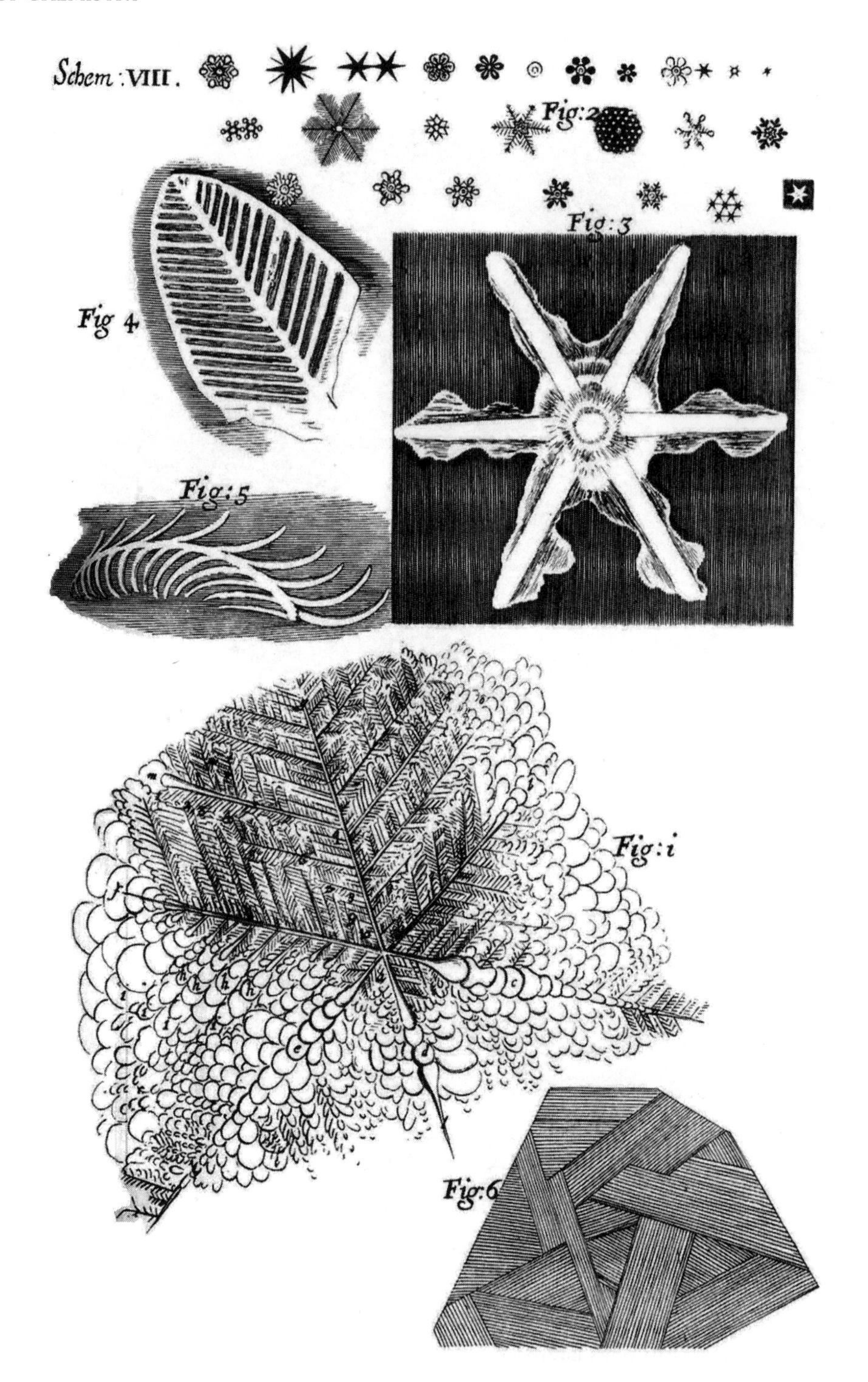

FIGURE 81. ■ Ice crystals viewed by Hooke under the microscope. He understood that their sixfold symmetry derived from packing of spheres (Figure 80). Hooke found that the ice crystals derived from urine were pure water (they lacked the "urinous" taste).

starting to freeze. Hooke referenced these observations to Observation XIII (Figure 80) and specifically to the hexagonal closest packing of spheres to form crystals. While he could assign no chemical meaning to these spheres, some 140 years later John Dalton used his atomic theory to explain the crystalline structure of ice with similar illustrations employing spheres to represent molecules of water.[10]

Chemically speaking, Observation XVI (Of Charcoal, or *burnt* Vegetables)[11] is the most exciting essay in *Micrographia*. Vegetable matter may be placed in a crucible, thoroughly surrounded and covered with sand and heated by fire. Once the heating is stopped and the sand is allowed to cool, charcoal may be recovered. However, if the sand is still hot (or even warm), the uncovered charcoal will burst into flames and be completely consumed. Other oxygen-deficient environments (including vacuum) did not support combustion of charcoal. However, charcoal heated in vacuo inflamed as soon as atmospheric air was introduced. Of course, gunpowder (charcoal, sulfur, and saltpetre) had been known for centuries. Heating charcoal with saltpeter produced very vigorous and complete combustion in a closed vessel (as well as under water). In contrast, combustion of charcoal in a closed vessel containing atmospheric air soon petered out. Similar observations were also made by Hooke using sulfur instead of charcoal. He posited that air is "a menstruum" capable of "dissolving" "sulphureous" (i.e., combustible) bodies. Furthermore:[11]

> the dissolution of sulphureous bodies is made by a substance inherent, and mixt with the Air, that is like, if not the very same, with that which is fixed in Salt-petre, which by multitudes of Experiments that may be made with Saltpetre, will, I think, most evidently be demonstrated.

We will shortly speak of Dr. John Mayow, a friend and companion of Hooke. Why is Mayow, rather than Hooke, generally credited with the discovery that a component of air supports both combustion and respiration? Partington[12] points out that Hooke postulated that the substance "mixt" in air was a "saline substance" (finely divided, suspended saltpetre or niter, perhaps?) that might be somehow "strained out." It was Mayow who correctly proved that the active substance was a gaseous component of atmospheric air.

Micrographia illustrated numerous small objects microscopically. The enlarged image of the stinger of a bee,[13] for example, actually provided very useful insights into its mode of action. However, for collectors of old tomes, far worse than fierce stinging bees and plague-carrying fleas is the fearsome bookworm (Figure 82)![14] Let us hear Hooke:

> And indeed, when I consider what a heap of Saw-dust or chips this little creature (which is one of the teeth of Time) conveys into its entrails, I cannot chuse but remember and admire the excellent contrivance of Nature, in placing in Animals such a fire, as it is continually nourished and supply'd by the materials convey'd into the stomach, and fomented by the bellows of the lungs; and in so contriving the most admirable fabrick of Animals, as to make the very spending and wasting of that fire, to be instrumental to the procuring and collecting more materials to augment and cherish it self, which indeed seems to be the principal end of all the contrivances observable in bruit animals.

WANTED DEAD OR ALIVE

"The Bookworm"

aka "Silverfish"

aka *Tysanura*

Reward Offered by Antiquarian Book Collectors Anonymous

FIGURE 82. ■ No book collector or librarian will protest this "wanted" poster.

And this some 120 years before Lavoisier proved with balance and calorimeter that respiration is combustion!

1. A. Greenberg, *A Chemical History Tour*, John Wiley & Sons, New York, 2000, pp. 87–90.
2. Greenberg, op. cit., pp. 91–92.
3. *The New Encyclopedia Britannica*, Encyclopedia Britannica Inc., Chicago, 1986, Vol. 6, p. 44.
4. J.R. Partington, *A History of Chemistry*, Macmillan & Co. Ltd., London, Vol. 2, 1961, pp. 550–570.
5. Partington, op. cit., pp. 551–552.
6. R. Hooke, *Micrographia: Or Some Physiological Descriptions of Minute Bodies Made by Magnifying Glasses. With Observations and Inquiries thereupon*, Jo. Martyn and Ja. Allestry, Printers to the Royal Society, London, 1665. See also the facsimile reprint published by Culture Et Civilisation, Brussels, 1966.
7. Hooke, op. cit., Preface
8. Hooke, op. cit., pp. 82–88.
9. Hooke, op. cit., pp. 88–93.
10. Greenberg, op. cit., pp. 173–175.

11. Hooke, op. cit., pp. 100–106.
12. Partington, op. cit., p. 558.
13. Hooke, op. cit., pp. 163–165.
14. Hooke, op. cit., pp. 208–210.

A FLEETING WHIFF OF OXYGEN?

John Mayow (1641–1679)[1] came very close to "nipping" phlogiston theory "in the bud" almost immediately after Becher first proposed its original form in 1669. He first entered Oxford in 1658, was admitted as a scholar in 1659, and was elected a fellow of All Souls College in 1660. Mayow became a "profess'd physician" around 1670, although Partington could find no evidence for a formal medical degree.[1] It is not exactly clear when Mayow and Hooke met or whether Mayow ever met Boyle. He does appear to have been given access to the Boylean vacuum pump in Oxford during the 1660s.[2]

In 1668 Mayow published two tracts dealing with respiration and rickets. These were revised in 1674 and published with three additional tracts to constitute the *Tractatus Quinque Medico-Physici. . . .*[3] It is in this work that Mayow identifies his "*Spiritu Nitro-Aereo*" or "nitro-aerial spirit." Just as Hooke had done in his 1665 book *Micrographia*, Mayow identifies a substance in the air that is required to support the combustion of "sulphureous" matter just as niter or saltpetre was known to. Flammable substances were said to contain "sulphureous matter," and this was, of course, strongly related to phlogiston theory. Indeed, one can trace these concepts back to Paracelsus' three principles—sulfur, mercury, and salt. However, Mayow's powerful contributions[1] to chemistry were the realizations that (1) a component of air supported combustion; (2) this component of air had the same effect as nitre or saltpetre; (3) this component of air also supported respiration; and, most uniquely, (4) this was a specific gaseous component of air. Thus, "atmospheric air" contained a gaseous component capable of supporting combustion and respiration and another gaseous component that could not. We have described Mayow's apparatus and experiments elsewhere.[4] He correctly explained a curious observation of Boyle's concerning gunpowder. It was known that the saltpetre in gunpowder provided a much greater amount of "nitro-aerial spirit" than did atmospheric air. Furthermore, the "fuel" components of gunpowder, carbon, and sulfur could each burn in a closed vessel up to a point and then would extinguish. In contrast, a closed vessel containing carbon or sulfur combined with saltpetre would burn to completion just as gunpowder would under these circumstances. However, Boyle placed some gunpowder in a circle on a surface under vacuum.[5] Under a heating lens he observed slow, localized ignition of the particles of gunpowder directly exposed to the intense light. Removal of the lens stopped the burning. However, if the burning lens was trained on some crystals of gunpowder in the powder circle and the system was then opened to the atmosphere, full conflagration occurred instantly. Mayow correctly reasoned that "nitro-aerial" particles had to be in direct contact with charcoal or sulfur to produce combustion.[6]

FIGURE 83. ■ A water spout depicted in John Mayow's *Tractus Quinque Medico Physici* (1681 edition, published in The Hague; the first edition was published in Oxford in 1674).

Partington[7] remarks on Mayow's claim to have heated niter and collected the resulting nitric acid. He notes "if he had actually tried the experiment he could have discovered oxygen." However, as Partington and others remarked, Mayow did not have the means to capture, manipulate, and study gases. These techniques awaited development by Stephen Hales[8] in 1727 and subsequent improvements by William Brownrigg, Joseph Black, Henry Cavendish, and Joseph Priestley.

The *Tractatus Quinque* concerned itself with other scientific questions beyond the medical, physiological, and chemical. For example, Mayow discussed the origins of water spouts as due to air turbulence (see Figures 83 and 84; see also Benjamin Franklin's studies of these phenomena and Figure 119 later in this book). Mayow's explanation of lightning and thunder are reminiscent of those of Paracelsus[9] and imagine explosions between "nitro-aerial" spirit and "sulphureous" matter in the atmosphere.

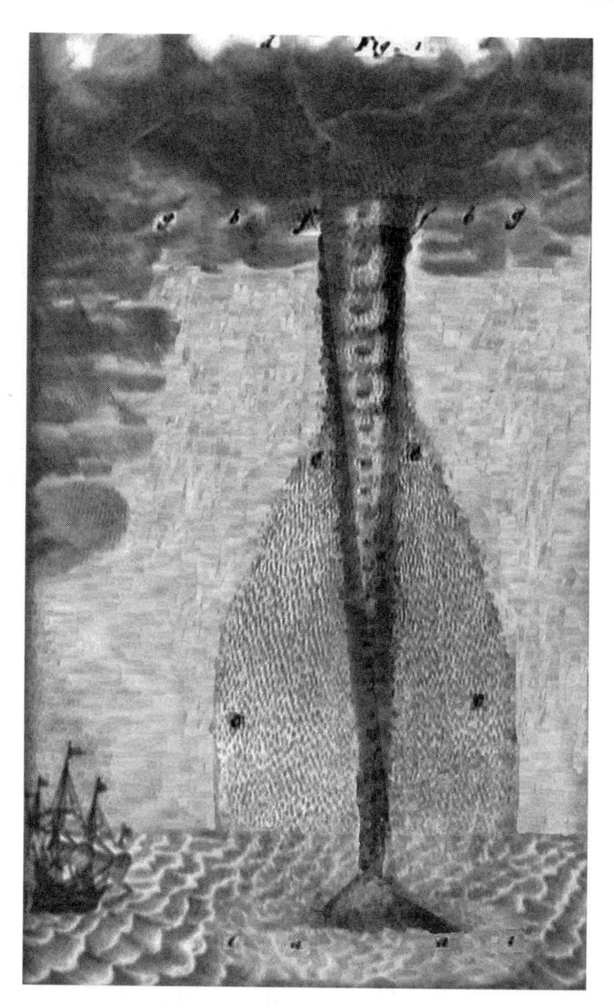

FIGURE 84. ■ A second water spout depicted in the 1681 edition of Mayow's *Tractus Quinque Medico Physici*. Mayow came quite close to discovering that saltpetre contained oxygen, which could support combustion. Robert Boyle and his assistant Robert Hooke, who was friendly with Mayow, also investigated the ability of saltpetre to sustain combustion.

---

1. J.R. Partington, *A History of Chemistry*, Macmillan & Co. Ltd., London, 1961, Vol. 2, pp. 577–614.
2. Partington, op. cit., p. 604.
3. J. Mayow, *Tractatus Quinque Medico-Physici. Quorum primus agit de Sal-Nitro, et Spiritu Nitro-Aereo. Secondus de Respiratione. Tertius de Respiratione Foetus in Utero, et Ovo. Quartus de Motu Musculari, et Spiritibus Animalibus. Ultimus de Rhachitide*, Sheldonian Theatre, Oxford, 1674. (A second Latin edition was published in The Hague, 1681; an English translation was published by the Alembic Club, Edinburgh, in 1907.)
4. A. Greenberg, *A Chemical History Tour*, John Wiley & Sons, New York, 2000, pp. 105–106.
5. Partington, op. cit., p. 527.
6. Partington, op. cit., p. 589.
7. Partington, op. cit., p. 588.
8. Greenberg, op. cit., pp. 122–130.
9. Partington, op. cit., p. 133.

LUCIFER'S ELEMENT AND KUNCKEL'S PILLS

Urine, the golden liquid endowed with vital and mystical properties, was used for centuries in thousands of alchemical preparations. Undoubtedly it has been distilled to dryness countless times. However, in 1669 a competent but obscure alchemist, Hennig Brand, boiled urine, concentrating it to a thick syrup, from which a red oil was distilled. The retort's black carbon residue was added to this oil and heated in an earthen retort. The scene is imagined by John Emsley in his wonderful book *The 13th Element*.[1] We "witness" Brand observe the distillation of a heavy glowing liquid that bursts into flames as it contacts air.[2] Once Brand isolates the liquid in his receiver, it solidifies but continues to glow. Imagine the wonder that this provoked—a glowing, fiery secret concealed in our own bodies and our excreta! Brand had discovered the element phosphorus ("bringing light"). The melting point (44°C) and boiling point (280°C) of white phosphorus are quite low. This explains its ease of distillation and the fact that impurities may lower the melting point sufficiently to make "liquid phosphorus".

Rather than publishing an epistle on urine and its transformation to phosphorus, Brand kept his work in Hamburg secret for about six years, hoping to make money as soon as he discovered what it was good for.[2] However, exhausting the fortune of his second wife on this fruitless search, he eventually publicized the discovery, and his work came to the attention of Johann Kunckel, recently retired unsuccessful gold-maker to the Elector of Saxony,[3] and now professor at the University of Wittenberg. Kunckel visited Hamburg but could learn no details from Brand. Kunckel, in turn, informed another alchemical colleague, Johann Daniel Kraft in Dresden, about the new substance. Sensing opportunity, Kraft hastened immediately to Hamburg and purchased all of Brand's phosphorus supply along with exclusive rights and a pledge by Brand to secrecy. This bit of business chicanery occurred right under Kunckel's nose, as it were, and the best the distinguished academician could get from Brand was the hint that urine was the source.[2]

Kunckel (1630–1703),[4] the son of an alchemist, had received no formal academic training but was an able and respected scientific investigator. With only the vague hint provided by Brand, Kunckel independently discovered the process and published on the properties of his *noctiluca constans* ("unending nightlight"), although not the method of its preparation, in 1678. Robert Boyle obtained samples from Kraft and Kunckel during this period and published work in 1680 on the preparation and properties of liquid phosphorus, the "Aerial Noctiluca" ("nightlight spirit"), and another in 1682 on solid phosphorus, the "Icy Noctiluca."[4] Even the eminent mathematician Gottfried Wilhelm Leibniz (1646–1716), who independently of Sir Isaac Newton developed the calculus, maintained a lifelong interest in alchemy, as did arch rival Newton, and wrote on the experimental investigations of phosphorus.[4,5] The magic imagery associated with phosphorus is apparent in Figure 85, the fanciful frontispiece to Johann Heinrich Cohausen's 1717 book[6] on phosphorus, where we see both Hermes and a flying dragon as sources of light and fire. Figure 86[7] apparently shows Johann Daniel Kraft "pitching" phosphorus to Leopold I early in his long reign (1658–1705) as

FIGURE 85. ▪ The mystical title page of Johann Heinrich Cohausen's 1717 treatise on phosphorus. Hermes and the flying dragon are sources of fire and light—properties of white phosphorus.

Holy Roman Emperor. In this figure, *1* depicts flashing phosphorus (solid); *2*, liquid phosphorus giving off smoky fumes as it sits motionless; and *3*, a shine-in-the-dark barometer. Item 5 is a kind of "flashing sign"[8] with the Emperor's name printed neatly in solid phosphorus—a very modern bit of slick salesmanship.

Kunckel was certainly an enthusiastic alchemist, if unsuccessful gold-maker. Believing that mercury was the spirit of metallicity retained in the transmutation of metals, he reported extracting mercury from all metals.[9] However, his quantitative studies indicated that antimony gains weight upon calcination and also included measuring the strength of *aqua fortis* (nitric acid) by saturating it with silver, evaporating the solution to dryness, and weighing the remaining salt.[10] Kunckel also contributed to the art of glass-making and his 1679 book *Ars Vitraria Experimentalis*, included the 1612 work (in seven books) by Antonio

FIGURE 86. ▪ An illustration from Cohausen's 1717 *Lumen Novum Phosphoro Accensi*. How does one get the Holy Roman Empire phosphorus contract? Simple; devise a sign in which white phosphorus spells out the Emperor's name "LEOPOLDUS" so that it shines boldly in the dark.

Neri, updated with three of his own works.[11] Figure 87, from the 1679 book, depicts a glass-making furnace with workers fashioning bottles. In Figure 88 we see a pedal-operated bellows used to fabricate small glass toys.[4]

Emsley's book[1] traces the development and uses of phosphorus, which he suggests might be termed "The Devil's Element."[12] The basis for the destructive distillation of urine (or bone) to phosphorus is described. Organic matter, such as creatine, in urine decomposes under oxygen-poor conditions to form elemental carbon (e.g., charcoal). Under the high heat the carbon strips oxygen atoms from phosphate salts also present in the urine residue to form gaseous carbon monoxide. This is really not very different from the industrial process that produces white phosphorus from rock phosphate in the presence of coke and silica:[13]

$$2\,Ca_3(PO_4)_2 + 6\,SiO_2 + 10\,C \rightarrow P_4 + 6\,CaSiO_3 + 10\,CO$$

In his chapter[14] titled "The Toxic Tonic," Emsley notes the marketing of pills made of this exceedingly toxic element. Coated with a thin film of gold or silver for physical safety, these were marketed as "Kunckel's Pills" shortly after the famous chemist died. There are 60 pages covering the chemical, practical,

FIGURE 87. ■ A glass-making furnace depicted in Johann Kunckel's 1679 *Ars Vitraria Experimentalis* (courtesy Chemical Heritage Foundation). Kunckel was cheated by his "friend" Johann Daniel Kraft (he's the "pitchman" in Figure 86) who monopolized Hennig Brand's discovery of phosphorus. Nonetheless, the inadvertent hint from Brand that phosphorus came from urine was enough for the clever Kunckel to independently discover how to make it.

FIGURE 88. ■ Fabricating small glass toys with the aid of a pedal-operated bellows (from Kunckel's 1679 *Ars Vitraria Experimentalis*, courtesy Chemical Heritage Foundation).

business, and sociological history of matches, and this very interesting section almost reads like a novel. (Red phosphorus, the polymeric allotrope, was discovered in the nineteenth century by heating white phosphorus to 400°C in a closed vessel. It helped make the match industry safer.) Here, we first meet the late nineteenth-century English social reformer Annie Besant, who forms a union for women in the dangerous and exploitative match fabrication industry. We will meet her later in our book—some 20 years hence "divining" the internal structures of atoms (see p. 293). And Emsley describes the origin of the glow of white phosphorus—fully understood only in 1974. At the surface of white phosphorus, a solid composed of tetrahedral P_4 molecules, reaction with oxygen produces highly unstable molecules of HPO and P_2O_2, which luminesce close to the surface that formed them just before they "die."[15]

---

1. J. Emsley, *The 13th Element—the Sordid Tale of Murder, Fire, and Phosphorus*, John Wiley & Sons, Inc., New York, 2000. This book was first published in Great Britain as *The Shocking History of Phosphorus*, Macmillan Publishers Ltd. in 2000. It is a truly admirable book—a "page-turner," possibly a "barn-burner." The scope of the book can be imagined as author Emsley relates sadly and ironically that phosphorus was discovered in Hamburg and used in its horrific fire-bombing almost three centuries later. He notes that phosphorus was "the thirteenth chemical element to be isolated in its pure form." Aaron J. Ihde might have contested that since he lists zinc among the elements to have been isolated before 1600 (see A.J. Ihde, *The Development of Modern Chemistry*, Harper & Row, New York, 1964, p. 747). However, the separation of zinc from its oxide, a high-temperature process, was scientifically reported in the mid-eighteenth century, so Emsley's appellation appears to be "kosher."
2. Emsley, op. cit., pp. 3–24.
3. An "Elector" was a prince in the Holy Roman Empire who could participate in the election of an emperor.
4. J.R. Partington, *A History of Chemistry*, Macmillan & Co. Ltd., 1961, Vol. 2, pp. 361–377.
5. Partington, op. cit., p. 485.
6. J.H. Cohausen, *Lumen Novem Phosphoris Accensum, sive Exercitatio Physico-Chymica De causa lucis in Phosphoris tam naturalibus quam artificialibus*, Joannem Oosterwye, Amsterdam, 1717.
7. Cohausen, op. cit., p. 203.
8. Actually, the "*pomum*" was a globular hand warmer for clerics.
9. Partington, op. cit., p. 362.
10. Partington, op. cit., p. 375.
11. Partington, op. cit., p. 368.
12. Emsley, op. cit., pp. 299–302.
13. F.A. Cotton and G. Wilkinson, *Advanced Inorganic Chemistry*, fifth edition, John Wiley & Sons, New York, 1988, p. 386.
14. Emsley, op. cit., pp. 47–63.
15. Emsley, op. cit., p. 16.

THE EMPEROR'S MERCANTILE ALCHEMIST

Johann Joachim Becher (1635–1682)[1–3] has been gone for well over three centuries, and we usually think of him only as the *ur*-father of chemistry's first comprehensive theory: phlogiston. What we generally miss is that Becher may well have been the greatest mercantilist of the seventeenth-century Holy Roman Empire.[3] In this, he would share some common ground with Antoine Laurent Lavoisier, who, during the late eighteenth century, will become the Father of Modern Chemistry even as he functions as one of France's greatest economists.[4] In 1666, the 31-year-old Becher was appointed economic advisor to Leopold I and titled himself "Advisor on Commerce to His Majesty, the Emperor of the Holy Roman Empire."[5] Sixteen years later he would die in London and leave his family in such poor circumstances that one of his daughters was forced to become a domestic.[6]

The Holy Roman Empire lasted, in name at least, for over a thousand years following the conferral of the imperial title to Charlemagne by Pope Leo III in 800.[7] It consisted of a vast realm in central Europe with Germanic people at its core who furnished most of its traditional rulers. The Reformation in the sixteenth century created rebellious centers of power, notably among German princes who adopted Protestantism and rebelled against the Emperor. These religious tensions reinforced a bewildering "cat's cradle" of territorial conflicts and alliances leading to the start of the disastrous Thirty Years War (1618–1648).[8] By the time the war was settled with the Treaty of Westphalia in 1648, Spain had lost the Netherlands and its preeminence on the Continent, France had emerged as the major western European power, and many Germanic towns were ruined economically and the Empire irrevocably weakened. A century later, the famous French author and satirist Voltaire would quip that the Holy Roman Empire was "neither holy, nor Roman, nor an empire."[7] It would inauspiciously pass out of existence in 1806, two years after Napoleon declared himself emperor of France.

Becher came of age in the aftermath of the Thirty-Years War and like some other prominent chemists of the period, including Kunckel and Johann Rudolph Glauber, devoted himself to the prosperity of Germany. Becher was self-educated, developed an early interest in technology, and published his first work in 1654 on alchemy using the pseudonym Solinus Saltzthal.[1,2] By 1655 Becher had established himself as mathematical advisor to the Holy Roman Emperor Ferdinand III in Vienna and was advising him on alchemical processes.[9] Becher's first book on metal chemistry and iatrochemistry, *Natur-Kündigung der Metallen* (Figure 89), was published in 1660.[1] He believed in a "vitalist theory" in which minerals, as well as animals and plants, "have a sort of life and grow in the earth from seeds."[1] He obtained the M.D. degree at Mainz in 1661 and was appointed to the university's medical faculty in 1663 and became physician to the Elector at Mainz. Married in 1662 to a woman from a prosperous and well-respected family, the restless Becher moved to Munich in 1664 and became Medical and Mathematical Advisor to the Elector of Bavaria.[1–3] It was during this period that he became very much involved in commerce, organized the Eastern Trading Company and tried to establish for his patron a commercial colony in South America. As noted above, in 1666 he joined Leopold I in Vienna.

In his 1664 *Oedipus chimicus*,[10] Becher describes his early concepts of the elementary composition of matter.[2] His most famous work is his 1669 book commonly referred to as the *Physica subterranea*.[1–3,11] In this work, Becher argued[1,2] that air, water, and earth constituted the true elements with air being "an instrument of mixing." Metals and stones were said by Becher to be composed of three earths: *terra vitrescible* (glassy earth—the substance of subterranean matter), *terra pinguis* (fatty earth—combustibility) and *terra fluida* (odor, volatility, and other subtle properties). Becher concluded that all substances that burn, including metals such as tin and zinc, must contain *terra pinguis*, although Partington[1] notes that the fate of *terra pinguis* is never described by Becher. Indeed, Becher was well aware that metals *increased* in weight upon forming calxes. He attributed this to the accretion of fiery effluvia onto the metal as proposed earlier by Boyle.[12]

Becher's transformation[13] from a "purveyor of alchemical secrets" before 1655 into the trusted technical advisor to nobles and emperors over the next 15 years relied on his mastery of mechanics and science, amplified by his ability to

FIGURE 89. ■ Title page from the 1660 Frankfort edition of *Natur-Kündigung der Metallen* by Johann Joachim Becher. Becher is known to chemists as the father of phlogiston theory. However, he was at least as famous for his knowledge of economics and his status as Advisor on Commerce to Leopold I, Emperor of the Holy Roman Empire. (Courtesy The Roy G. Neville Historical Chemical Library.)

market himself as the expert to consult in a world full of unscrupulous pretenders. Here is a segment from a letter to Emperor Leopold authored during the 1670s:[13]

> Above all, however, because Your Imperial Majesty has a desire to have some trials made in these things, it would be necessary to take into service a loyal, honest, and knowledgeable subject, whom Your Imperial Majesty could trust with the processes of such worthless vagabonds, and who, privately and se-

> cretly, could in silence faithfully work out the processes and report on them to Your Majesty. If this does not happen, Your Imperial Majesty will never get to the bottom of this, nor understand the nature of these things, but instead will always be duped by these scoundrels.

Now, who does Doktor Becher have in mind as the Emeror's expert? Becher biographer Pamela H. Smith notes ironically that "Becher's portrayal of the selfish and gain-seeking projector resembles remarkably his own situation ten years earlier."[13]

Becher's involvement with German mercantile interests led to his design of a factory for manufacture of glassware and crafts, complete with laboratory and library. His edicts, in 1677, against French imports into southern Germany failed and led to his brief imprisonment in 1678.[2] In 1678, he was also involved in an unsuccessful attempt to commercialize Henning Brand's technology for phosphorus manufacture.[14] However, another "syndicate," headed by Gottfried Wilhelm Leibniz, the famous mathematician who was "in cahoots" with the shadowy industrial

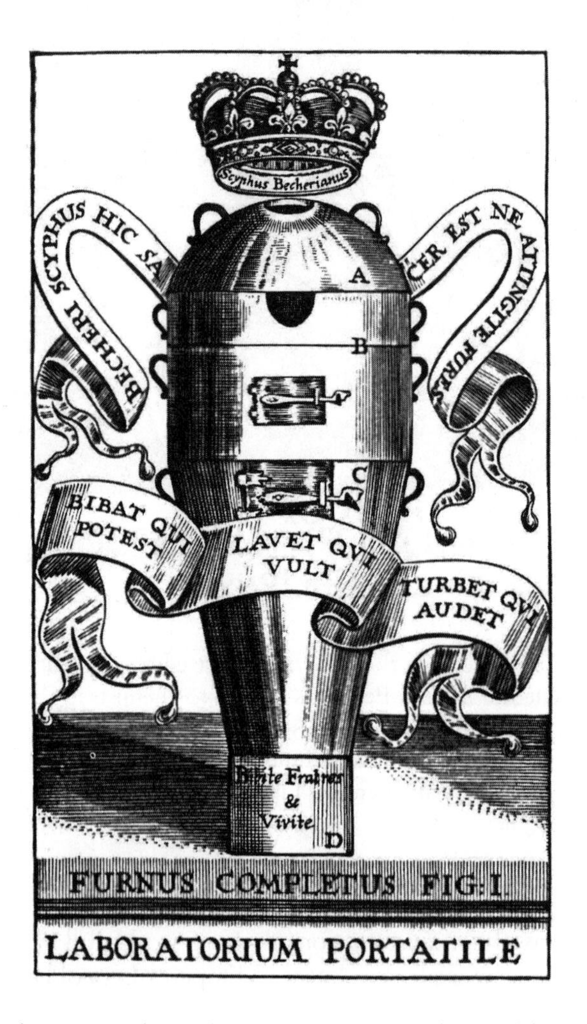

FIGURE 90. ■ Becher was nothing if not a venture capitalist, and here is the portable furnace he invented and marketed. One of these was purchased for twelve pounds by Robert Boyle. (From Becher's 1660 *Natur-Kündigung der Metallen*; courtesy The Roy G. Neville Historical Chemical Library.)

spy Johann Daniel Kraft, succeeded in bringing Brand and his technology to Hanover. You will remember our earlier discussion, gentle reader, of Kraft's swift appropriation of Kunckel's hint about Brand's discovery of phosphorus, followed by his attempt to "corner the market" and "shut Kunckel out." Brand and his wife Margaretha, themselves, were not above using the threat of joining Becher to extort additional funds from the Leibniz "syndicate." Frau Brand's letter to Leibniz is not very subtle: "Dr. Becher is ever so honest and four weeks ago, as he left Hamburg for Amsterdam, he honored my husband with ninety-four Reichsthaler."[14]

Anticipating the likely failure of a large-scale demonstration of his process for extraction of gold from sea sand scheduled in Holland in March, 1680,[2] Becher-

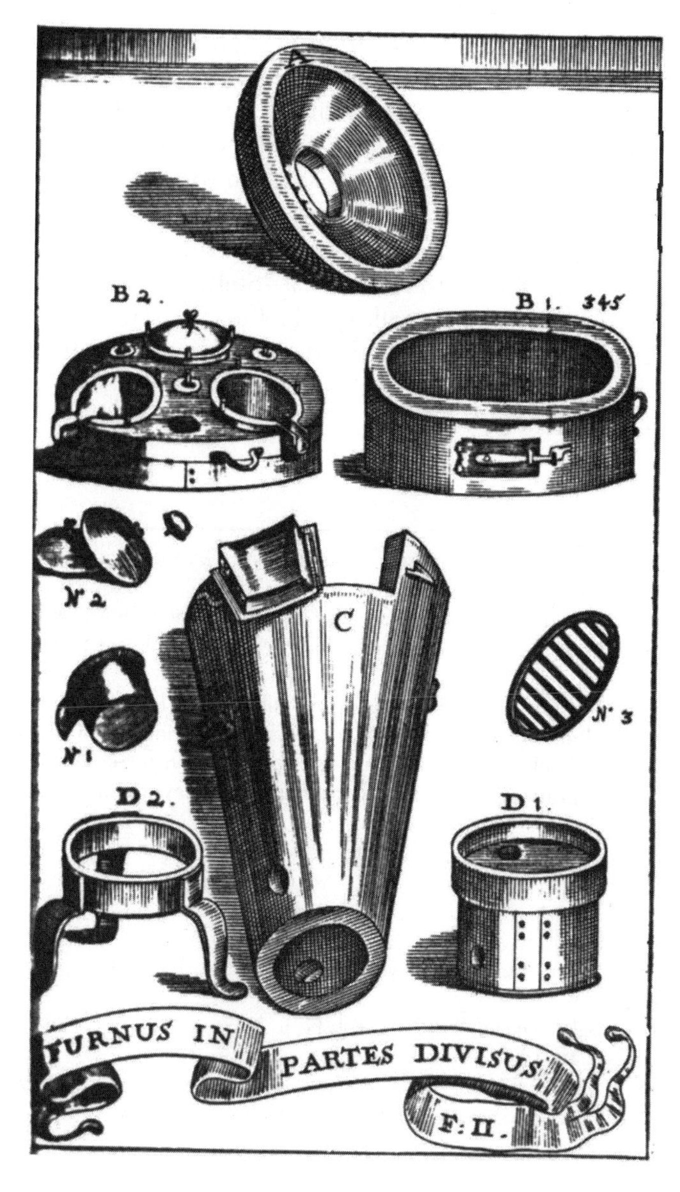

FIGURE 91. ■ Disassembled view of Becher's portable furnace (Figure 90). (From Becher's 1660 *Natur-Kündigung der Metallen*, courtesy The Roy G. Neville Historical Chemical Library).

er abruptly left for London, without his family. Although Robert Boyle was one his patrons in England, Becher was unsuccessful in his entreaties to the Royal Society for election to its membership. He did, however, sell three of his portable furnaces (Figures 90 and 91) at 12 pounds each. One of these was purchased by Boyle. I confess that I would love to read a play or short story reconstructing the interplay between the aristocratic Englishman Boyle and the eight-year-younger Becher, very possibly "burned out" from his Continental intrigues, close scrapes with the law, and abandonment of his family, in the final two years of his brief, adventurous life.

Becher's theory was largely unrecognized in his time and was embraced some three decades later by the famous physician Georg Ernst Stahl (1660–1734).[15–17] Although one frequently reads that Stahl was a "student" or a "disciple" of Becher, it is worthwhile to note explicitly that Stahl had just attained the age of 22 and was studying medicine in Jena (Germany) when Becher died in London in 1682. Although Stahl's interest in chemistry started very early, there is no mention of the two having ever met. Nonetheless, Stahl's reading of Becher's work and adoption of his theory led him to republish Becher's *Physica subterranea* in 1703. It is Stahl who coined the term "phlogiston" and developed the concept that this essence of fire was lost to the surroundings during combustion and calcinations.[15] Partington notes that "Stahl was proud, morose, atrabilious, . . . quarreled with his senior colleague Hoffmann, to whom he owed his appointment at Halle, . . . rarely answered letters, . . . showed contempt for all who differed from his views and reacted violently to criticism. These qualities . . . greatly enhanced his reputation."[17]

---

1. J.R. Partington, *A History of Chemistry*, Macmillan and Co. Inc., London, 1962, Vol. 2, pp. 637–652. I am grateful to The Roy G. Neville Historical Chemical Library (California) for supplying the three figures shown from Becher's 1660 *K~undigung der Metallen*.
2. C.C. Gillispie (ed.), *Dictionary of Scientific Biography*, Charles Scribner's Sons, New York, 1970, Vol. I, pp. 548–551.
3. P.H. Smith, *The Business of Alchemy—Science and Culture in the Holy Roman Empire*, Princeton University Press, Princeton, 1994. It is interesting that the international financier George Soros has written a book titled *The Alchemy of Finance*, Simon & Schuster, New York, 1987. Soros, a protégé of philosopher of science Karl Popper, employs finance and philanthropy to promote open societies.
4. J.-P. Poirier, *Lavoisier—Chemist, Biologist, Economist*, University of Pennsylvania Press, Philadelphia, 1996.
5. Smith, op. cit., p. 18.
6. Partington, op. cit., p. 638.
7. *The New Encyclopedia Britannica*, Encyclopedia Britannica, Inc., Chicago, 1986, Vol. 6, pp. 21–22.
8. *The New Encyclopedia Britannica*, op. cit., Vol. 11, p. 711.
9. Smith, op. cit., pp. 16–17.
10. A. Greenberg, *A Chemical History Tour*, John Wiley and Sons, New York, 2000, pp. 102–104.
11. Partington (see above) also cites an earlier 1667 version of this work.
12. Partington, op. cit., p. 650.
13. Smith, op. cit., pp. 76–80.
14. Smith, op. cit., pp. 248–255.
15. Partington, op. cit., pp. 653–686.
16. C.C. Gillispie, op. cit., 1975, Vol. XII, pp. 599–606.
17. A. Greenberg, op. cit., pp. 106–108.

THE HUMBLE GIFT OF CHARCOAL

Charcoal is hardly an awe-inspiring substance, yet it has played a critical role in human history. As noted in an earlier essay, charcoal's ability to strip phosphate of its oxygens at high temperature provided the surprised Brant with elemental (white) phosphorus. In modern terms we understand that the driving forces are thermodynamic. Carbon monoxide has the strongest covalent chemical bond

FIGURE 92. ▪ Oil-on-porcelain painting by artist L. Sturm, very likely the porcelain painter Ludwig Sturm (source: Dr. Alfred Bader). Although the painting is titled "The Alchemist," rational chemistry is occurring. The key to the picture is the pan of charcoal. It is likely that a metal oxide is being reduced to the metal by charcoal in the red-hot crucible. See color plates. (I am grateful to the Art Museum of the State University of New York at Binghamton for permission to use this image.)

in nature.[1] Energetically, the creation of strong bonds at the sacrifice of weaker bonds is a powerful driver of chemical reactions. Moreover, we understand that entropy (the degree of disorder in a system) can be a strong driving force as well. Production of a gas (carbon monoxide), which increases disorder and therefore entropy, is a potent driving force in reductions by charcoal.[2] Escape of the gas into the open atmosphere prevents recombination of carbon monoxide's lone oxygen to re-form its "parent" substance, and this further drives the reaction. Through the ages charcoal has been heated with various metal calxes (powdery oxides) to reduce them to the corresponding pure metals.[3] The by-product, carbon monoxide, simply disappears into thin air. What a magical effect! Indeed, freshly made activated charcoal is also an incredibly powerful absorbent that can remove odors, decolorize liquids, and even make red wine look like water. What a *terrible* effect!

Figure 92 shows an oil painting on porcelain by an L. Sturm and given the title "The Alchemist."[4] The artist is likely to be a porcelain painter in Bamberg and Munich named Ludwig Sturm (1844–1926).[5] The title *could* be an apt one in the sense that the central figure appears to be performing an operation for the benefit of two wealthy clients in eighteenth-century dress. However, alchemy had reached its apex during the midseventeenth century, and by the mid-eighteenth century chemistry was firmly on its way to becoming a precise science. The gullible wealthy had also "wised up" by this period. What *is* happening in this painting? Clearly, the chemist's assistant has provided a red-hot crucible using his furnace and bellows and the chemist is adding a powdery calx to the crucible. The key to the figure is the pan of powdered charcoal that appears just in front of the wealthy client on the right. One might imagine adding the charcoal to the crucible just prior to addition of the calx. Were the calx black oxide of copper, the result would be particularly thrilling—a hissing of gas from the dark mass and appearance of a reddish, golden drop of liquid metal that would soon solidify into copper.

A phlogistonist's view would be somewhat different. Charcoal is "fatty" and loaded with phlogiston—the essence of fire. The copper calx would actually be copper devoid of phlogiston. The chemical operation shown would reduce the calx back to the metal by returning its full complement of phlogiston. The ashes remaining in the crucible would then be charcoal devoid of its phlogiston. Now was the central figure an alchemist ("puffer" or charlatan?) seducing two well-heeled investors who wished to multiply their fortunes? I suspect not. He was probably a competent early chemist or metallurgist seeking support from some wealthy investors.

---

1. Professor Joel F. Liebman, personal communication.
2. The thermodynamics of this reaction can be calculated using standard enthalpy and entropy of formation data in M.W. Chase, Jr., *NIST-JANAF Thermochemical Tables*, fourth edition, *Journal of Physical and Chemical Reference Data, Monograph* 9, 1998 (see also the NIST Website at *http://nist.gov*). For a metal having weaker affinity for oxygen, such as in mercury(II) oxide, both enthalpy and entropy favor this reaction. For a metal having somewhat stronger affinity, as in copper(II) oxide, enthalpy disfavors the reaction but is overwhelmed by entropy.
3. The most important pyrometallurgical operation is the reduction of the iron ores hematite (Fe_2O_3) and magnetite (Fe_3O_4). Although carbon (in the form of coke) is added to these ores at high heat, the chemistry is more complex than simply passing oxygen from iron to carbon. The

blast furnace housing this operation provides hot air. Oxygen in the air forms carbon monoxide from the coke and it is the carbon monoxide that reduces the iron ores by stripping them of oxygen to produce carbon dioxide. This is complemented by water also present in the blast air that likewise converts coke to carbon monoxide. Water's by-product, hydrogen, similarly strips the iron ore of oxygen to produce water. (See T.L. Brown, H.E. LeMay, Jr., and B.E. Bursten, *Chemistry—the Central Science,* seventh edition, Prentice-Hall, Upper Saddle River, NJ, 1997, pp. 872–875.

4. I am grateful to Dr. Lynn Gamwell, Director, University Art Museum, State University of New York at Binghamton for kindly providing this image from the museum collection.
5. H. Vollmer (ed.), *Allgemeines Lexikon Der Bilden den Künster von der Antike Bis Zur Gegenwart,* Verlag Von E.A. Seeman, Leipzig, 1938, Vol. XXXII, p. 257. I am grateful to Dr. Alfred Bader for commenting on this painting and making me aware of the artist.

THE SURPRISING *CHEMICAL* TAXONOMIES OF MINERALS AND MOLLUSKS

Eighteenth-century Sweden[1] enjoyed a powerful mining and metallurgy industry, becoming the main source of iron for the rest of Europe. Its other abundant natural resource—virgin forests—made Sweden a center for the lumber industry and furniture manufacture. This may explain why detailed scientific classification of both the plant and mineral kingdoms originated in Sweden during this period. On the other hand, it may simply have been the Lutheran yearning for order and harmony.

In 1753, Carolus Linnaeus (1707–1778)[2] published the *Species Plantarum,* providing the first systematic taxonomy of flowering plants and ferns. It was based largely on the external structures (morphologies) of flower parts. External appearance also played a major role in mineral classification. For example, gemstones such as diamonds and rubies were assumed to be closely related. In 1758, another Swede, Axel Frederic Cronstedt [1722 (or 1702)–1765],[3] published (anonymously) his *Försök till Mineralogie.* In this book, Cronstedt placed all minerals into four chemical groups: earths, salts, bitumens, and metals.[3] However, this crude classification was published only two years after Black reported "fixed air" and predated the chemical revolution by about two decades.

In order to illustrate the chemical confusion prevalent in eighteenth-century mineralogy, a couple of examples will suffice. Plumbago or Flanders stone is a slippery grayish mineral that darkens the hands.[4] It was also called "black lead," due to the superficial resemblance to the grayish, soft metal. Plumbago was employed in seventeenth-century pencils;[5] hence its more modern name *graphite.* Of course we still commonly employ "lead" pencils.[6] Johan Gottschalk Wallerius (1709–1785), another Swede, classified graphite as a kind of talc.[7] In 1779, the great Swedish chemist Carl Wilhem Scheele oxidized graphite with niter, collected the "fixed air" produced, and concluded that the mineral consisted of pure carbon.[8] Diamond, on the other hand, appears to be as distinct from graphite, as one can imagine. It is crystalline, clear and harder than rock. Indeed, diamond would appear to be a close relative of other rare, beautiful gems, including ruby, garnet, and sapphire. However, rumors about experiments in which diamonds were burned gained increasing credibility during the seventeenth and eighteenth centuries.[9] In 1760, François I, Emperor of Austria, described an experiment wherein diamonds and rubies were burned for 24 hours in crucibles.[9] Upon opening the crucibles, the rubies remained unchanged but the diamonds had disappeared without a trace. In 1797, Smithson Tennant oxidized diamond with nitre and proved conclusively that diamonds are also pure carbon.[10]

The blowpipe was an effective early instrument for analyzing the composition and chemistry of minerals. Although its origins are ancient, blowpipes were perfected in Sweden during the eighteenth century.[11] Scheele and Torbern Bergman used the instrument extensively. Figure 93 is from Bergman's essays and describes in detail the construction and use of a solid silver blowpipe.[12] Section A (upper right), starts with the mouthpiece and tapers to a tight fit with unit *B*, which forces the breath to make a 90° turn and collects in a trough the mist and droplets of water in the operator's exhalations. Unit *B* attaches to pipe C that ends in a smooth, small round orifice g that trains the exhalation on the candle flame. The flame is blown horizontally and its reducing or oxidizing regions may be focused on samples as desired. Tiny mineral samples may be exposed to the flame while sitting in spoon *E*, made of silver or gold, or, if the mineral is nonflammable, seated in an indentation in a piece of charcoal. Hammer *F* pounds pieces of minerals within metal ring *H* on metal plate G. Pieces of mineral are handled using forceps *I*.

Now here's the tricky part. The flow of exhaled air onto the flame must be smooth and continuous, sometimes for minutes at a time. Try *that*. However, Bergman[12] assures us that with practice the technique can be mastered—fill your cheeks with air and as you inhale and exhale through your nostrils, keep your cheeks replenished with air, and keep squeezing them steadily with the fingers of one hand so that the exhalation remains steady and continuous. Esteemed reader, you have my leave to take the day off and practice.

There are several protocols to follow. First, expose the sample to the outer (fuel-deficient), oxidizing part of the flame. If the sample survives, expose it to the tip of the blue, reducing, hotter section of the flame. If the sample will not melt in the flame, then Bergman offers three "fluxes," substances that aid fusion (melting) of samples: an acidic phosphate salt, an alkaline salt (sodium hydroxide), and a neutral salt (borax). The flux is melted; then a finely ground mineral is added and brought to reaction with the flux, and the results are recorded. Such blowpipe analyses were both highly specific and extremely sensitive.

As the eighteenth century came to a close, advances in mineral analyses and chemical theory merged to set the stage for a revolution in mineral taxonomy. Figure 94 is from the 1801 *Traité de Mineralogie* by abbé René Just Haüy (1743–1822).[13] In this book the amorphous yellow crystals of sulfur and the clear crystalline diamond, as dissimilar as they are, appear together in Plate LXII as flammable chemical elements. Haüy was the first to understand that crystals cleaved along specific faces fixed by underlying crystal symmetries. He is perhaps the principle founder of crystallography.[14] William Hyde Wollaston (1766–1828), carefully measured the angles of crystalline faces, unified the ideas of Hooke (see Figure 80) and Dalton[15] about crystal packing and applied them to understanding crystal symmetry and cleavage (see Figure 95).[14] The mineralogical studies started in Sweden combined with the "new chemistry" enabled American mineralogist James Dwight Dana to propose, in 1837, the chemical classification of minerals that remains functional today.[16]

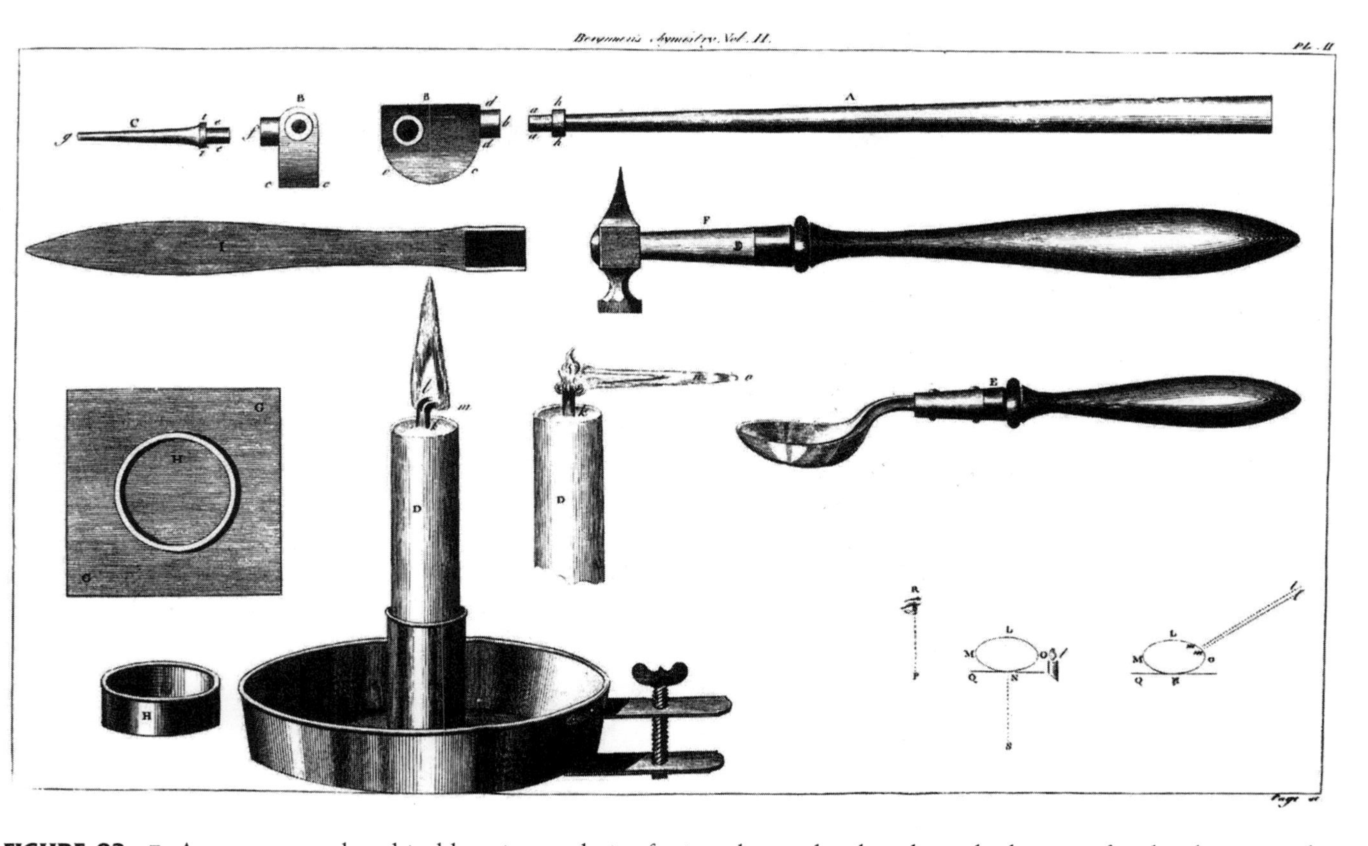

FIGURE 93. ▪ Apparatus employed in blowpipe analysis of minerals was developed to a high state of technology in eighteenth-century Sweden (from Torbern Bergman's 1788 *Physical and Chemical Essays*).

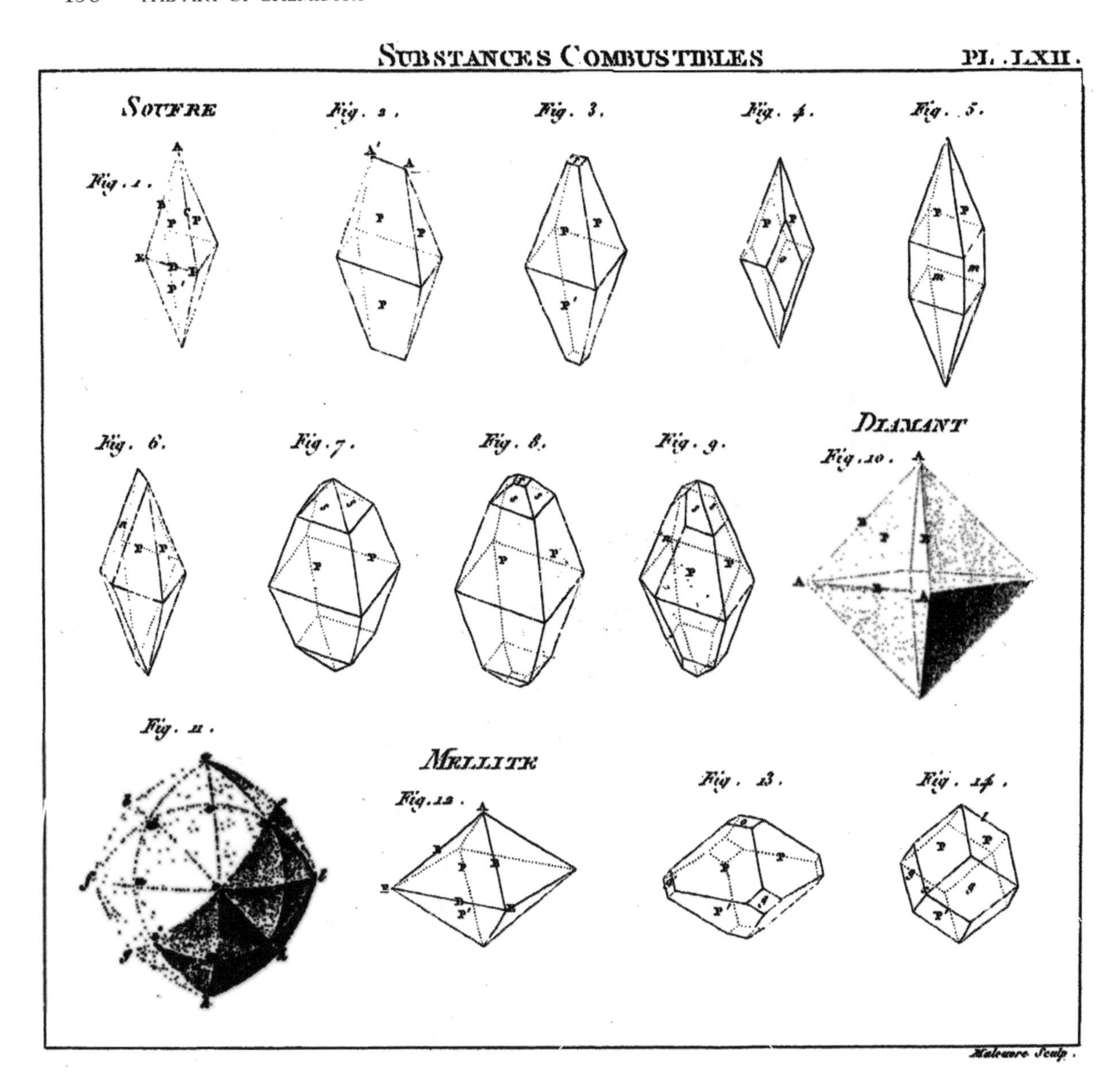

FIGURE 94. ■ Chemical classification of minerals illustrated in René Haüy, *Traité de Minéralogie*, Paris, 1801 (courtesy Chemical Heritage Foundation).

Now, how do these discussions of minerals relate to mollusks? The systematics of the Linnaeus taxonomy were based on external structure (morphology). Developed a century before Darwin's discovery of evolution, it lacked the insights derived from the theory of natural selection. Thus, we now better understand the external similarity between sharks and dolphins (frequently confused by nervous swimmers) by understanding the parallel evolutionary paths that allow them to occupy similar environments and niches. Morphology here is very misleading—the shark, a fish, and the dolphin, a mammal, are about as

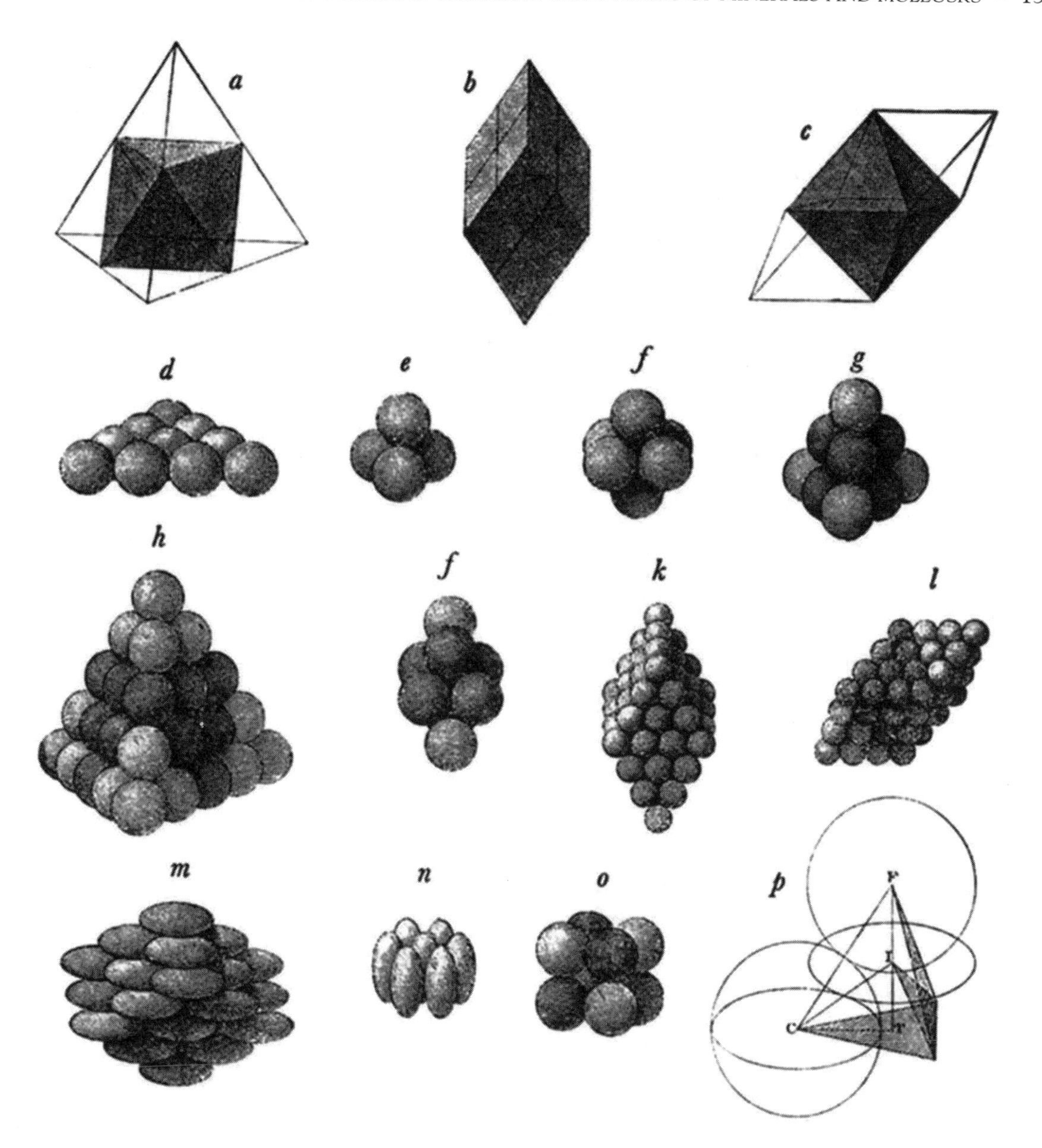

FIGURE 95. ▪ William Hyde Wollaston's explanation of crystalline faces (from C. Singer, *The Earliest Chemical Industry*, The Folio Society, with permission from The Folio Society).

different as a ruby and a diamond. In contrast, dolphins and horses seem to be as morphologically unrelated as graphite and diamond. However, they both are warm-blooded, give birth to live young free of eggs, and nurse their young. The hidden chemical reality of graphite and diamond is that they are both pure carbon.

There are six species of clams shown schematically in Figure 96.[17] Four of these are species in the genus *Calyptogena*, and the remaining two are from genera *Vesicomya* and *Ectenagena*. These classifications have been based largely on the morphologies of the clams' shells since this is the portion that survives once the clam dies and also permits linkages with fossil ancestors (see the essay on Lamarck later in this book, see p. 194). However, during the latter part of the twentieth century, new chemical tools, including protein sequencing and later DNA sequencing, were developed to study the hidden dimensions in such phylogenetic relationships. Specifically, each characteristic protein in an organism is coded for by a gene. The hemoglobin of a horse more closely resembles that of a cow than that of a mouse. In this case, at least, the obvious morphological relationship reflects the underlying genetic differences. A new field termed "genomics" arose toward the end of the twentieth century as advances in chemical analysis, automation, and bioinformatics allowed direct comparisons of the huge genetic sequences of different organisms. The results are occasionally quite surprising.

Figure 96 depicts the systematic relationships between the six clam species explored based not on the morphologies of the shells but rather on the DNA sequences pertaining to mitochondrial oxidase subunit I.[17] The phylogenetic systematics shown are displayed as a cladogram[18] in which each branch in the tree represents a distinct modification to form a new species. Genomically, the relationships between the clams are quite different from the classification based on morphology (Figure 96). The hidden chemical reality is considerable different from the conclusions based on structure. Thus, on the basis of this chemi-

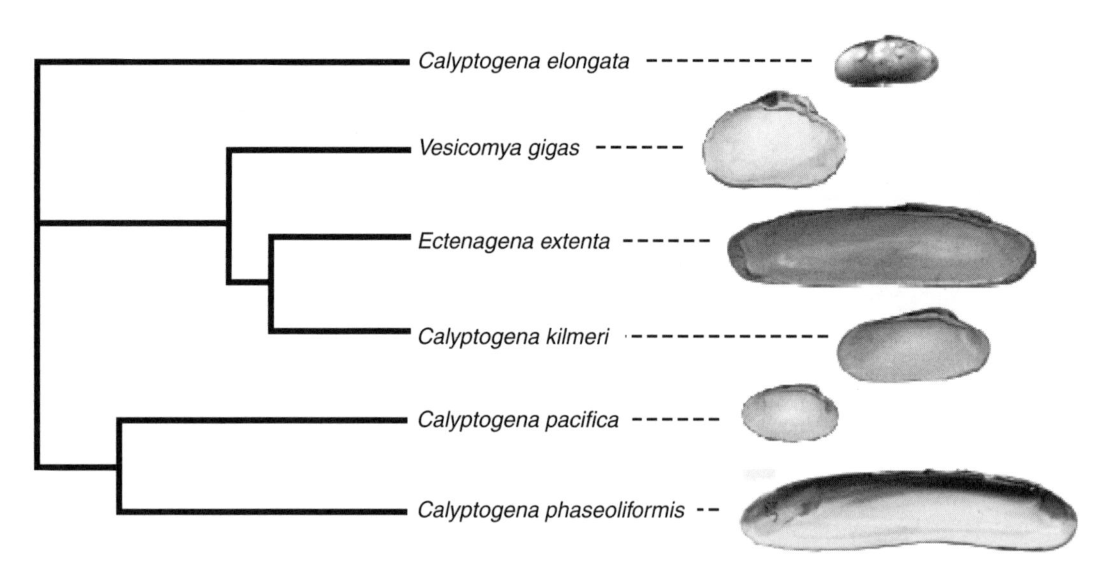

FIGURE 96. ■ Modern cladistic display of phylogenetic relationships among selected clams. These classifications are based on chemical criteria (DNA sequences) rather than shell structure (morphology). These biochemical studies indicate that C. *kilmeri* is genetically more similar to V. *gigas* and E. *extenta* than it is to the more morphologically similar C. *elongata*, C. *pacifica*, and C. *phaseoliformis*. Mineralogy underwent a similar evolution from morphological to chemical taxonomy more than two centuries ago. Prior to 1770, diamond was thought to be closely related to ruby and totally unrelated to plumbago (graphite). However, chemistry showed that diamond and plumbago are each pure carbon and are both totally unrelated to ruby. (I am grateful to Professor Robert Vrijenhoek for introducing me to cladistic concepts, discussing his studies on clams, and furnishing the figures shown.)

cal classification, *V. gigas*, *E. extenta*, and *C. kilmeri* are representatives of three different genera but could be assigned to one genus since they are "monophyletic" (trace to a common ancestor). The cladogram indicates that *C. pacifica* and *C. kilmeri*, currently classified in the same genus, have less in common than do *C. kilmeri* and *E. extenta* (*E. extenta* appears to be morphologically distant from *C. kilmeri*).[17] Professor Robert Vrijenhoek discovered that clams in Monterey Bay called *Calyptogena pacifica* in fact represent three morphologically similar but genetically distinct species occupying different depths. The true *Calyptogena pacifica* is not even found in Monterey Bay but in the vicinity of Washington State.[17]

As Genomics passes through its infancy it begins to raise many difficult questions. For example, not all proteins reflect identical phylogenetic relationships between closely related species. The use of the specific DNA segment noted above to determine the phylogenetic relationships between the six clams depicted in Figure 96 is not the only possibility. As more knowledge is gained in this revolutionary field, decisions on most appropriate sequences as well as weighting factors may emerge. Does this return us to the subjectivity of morphological classification, or can certain "critical" genes be accepted as the determining indices? These are certainly much more complex questions than those raised two centuries earlier about minerals that are chemically simpler, lack functional tissues and organs, do not metabolize other minerals, and, so far as we know, neither mate nor evolve.

---

1. *The New Encyclopedia Britannica*, Encyclopedia Britannica Inc., Chicago, 1986, Vol. 28, pp. 332–350.
2. *The New Encyclopedia Britannica*, op. cit., Vol. 7, pp. 379–380.
3. J.R. Partington, *A History of Chemistry*, Macmillan & Co., Ltd., 1961, Vol. 2, pp. 173–175.
4. Partington, op. cit., p. 91.
5. Partington, op. cit., p. 104.
6. Long-time friend Professor Joel F. Liebman is a theoretician who has not done an experiment for at least 35 years. He quips that the only chemical research hazard he faces is "lead poisoning" from an inadvertent jab from his own pencil.
7. Partington, *A History of Chemistry*, Macmillan & Co., Ltd., London, 1962, Vol. 3, p. 170.
8. Partington (1962), op. cit., pp. 216–217.
9. J.-P. Poirier, *Lavoisier—Chemist, Biologist, Economist*, Engl. transl., University of Pennsylvania Press, Philadelphia, 1996, pp. 47–50.
10. Partington (1962), op. cit., pp. 703–705.
11. J.J. Berzelius, *The Use of the Blowpipe in Chemical Analysis, and in the Examination of Minerals* (transl. J.G. Children), London, 1822.
12. T. Bergman, *Physical and Chemical Essays* (transl. E. Cullen, J. Murray, London, 1788, pp. 471–529.
13. R. Haòy, *Traité de Minéralogie*, Louis, Paris, 1801.
14. C. Singer, *The Earliest Chemical Industry*, The Folio Society, London, 1948, pp. 291–307. We are grateful to the Folio Society for permission to reproduce this figure.
15. A. Greenberg, *A Chemical History Tour*, John Wiley & Sons, New York, 2000, pp. 173–175.
16. *The New Encyclopedia Britannica*, op. cit., Vol. 24, pp. 129–138.
17. I am grateful to Professor Robert Vrijenhoek, who helped inspire this essay and who supplied the drawings in Figure 96 as well as extremely helpful discussions. I also wish to acknowledge helpful conversations with Professors Judith Weis and Will Clyde.
18. I.J. Kitching, P.L. Forey and D.M. Williams, in *Encyclopedia of Diversity*, S.A. Levin (ed.), Vol. 1, pp. 677–707.

THERE IS TRUTH IN CHALK

In vino veritas ("there is truth in wine") implies that, suitably "lubricated," a person may be more likely to confess all. However, the German apothecary Johann Friedrich Meyer (1705–1765)[1] might well have said "*In calcis veritas*" ("there is truth in chalk")[2] since it is through calcium carbonate (chalk; limestone) that he discovered his *acidum pingue*—said to be the general principle innate in all bodies, the principle in fire, and the component of all acids (see Figure 97).[3] And *well* might he have said it since he confessed to consuming 1200 pounds of chalk over eight years to cure his own violent stomach acidity.[4]

Until 1756, the only known gas or "air" was indeed common air. In that year, Dr. Joseph Black published his paper on the isolation and properties of "fixed air" (carbon dioxide or CO_2).[5] He had used the pneumatic techniques of Stephen Hales and William Brownrigg to capture the "air" that was "fixed" in chalk ($CaCO_3$). Although Van Helmont had worked with this "air" over 100 years earlier, neither he nor other contemporaries truly characterized it.

In Black's time only three major alkalis were recognized:[5] vegetable, marine, and volatile. Each of these was found in "mild" and "caustic" forms. Black's careful quantitative studies correctly convinced him that loss of "fixed air" is what converted mild alkalis to caustic alkalis. In modern terms, this can be summarized as follows:

Alkali Family	"Mild" Alkali	$\xrightarrow{-CO_2}$ $+H_2O$	"Caustic" Alkali
Vegetable	K_2CO_3		KOH
Marine	Na_2CO_3		NaOH
Volatile	$(NH_4)_2CO_3$		NH_4OH

Meyer's theory was essentially the reverse of Black's.[1] Meyer's *acidum pingue* ("fatty" or "oily acid") was said to be a component of all acids. When the mild alkalis (which Black understood to be carbonates) were reacted with acids, the effervescence indicated absorption of the *acidum pingue* found in all acids. Caustic alkalis were, as noted above, saturated with *acidum pingue* and thus did not effervesce when reacted with acids. The slippery feeling of caustic alkalies arose from the "oily acid" saturating them. Figure 97 depicts a table of affinities with *acidum pingue*.

At this point, parallels with phlogiston theory become all too apparent.[1] Metal calxes were said to gain phlogiston and become metals when heated with charcoal, a substance laden with phlogiston. Lavoisier established that, to the contrary, the calxes actually *lost* oxygen to the charcoal to form CO_2. According to Meyer, mild alkalis were said to gain *acidum pingue* from the fire to become strong alkalis. Black established that, in fact, they *lost* CO_2 in these transformations.

To add to this confusion, however, conversion of a metal such as calcium to its calx in the fire would require adding *acidum pingue* from the fire. The calx is

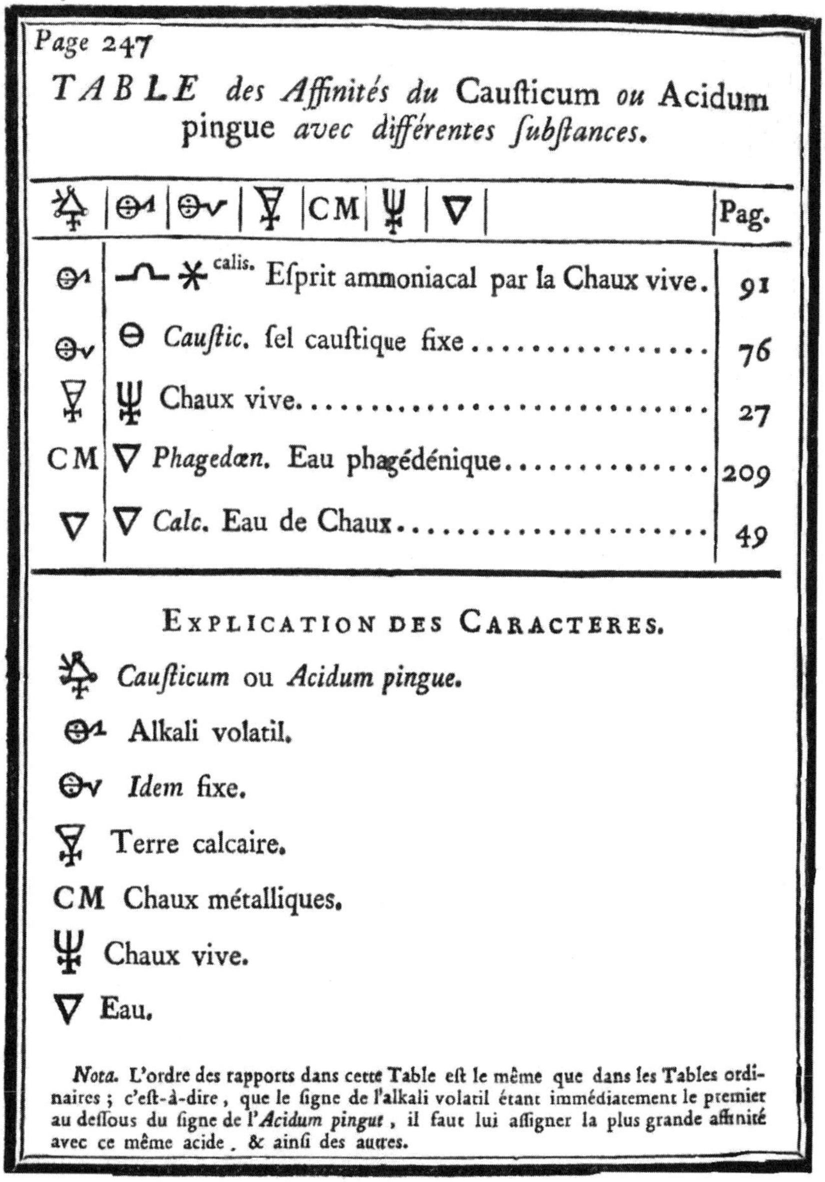

Page 247

TABLE des Affinités du Caufticum *ou* Acidum pingue *avec différentes fubftances.*

	⊖1	⊖v		CM		▽	Pag.

		Pag.
⊖1	calis. Efprit ammoniacal par la Chaux vive.	91
⊖v	⊖ *Cauftic.* fel cauftique fixe................	76
	Chaux vive.........................	27
CM	▽ *Phagedæn.* Eau phagédénique..............	209
▽	▽ *Calc.* Eau de Chaux....................	49

EXPLICATION DES CARACTERES.

Caufticum ou *Acidum pingue.*

⊖1 Alkali volatil.

⊖v *Idem* fixe.

Terre calcaire.

CM Chaux métalliques.

Chaux vive.

▽ Eau.

Nota. L'ordre des rapports dans cette Table eft le même que dans les Tables ordinaires ; c'eft-à-dire, que le figne de l'alkali volatil étant immédiatement le premier au deffous du figne de l'*Acidum pingut*, il faut lui affigner la plus grande affinité avec ce même acide, & ainfi des autres.

FIGURE 97. ▪ A table from Johann Friedrich Meyer's 1766 *Essais de Chymie*. He believed in an *acidum pingue* ("fatty" or "oily" acid) present in strong ("caustic") alkali (e.g., KOH) and absent in mild alkali (e.g., K_2CO_3). Addition of *acidum pingue* actually corresponded to loss of CO_2 much as the addition of phlogiston corresponded to the loss of another invisible gas—oxygen.

indeed heavier. However, addition of *acidum pingue* to mild alkalis to form caustic alkalis would suggest that the latter should also be heavier. They are, however, lighter because of loss of carbon dioxide during the transformation. The source for this "confusion within confusion" is that phlogiston (*terra pinguis* or "fatty earth") is the essence of fire internal to a combustible substance. In contrast, *acidum pinguis* is a component of fire that acts as an external agent to the combustible material. Black and Lavoisier "trimmed the fat" from both *acidum pingue* and *terra pinguis* or, perhaps, emulsified these theories.

---

1. J.R. Partington, *A History of Chemistry*, Vol. 3, Macmillan & Co. Ltd., London, 1962, pp. 145–146.
2. "There is truth in chalk" is self-evident to any teacher who has not been fully converted to the computerized classroom.
3. (J.)F. Meyer, *Essais de Chymie, Sur La Chaux Vive, La Matiere Elastique et Electricque, Le Feu, Et L'Acide Universel Primitif, Avec un Supplement Sur Les élements* (M.P.F. Dreux, transl.), Vol. 1, Chez G. Cavalier, Paris, 1766, p. xv.
4. J. Ferguson, *Bibliotheca Chemica*, Vol. II, Derek Verschoyle Academic and Bibliographical Publications Ltd., London, 1954, p. 93.
5. Partington, op. cit., pp. 135–143.

SECTION V
TWO REVOLUTIONS IN FRANCE

IN THE EARLY MORNING HOURS OF THE CHEMICAL REVOLUTION

Revolutions usually begin quietly, build momentum, and then explode with a cataclysmic event such as the storming of the Bastille. So, too, did the chemical revolution begin with early rumbles of discontent with phlogiston theory. The observed increases in weight as metals *lost* phlogiston to become their calxes were explicable only if chemists repressed all their experiences and common sense and accepted negative mass. Often, early acts of rebellion happen quite innocently, and the consequences are understood much later. Stephen Hales' publication in 1727 of his methods for collecting gases led to Joseph Black's report of carbon dioxide in 1756, the isolation of hydrogen gas by Henry Cavendish in 1766, and ultimately the isolation of oxygen gas by Carl Wilhelm Scheele in 1771 or 1772 and Joseph Priestley in 1774. Cavendish initially thought that he had actually captured the phlogiston released from metals by the aqueous acids that converted them to calxes. He could not possibly have understood that the gas actually came from the "element" water and not the metal. Scheele and Priestley imagined the "air" they captured as strongly attracting phlogiston, the essence of fire, from combustibles and metals. Perhaps *the* cataclysmic event in the chemical revolution was the full realization in 1783 that Cavendish's phlogiston and Priestley's dephlogisticated air combined to give water and did not simply disappear with a "poof" into "air" or simply *nothing*.

Figures 98–107 are from the fabulous 35-volume, folio encyclopedia of philosopher Denis Diderot (1713–1784) and mathematician Jean Le Rond D'Alembert (1717–1783), published between 1751 and 1772, with later supplements (1776–1780).[1] Diderot's progressive philosophy had earlier run afoul of the reactionary French Church and State and he spent three months in prison during 1749.[2] The encyclopedia, suffused with the more progressive thinking of the Enlightenment, is still highly treasured for its exquisite printing as well as content. The elegant symbols in Figures 98–101 illustrate the "Babel" of chemical nomenclature that existed prior to publication by De Morveau, Lavoisier, and their colleagues of the *Nomenclature Chimique* in 1787. These figures combine elements, alloys and other mixtures, compounds, chemical operations, quantities, glassware, and other apparatus. We will very briefly describe only a few of these.

The ancients likened the seven known metals to the sun, the moon and the five known planets (Mercury, Venus, Mars, Jupiter, and Saturn)[3]—see Figures 70 and 71. The first symbol in column 1 of Figure 98 represents the alloy steel (*Acier*), and its resemblance to iron (*Fer*—Figure 99, column 1) is obvious. *Limaille d'Acier* (Figure 99, column 2) refers to steel shavings or powder. The shield and spear of the god of war Mars symbolized the metal widely used in weaponry, whose red calx suggested the red planet. *Airain brulé* (Figure 98, column 1) is roasted brass (or bronze), an alloy of copper (*Cuivre* or Venus, Figure 101, column 2—the mirror of Venus) and tin (*Estain*, Figure 99, column 1—Jupiter—the

Pl. 1

Acier. Azur. Chaux d'Œufs. Cucurbite.
Airain brulé. Bain Marie. Chaux d'Or. Cuillerée.
Air. Bain de Vapeurs. Chaux de Vitriol. Demie dragme.
Alembic. Blanc d'Espagne. Chaux vive. Demie livre.
Alun. Bol Armenien. Chopine. Demie once.
Alun de Plume. Baume. Cinabre. Digérer.
Amalgame. Borax. Cinabre d'Antimoine. Distiller.
An. Brique. Cire. Eau.
Ana. Brique pulvérisée. Coaguler. Eau commune.
Antimoine. Calciner. Congeller. Eau forte.
Argent. Camphre. Corail. Eau regale.
Regule d'Arsenic. Cimenter. Corne de Cerf. Eau de Vie.
Athanor. Cendre. Crane Humain. Ecorce de grenade.
Vitriol Rouge. Cendre Clavelée. Creuset. Ecume de Nitre.
Vitriol blanc. Céruse. Cristal. Esprit.
Aimant. Chaux. Cristal de Saturne. Esprit de Vin.

Caracteres de Chymie.

FIGURE 98. ■ Chemical symbols (see text) from the eighteenth-century encyclopedia published by philosopher Denis Diderot and mathematician Jean Le Rond D'Alembert.

FIGURE 99. ■ Chemical symbols (see text) from the eighteenth-century encyclopedia published by Diderot and D'Alembert.

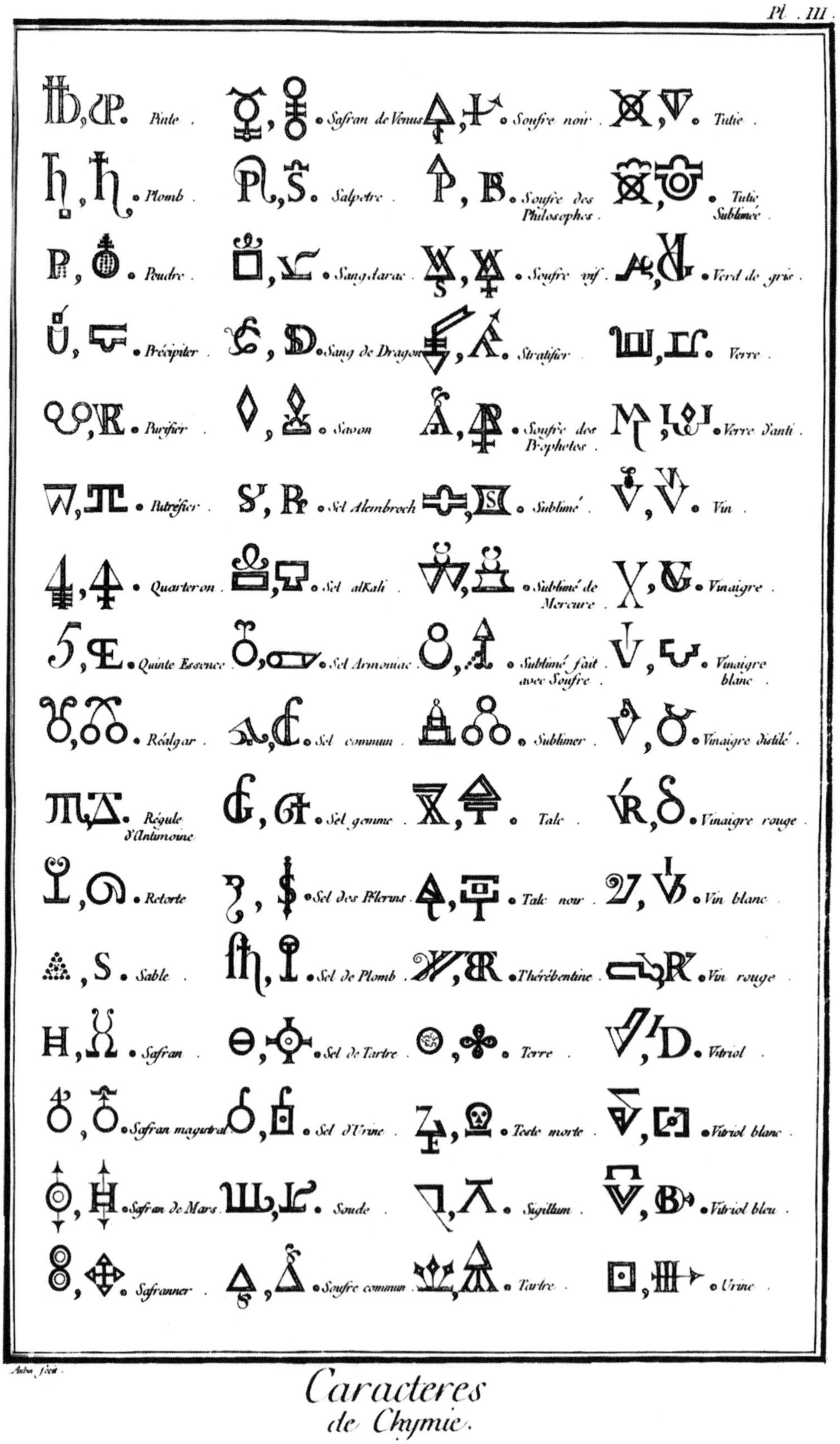

Pl. III.

Pinte. Safran de Venus. Soufre noir. Tutie.
Plomb. Salpetre. Soufre des Philosophes. Tutie Sublimée.
Poudre. Sangdarac. Soufre vif. Verd de gris.
Précipiter. Sang de Dragon. Stratifier. Verre.
Purifier. Savon. Soufre des Prophetes. Verre d'anti.
Putréfier. Sel Alembroch. Sublimé. Vin.
Quarteron. Sel alkali. Sublimé de Mercure. Vinaigre.
Quinte Essence. Sel Armoniac. Sublimé fait avec Soufre. Vinaigre blanc.
Réalgar. Sel commun. Sublimer. Vinaigre distilé.
Régule d'Antimoine. Sel gemme. Talc. Vinaigre rouge.
Retorte. Sel des Pélerins. Talc noir. Vin blanc.
Sable. Sel de Plomb. Thérébentine. Vin rouge.
Safran. Sel de Tartre. Terre. Vitriol.
Safran magistral. Sel d'Urine. Teste morte. Vitriol blanc.
Safran de Mars. Soude. Sigillum. Vitriol bleu.
Safranner. Soufre commun. Tartre. Urine.

Aubin fecit.

Caracteres de Chymie.

FIGURE 100. ■ Chemical symbols (see text) from the eighteenth-century encyclopedia published by Diderot and D'Alembert.

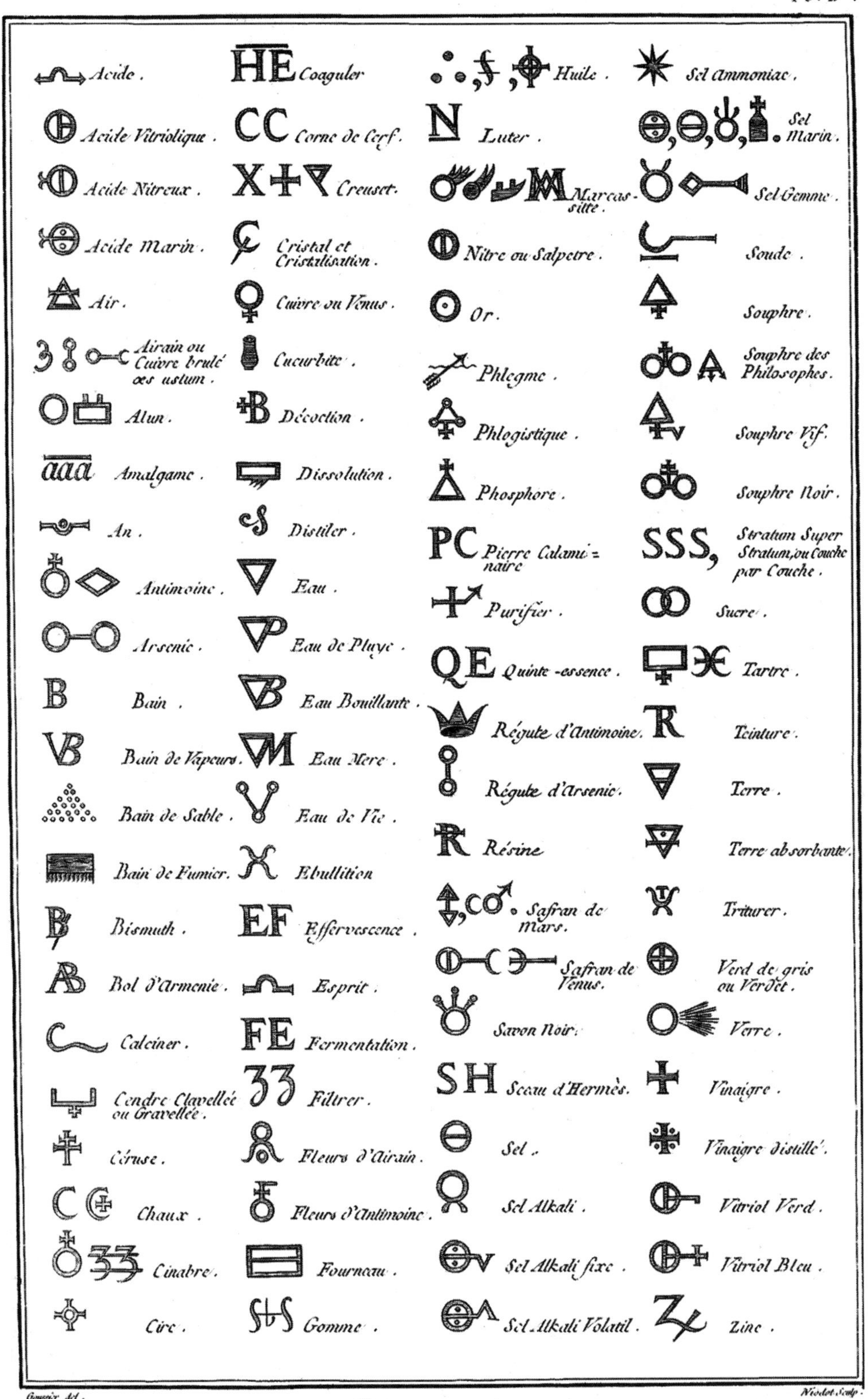

FIGURE 101. ■ Chemical symbols (see text) from the eighteenth-century encyclopedia published by Diderot and D'Alembert.

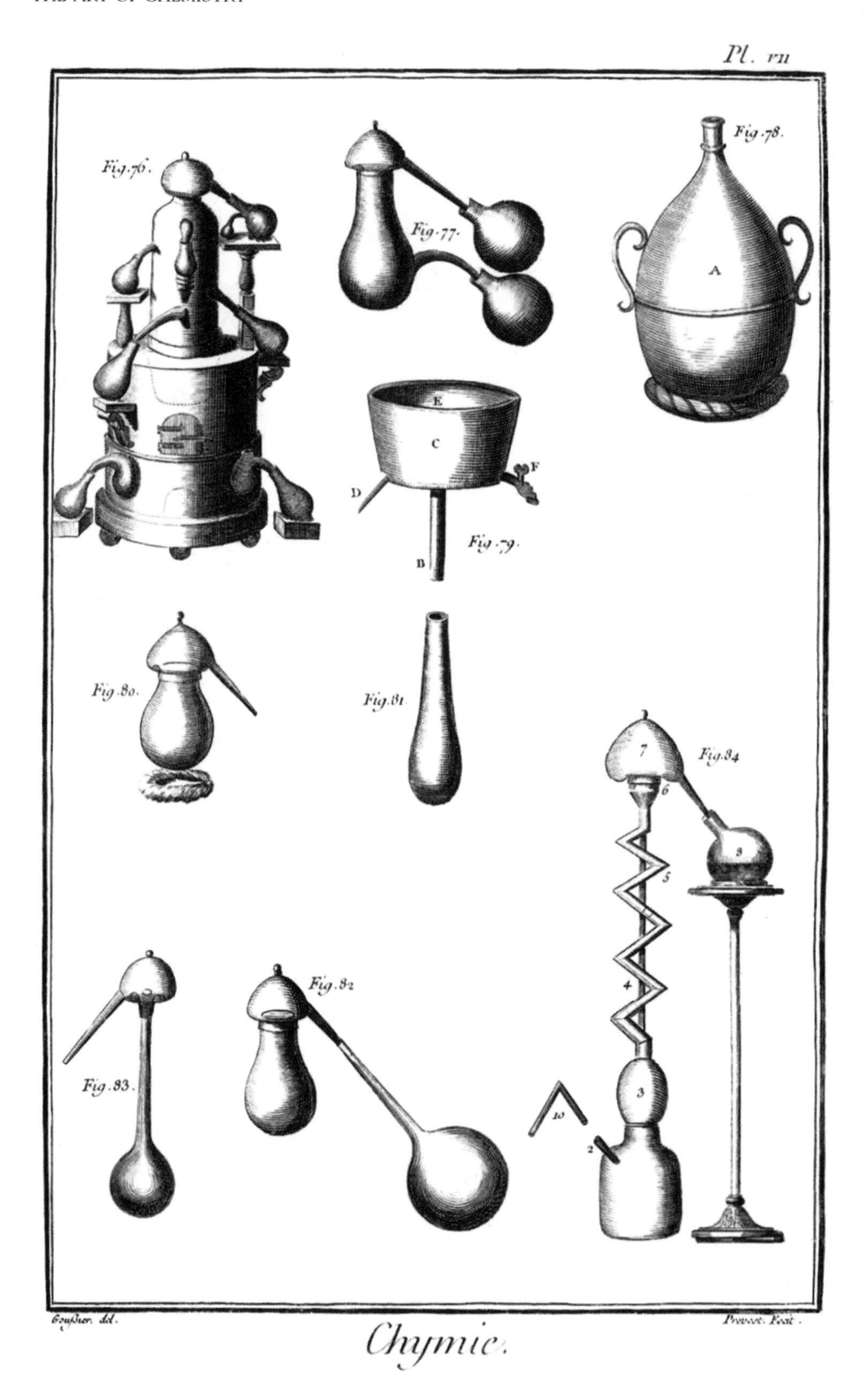

FIGURE 102. ▪ Distillation apparatus (see text) from the eighteenth-century encyclopedia published by Diderot and D'Alembert.

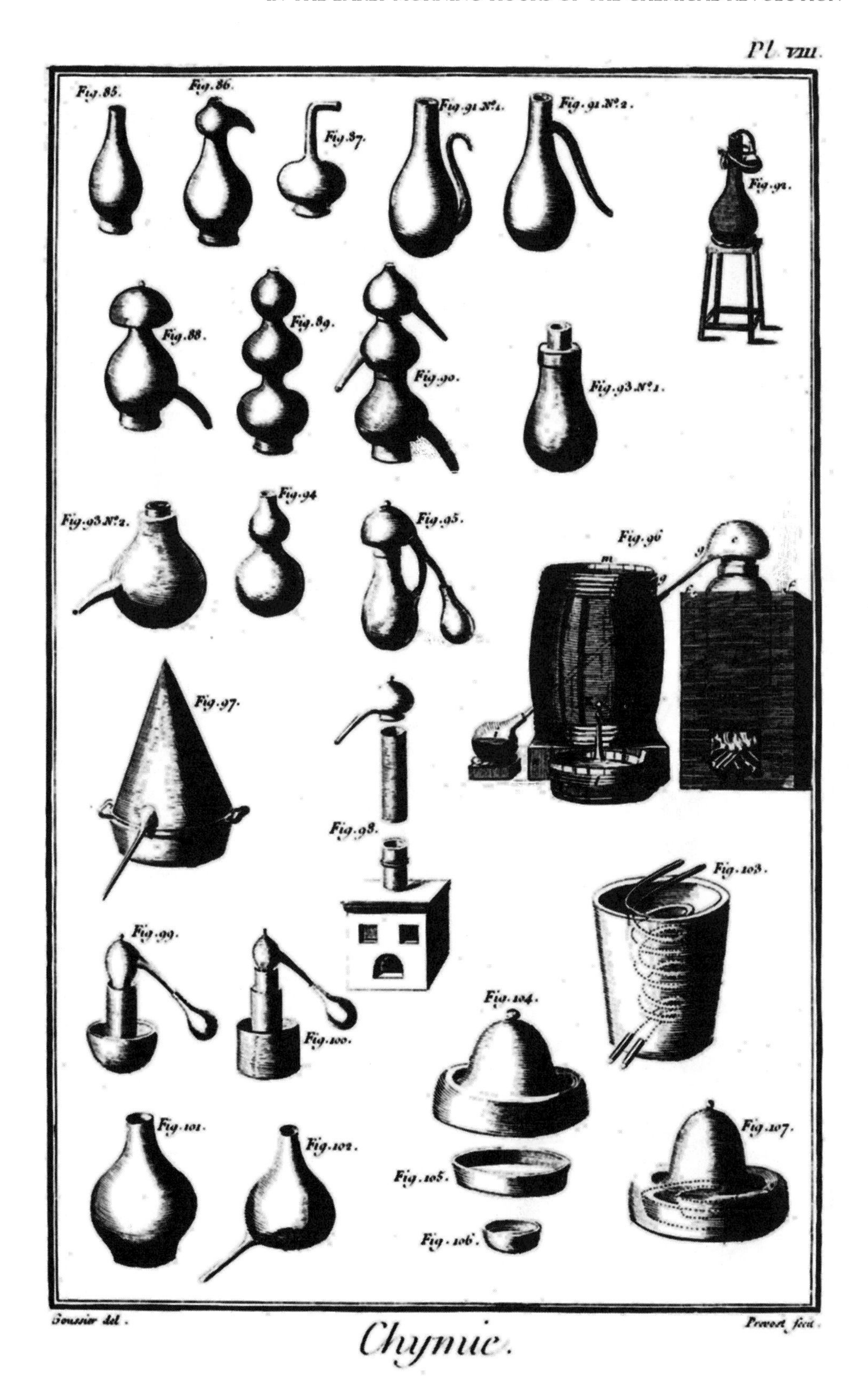

FIGURE 103. ■ Distillation apparatus (see text) from the eighteenth-century encyclopedia published by Diderot and D'Alembert.

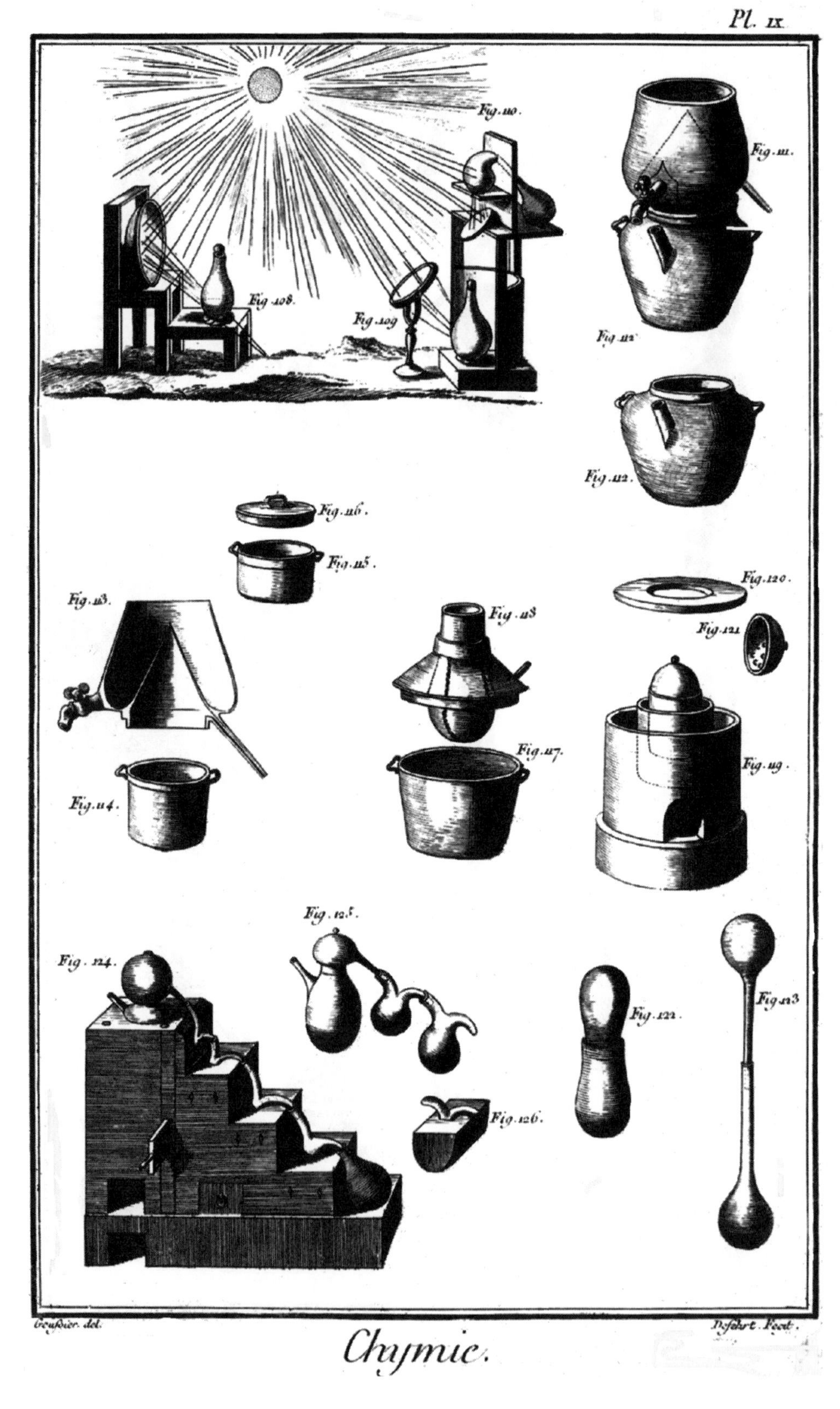

FIGURE 104. ■ Distillation and recirculation apparatus (see text) from the eighteenth-century encyclopedia published by Diderot and D'Alembert.

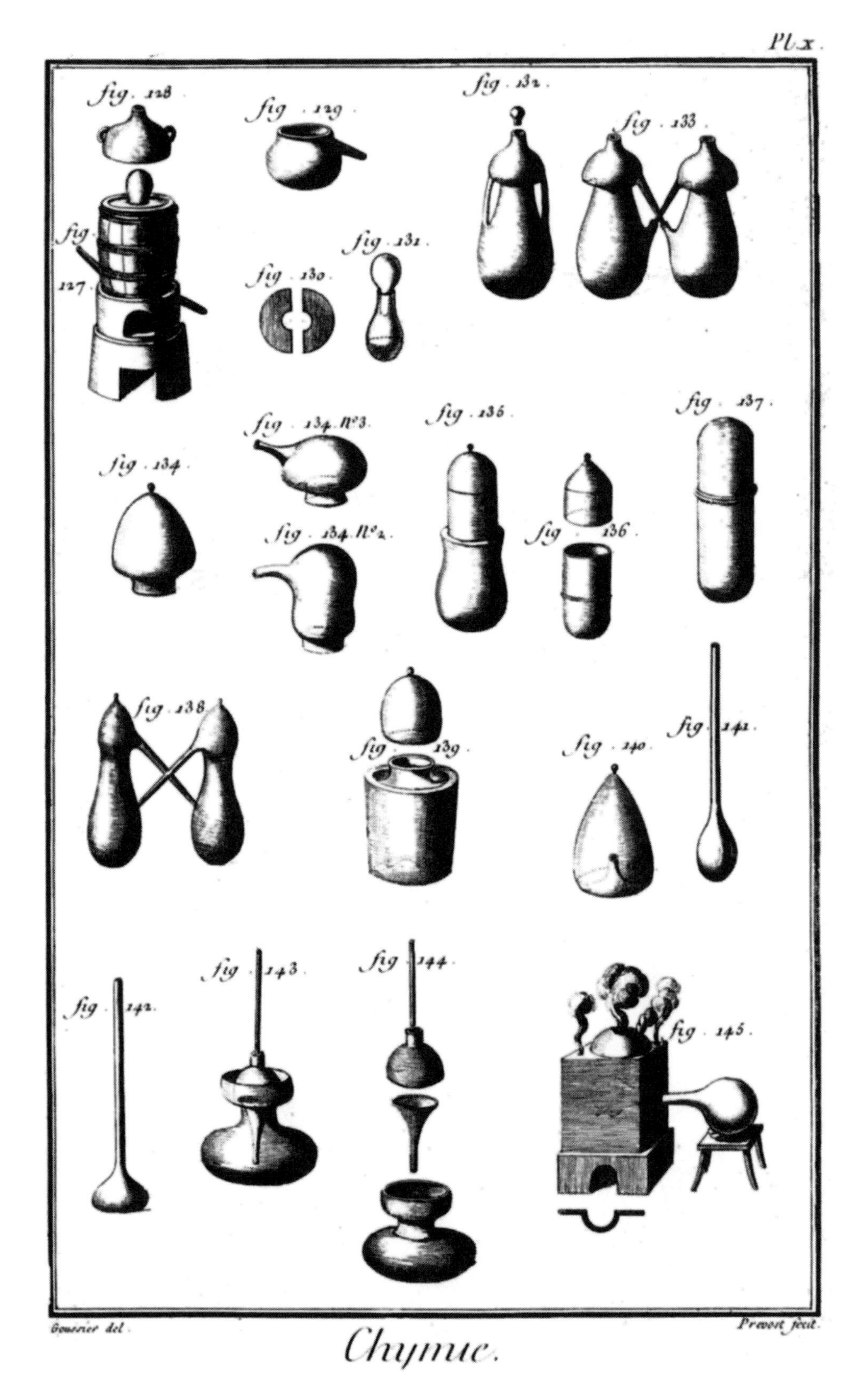

FIGURE 105. ■ Various types of recirculation apparatus (see text) from the eighteenth-century encyclopedia published by Diderot and D'Alembert.

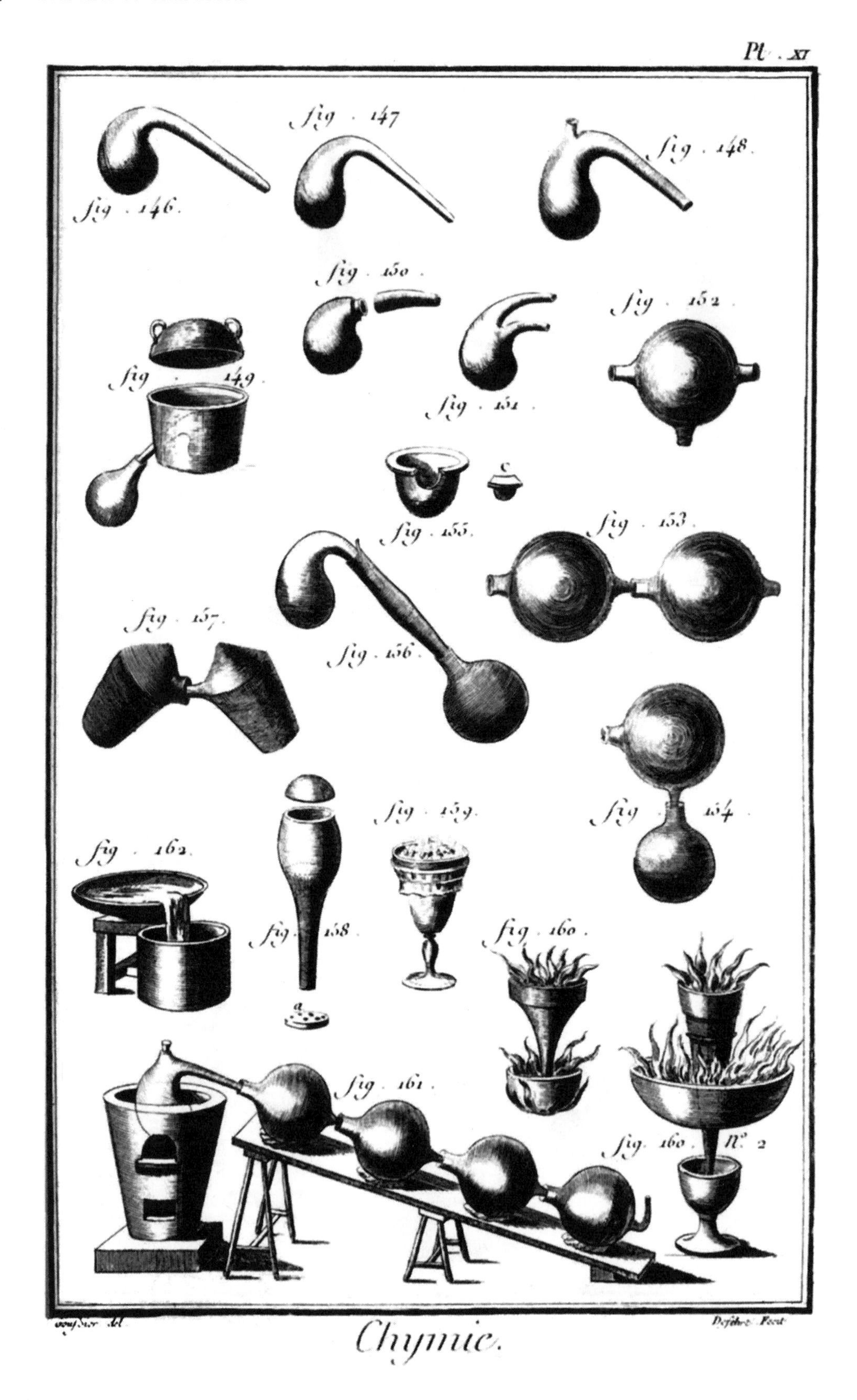

FIGURE 106. ■ Various retorts and "balloon" apparatus (see text) from the eighteenth-century encyclopedia published by Diderot and D'Alembert.

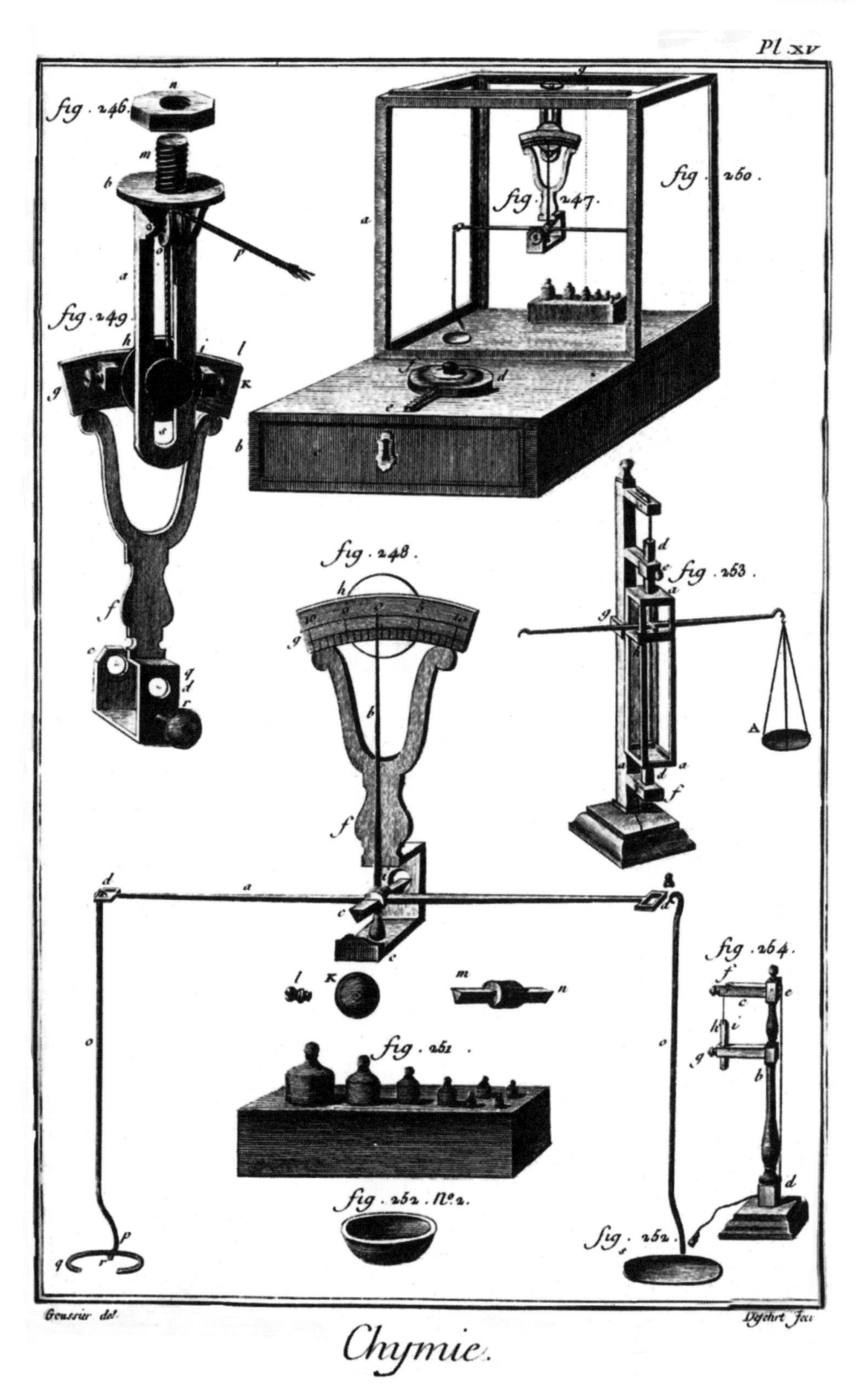

FIGURE 107. ▪ Fully assembled chemical balance belonging to Guillaume François Rouelle, demonstrator at the *Jardin du Roi* in Paris, depicted in the eighteenth-century encyclopedia published by Diderot and D'Alembert. G.F. Rouelle's demonstrations inspired Antoine Lavoisier to enter chemistry. Rouelle was a firm phlogistonist, and in less than two decades, his former student would demolish phlogiston with the modern theory of oxidation.

symbol may be derived from the Arabic four possibly because Jupiter is the fourth most distant planet from earth[3]). The outermost planet visible to the naked eye was Saturn—the slowest-moving according to earth-bound observers. Dark, dense lead (*Plomb*, Figure 100, column 1) or Saturn is symbolized by the scythe of the slow-moving god of sowing or seed. The caduceus of Mercury is itself a symbol consisting of male and female snakes (Figure 99, column 3) entwined about the god's staff. Silver (*Argent*) is the moon (Figure 98, column 1) and gold (*Or*) the sun (Figure 99, column 4).

Antimoine (Figure 98, column 1) actually represents stibnite (Sb_2S_3), while "regulus" (the refined state) of antimony (Figure 100, column 1) is actually the pure element.[4] "Flowers of antimony" (Figure 101, column 2, symbol 21) refers to what we now recognize as antimony oxide (Sb_2O_3) ("flowers" symbolize volatile salts that may be sublimed).[4] In Figure 100 there are no less than five listings for sulfur (*soufre*): common, black (*noir*), living or natural (*vif*), "philosophical sulfur" (one of Paracelsus' three principles "actually" derived alchemically from gold—but never mind) and "Sulfur of the Prophets." "Spanish white" (Figure 98, column 2, symbol 4)[4] can be bismuth oxychloride or bismuth oxynitrate but I see no resemblance to the symbol for bismuth itself (Figure 101, column 1). The *Bain Marie* (Figure 98, column 2, symbol 2) is the hot-water bath developed in ancient times by Maria Prophetessa (Mary the Jewess—a genuine historical alchemist[5]) for gentle controlled heating of chemical vessels. The following symbol represents a steam bath. *Aimant* (Figure 98, column 1) is a magnet. Phlogiston, of course, occupies an honored place (Figure 101, column 3). Incidentally, *Phlegme*, just above phlogiston, refers to an aqueous distillation fraction and not to the "yucchy" stuff the Brits call "catarrh." *Cornede Cerf* (Figure 98, column 3) is deerhorn. "Blood of the Dragon" (often symbolizing the Philosopher's Stone[6]—see Figure 20) is listed in Figure 100 (column 2, symbol 4).

Some of the glassware and other apparatus that preceded and ushered in the chemical revolution are depicted in Figures 102–107. Many of these predated the revolution by hundreds of years. In Figure 103 a complete alembic (*Fig.* 86) is accompanied by its cucurbit (*Fig.* 85) minus still head. *Figure* 89 and *Fig.* 90 each depict stacks of three aludels, often used for sublimation. *Figure* 97 is referred to as a "chapel of the ancients used to distill their rose water"; it is often referred to as a "rosenhut," which acts as a primitive air condenser.[7] *Fig.* 96 shows a complete distillation apparatus for obtaining spirit of wine, aromatic spirits, and essential oils. A condenser pipe passes through a barrel filled with ice water. An even more effective condenser system is the "double serpent" winding through a chamber filled with ice water. One serpent can be used to condense spirit of wine, the other for aromatic spirits or essential oils—drink and perfume, separate and not co-mingled. *Figure 107* depicts an eighteenth-century apparatus for distillation using the heat of the sun. The inner glass bowl containing the sample to be distilled is placed in a larger earthen bowl covered by a glass bell jar. The apparatus is set in a base, and volatiles from the inner vessel collect on the inside of the bell jar, flow down the sides, and collect in the larger bowl.

The top of Figure 104 depicts three different means for using the sun to provide heat for distillation. *Figures 122* and *123* show a double cucurbit and a double matras joined so as to allow reflux or "cohabation" (redistillation or "circulation"). The wonderful apparatus in *Fig. 124* also serves for cohobation. In Figure 105, the pelican (*Fig. 132*) and double pelicans (*Figs. 133 and 138*) also serve the purpose of cohobation. *Figure 127* (with *Fig. 128*) is an apparatus designed to provide a smoke bath for a reaction vessel. *Figure 143* depicts the "Enfer de Boyle" or Robert Boyle's "hell"—a piece of apparatus for fully sealing airtight a chemical mixture that will be strongly heated by fire.

Figure 106 shows five styles of retorts at the top. The glass balloons depicted in *Fig. 153* are combined to form an apparatus (*Fig. 161*) for distillation of vapors of antimony powder used in a canon powder also including charcoal and saltpetre. *Figure 159* is a glass *descensum* apparatus for filtration of melts. If a melt tends to resolidify before purification, modified, heated apparatus such as *160* and *160 No. 2* may be employed.

Figure 107 shows details as well as the fully assembled assaying balance belonging to Guillaume François Rouelle (1703–1770).[8] Rouelle was demonstrator in chemistry from 1742 through 1768 at the *Jardin du Roi* in Paris. He was, not surprisingly, a firm believer in phlogiston. His dynamic lecture style has been described thusly:[8] "On entering the laboratory for his lecture he was correctly dressed in velvet and with a powdered wig, holding his three-cornered hat under his arm. As he warmed to his subject he dispensed with hat, wig, coat, and waistcoat in turn." One of Rouelle's entranced students[8] was a certain Antoine Laurent Lavoisier, who would eventually become the "master of the chemical balance" and, in so doing, overthrow the phlogiston theory that captivated his teacher.

1. *The Haskell F. Norman Library of Science and Medicine*, Part II, Christie's, New York, 1998, pp. 124–125.
2. *The New Encyclopedia Britannica*, Encyclopedia Britannica, Inc., Chicago, 1986, pp. 79–81.
3. J. Read, *Prelude to Chemistry*, The Macmillan Co., New York, 1937, pp. 88–89.
4. J. Eklund, *The Incompleat Chymist—Being An Essay on the Eighteenth-Century Chemist in His Laboratory, with a Dictionary of Obsolete Chemical Terms of the Period*, Smithsonian Institution Press, Washington, DC, 1975.
5. Read, op. cit., p. 128.
6. Read, op. cit., pp. 15, 195.
7. Read, op. cit., pp. 76–77.
8. J.R. Partington, *A History of Chemistry*, Macmillan & Co. Ltd., Vol. 3, 1962, pp. 73–76.

AN EARLY BUT DISTANT RELATIVE OF THE PERIODIC TABLE

The earliest systematic attempt to organize pure substances according to chemical (and physical) properties was the table of affinities developed in 1718 by Étienne-François Geoffroy (1672–1731).[1,2] His 16-column table gathered acids in columns 1–4, bases in columns 6–8, and metals in columns 10–15. The substance immediately below the column header had the greatest affinity for it, the substance at the bottom, the lowest affinity. The final column in Geoffroy's table was headed by water followed by alcohol and then salt, having less affinity for water than alcohol. Addition of alcohol to saltwater displaces salt and precipitates it.[2] Solubility, of course, is a physical property rather than a chemical property. Geoffroy was a phlogistonist, and phlogiston appeared just below vitriolic acid in his table.[2]

Following Vannucio Biringuccio in the sixteenth century and Albaro Alonso Barba in the seventeenth century, Christlieb Ehregott Gellert was one of the first to employ amalgamation to separate gold from its ores. In 1750 he extended Geoffroy's table to 28 columns (Figure 108).[3] Let us take a brief glance at the organization of this table of "solubilities" (actually chemical *and* physical affinities or *rapports*). Most of the symbols can also be found in the first four figures from the previous essay. In contrast to Geoffroy's table, Gellert placed the substance having *greatest affinity* for the column header at the *bottom* of the column. (Substances having no affinity are listed below the main table.) Let us examine Gellert's organization—the column headers are grouped as follows:[4]

Earths (columns 1–5)

1. Vitrifiable (fusible under fire) or siliceous earth (silicon dioxide)
2. Fluors or fusible earths (low-melting minerals)
3. Clay (possibly alumina)
4. Gypsum (calcium sulfate dihydrate or $CaSO_4 \cdot 2H_2O$)
5. Calcareous earth [strong—$Ca(OH)_2$ or mild—$CaCO_3$)

Alkalis (columns 6 and 7)

6. Fixed alkali (potassium carbonate—K_2CO_3)
7. Volatile alkali (aqueous ammonia solution—NH_4OH)

Acids (columns 8–12)

8. Distilled vinegar (acetic acid—$HC_2H_3O_2$)
9. Marine acid (hydrochloric acid—HCl)
10. Nitrous acid (actually nitric acid—HNO_3)
11. Vitriolic acid (sulfuric acid—H_2SO_4)
12. *Aqua regia* (HCl/HNO_3—3:1)

A salt (column 13)

13. Nitre (sodium nitrate—$NaNO_3$)

Nonmetals and metals (columns 14–27)

14. Sulfur
15. Liver of sulfur
16. Cobalt
17. Arsenic
18. Regulus of antimony
19. Glass of antimony (possibly antimony oxysulfate—$Sb_2O_2SO_4$)
20. Bismuth
21. Zinc
22. Lead
23. Tin
24. Iron
25. Copper
26. Silver
27. Mercury

Glass (column 28—fused form of any substance—"C" connotes calx, so that the first symbol below the header is calx of mercury).

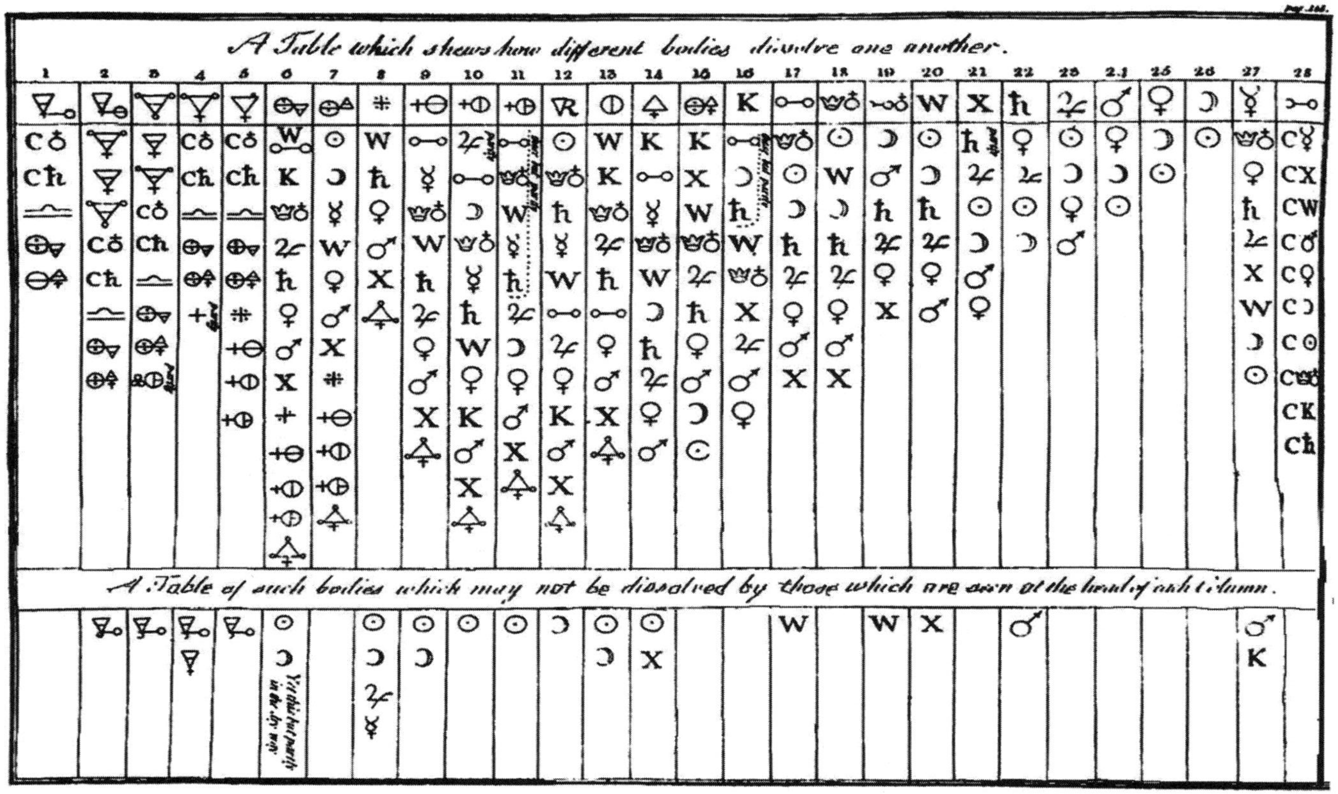

A Table which shews how different bodies dissolve one another.

A Table of such bodies which may not be dissolved by those which are seen at the head of each column.

FIGURE 108. ▪ Table of Affinities developed by Christlieb Ehregott Gellert in 1750. In this table, the highest affinity for the substance at the top of the table (the "header") is shown by the lowest substance in the table. The lowest affinity is exhibited by the substance just below the header. Substances in the lower table are unreactive with the header. Thus, in column 10 (nitric acid), phlogiston is seen to have the highest affinity because nitric acid is "dephlogisticated"—it supports calcination of metals. The reactivity of metals for nitric acid is as follows: zinc > iron > cobalt > copper>lead > mercury > silver > tin. Gold is the least reactive since it is actually unreactive with nitric acid. (From C.E. Gellert, *Metallurgic Chemistry*, 1776; courtesy Chemical Heritage Foundation.)

The unreactive nature of gold is readily apparent in its frequent appearance in the small bottom table of unreactive substances. Gold is not attacked by any acids (columns 8–11) except for *aqua regia* (column 12), where it is still seen to be the least reactive of the 12 substances listed. However, gold is the last of eight substances in column 27, thus indicating its high affinity for mercury—the very basis for Gellert's amalgamation process for isolating gold from its ores. Phlogiston appears as the last symbol in columns 8–12, indicating the highest affinity with acids. This derives, in part, from the fact that metals readily "dissolve" in acids, apparently losing their phlogiston to the surroundings and form calxes. Interestingly, phlogiston also has the highest affinity for the alkalis (columns 6 and 7) that, in turn, also have high affinities for acids. Note that distilled vinegar (acetic acid) is, as expected, the weakest of the acids since it "dissolves" the fewest substances. The highest affinity for niter (column 13) is also assigned to phlogiston since niter (as well as saltpetre) readily converts metals to calxes upon addition of heat. A century earlier, Boyle, Hooke, and Mayow had noted the ability of saltpetre to substitute for air in the combustion of sulfur and carbon as well as in the calcinations of metals. In the mid–1770s, oxygen was considered to be dephlogisticated air "hungry for phlogiston." In this sense, niter and saltpetre, which chemists soon would learn supplies oxygen, could also be considered "dephlogisticated."

The orders of affinities of metals for acids in Gellert's table very much reflect the modern Metal Activity Series (which indicates ease of oxidation). Here is one such comparison:

Gellert Column 10 (Nitric Acid)	Metal Activity Series[5]
Zinc	Zinc
Iron	Iron
Cobalt	Cobalt
Copper	Tin
Lead	Lead
Mercury	Copper
Silver	Silver
Tin	Mercury
Gold	Gold

If one takes silver metal and adds it to an aqueous solution of nitric acid, the metal slowly "dissolves," releasing bubbles of "air." (Some 15 years after Gellert's table was published, Henry Cavendish would capture these bubbles and determine that they were superlight and flammable and believe that his flammable air was phlogiston itself). We now recognize that silver has been oxidized and that the solution contains silver nitrate ($AgNO_3$) rather than the metal. If the more active metal copper is added (say, by placing a copper wire into the solution), the solution will begin to turn as blue as certain copper salts and silver will deposit on the wire. Thus, copper appears to have higher affinity for nitric acid than silver does, just as alcohol has a higher affinity for water than salt. In fact, there is considerable confusion here. For example, the metal oxidations are truly chemical changes; displacement of salt by alcohol is a purely physical change.

The organization of Geoffroy's or Gellert's tables into earths, alkalis, acids, salts, nonmetals, and metals has some hints of the future periodic table. The periodic table gathers metals, metalloids, and nonmetals in clusters and regions. Rankings within a chemical family reflect reactivity—fluorine is much more reactive than the other halogens; lithium is the least reactive of the alkali metals. The tables of affinities mix chemical and physical properties explicitly, while the periodic table is explicitly chemical in nature, although trends in physical properties also emerge. The periodic table includes elements only, while tables of affinities include elements and compounds. However, here it is well to remember that oxides of metals (left-hand side of the periodic table) are alkalis while oxides of nonmetals (right-hand side of the periodic table) are acids. What is chiefly missing from the tables of affinities is periodicity itself. Periodic behavior could occur only if there were some measurable scalar increase in a property such as atomic mass (early nineteenth century) or atomic number (early twentieth century). Nonetheless, these affinity tables forced chemists to explicitly imagine the systematic organization of their field—perhaps the first true step in becoming a modern science.

1. J.R. Partington, *A History of Chemistry*, Macmillan and Co. Ltd., London, Vol. 3, 1962, pp. 49–55.
2. A. Greenberg, *A Chemical History Tour*, John Wiley & Sons, New York, 2000, pp. 118–121.
3. C.E. Gellert, *Metallurgic Chymistry. Being a system of Mineralogy in General, and of all the arts arising from the science. To the great improvement of manufacturers, and the most capital branches of Trade and Commerce. Theoretical and Practical. In two parts, Translated from the original German of C.E. Gellert.* By I.S. London and T. Becker, 1776. (This is the English translation of the original German edition (1751–1755.) The author is grateful to Ms. Elizabeth Swan, Chemical Heritage Foundation, for supplying this image.
4. J. Eklund, *The Incompleat Chymist—Being an Essay on the Eighteenth-Century Chemist in His Laboratory, with a Dictionary of Obsolete Chemical Terms of the Period*, Smithsonian Institution Press, Washington, DC, 1975.
5. T.L. Brown, H.E. LeMay, Jr., and B.E. Bursten, *Chemistry—The Central Science*, seventh edition, Prentice-Hall, Upper Saddle River, NJ, 1997, pp. 128–132.

APOTHECARY'S ASSISTANT *AND* MEMBER OF THE ROYAL ACADEMY OF SCIENCES

Carl Wilhelm Scheele (1742–1786)[1,2] was born in Stralsund, Sweden, a Baltic Sea port that was ceded to Prussia in 1815 and is now in the northeastern part of Germany. The school that provided Scheele's solid elementary education did not offer a Gymnasium[2] (university-preparatory secondary education) course. Stimulated by his older brother's training, young Scheele was sent to the same pharmacy in Göteberg, under the supervision of Martin Bauch. Bauch was sophisticated and current in his chemistry, and Scheele began to bloom as a young chemist as he read the great texts of Johann Kunckel and Caspar Neumann and tried their experiments. When the pharmacy was sold in 1765, Scheele moved to Malmö, where he worked as apprentice in a pharmacy and met Anders Retzius, lecturer at the University of Lund.[1,2] He was given considerable freedom to continue his experimental interests at Malmö. Tempted by the scientific sophistication of Stockholm, in 1768 Scheele accepted an assistant's position in a pharmacy but found his duties limited to preparing prescriptions.[2] In 1770, he joined a pharmacy in Uppsala, where he was finally given his own workbench.[2] In Uppsala he developed a friendship with young Johan Gottlieb Gahn (1745–1818), who would discover manganese four years later. Gahn introduced Scheele to Sweden's foremost chemist Tobern Bergman (1735–1784).[3]

The Uppsala period (1770–1775) was one of incredible productivity for Scheele. He made numerous important discoveries, but his discovery of oxygen in 1771 or 1772 (possibly even earlier) is considered his single greatest work.[1] It is important to remark that Joseph Priestley independently discovered oxygen in 1774 and immediately published his findings. Scheele was apparently totally unaware of Priestley's work and submitted his book manuscript *Chemische Abhandlung von der Luft und dem Feuer* (eventual English title *Chemical Observations and Experiments on Air and Fire*[4]) to the printer in Uppsala in December 1775 as well as to Bergman in early 1776.[1] He first learned of Priestley's discovery of oxygen in August 1776, from a letter written to him by Bergman.[1] He was by that time already very frustrated with the slowness of Bergman's review of his manuscript as well as the printer's delay in publishing it.[1] Scheele's work, reporting his discovery of oxygen some six years earlier, was finally published in 1777 (Figure 109).[5] It would take over a century before Scheele's primacy in the discovery of oxygen would be widely accepted.[1,2]

Figure 110 is from the first English edition (1780) of Scheele's *On Fire and Air*.[4,6] In the upper left of Figure 110 (see *Fig. 1*) is shown a vial in which three teaspoonfuls of steel shavings and one ounce of water were placed and then followed by half an ounce of sulfuric acid. A cork with a hole containing a long, hollow glass tube was inserted tightly in the vial (*A*). The vial was placed in a vessel of hot water (*BB*, in order to accelerate solution). Flammable air (hydrogen) emerging from the top of the tube was then ignited by a candle and the flame positioned into the center of the 20-ounce retort C. When the fire self—extinguished, water from the vessel had climbed to level *D* in the retort. The space up to level *D* corresponded to four ounces of water. Thus, the volume of the original air in the retort had diminished by one-fifth. Scheele added limewater to the retort and noted that no fixed air (carbon dioxide) was produced from this combustion. He did not, unfortunately, detect the small quantity of water vapor formed. Scheele concluded that there was a distinct gaseous component in atmospheric air that supports combustion. Scheele believed that phlogiston was absorbed from the flammable substance by this gaseous substance. But where did it go? He postulated that the imponderable phlogiston and the new gas combined to form a substance that escaped by penetrating the pores of the glass vessel.

Figure 2 (see Figure 110) depicts a similar experiment involving a candle. A tough waxy mass is placed in the bottom of dish A. A strong iron wire is planted into this mass at one end and pressed into the bottom of a candle at the other end C. The candle is lit, and an inverted retort is placed over it and fastened tightly onto the waxy mass below. Water is added to the dish. By the time the candle had extinguished, only 2 of the original 160 ounces of the retort's capacity were occupied by the water. Scheele found that carbon

Carl Wilhelm Scheele's
d. Königl. Schwed. Acad. d. Wissenschaft. Mitgliedes,
Chemische Abhandlung
von der
Luft und dem Feuer.
Nebst einem Vorbericht
von
Torbern Bergman,
Chem. und Pharm. Prof. und Ritter; verschied. Societ. Mitglied.

G. H. B. M.r.

Upsala und Leipzig,
Verlegt von Magn. Swederus, Buchhändler; zu finden bey S. L. Crusius.
1777.

FIGURE 109. ■ Title page from the first (Upsala and Leipzig) edition of Carl Wilhelm Scheele's monumental work on air and fire. Although Scheele first discovered oxygen in 1771 or 1772, long delays in review by Bergman and publication allowed Joseph Priestley to be first in publishing oxygen's discovery (1774). (Courtesy Roy G. Neville Historical Chemical Library.)

dioxide was created in this combustion and correctly reasoned that production of fixed air would soon extinguish the flame. He could not, at the time, understand that for every molecule of gaseous oxygen lost, one molecule of gaseous carbon dioxide would be formed, although this gas also has significant water solubility.

Figure 3 shows the bladder employed to collect oxygen. Oil of vitriol (sulfuric acid) was added to saltpetre (KNO_3), and the gases produced included oxygen as well as red vapors of NO_2. These were collected in a bladder filled with milk of lime [$Ca(OH)_2$] solution that "fixed" the NO_2 vapors and left pure oxygen. *Fig-*

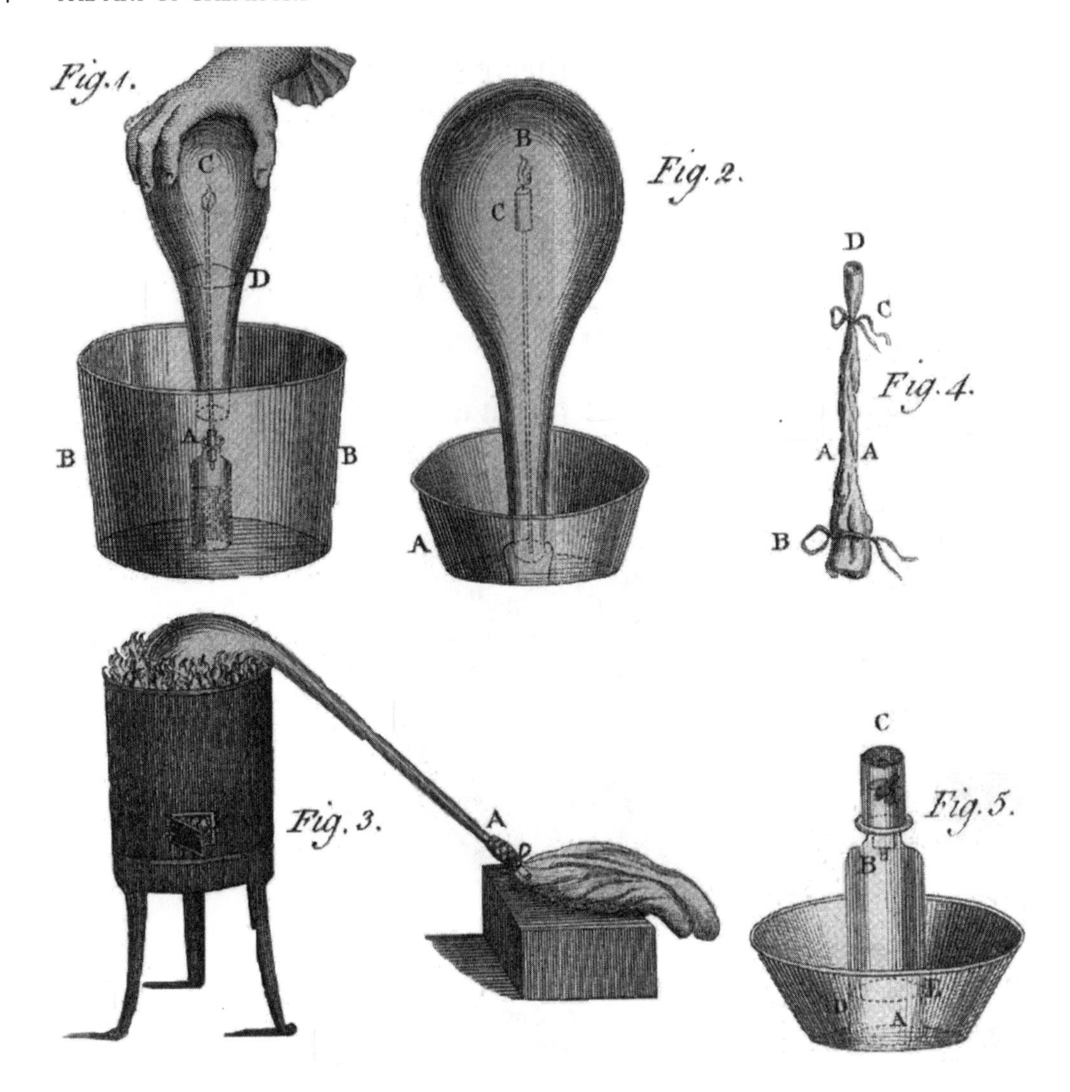

FIGURE 110. ▪ The plate from the first English edition (1780) of Scheele's *Chemical Observations and Experiments on Fire and Air*. In *Fig. 3*, oil of vitriol and saltpetre are heated together. The resulting gases include NO_2, trapped by milk of lime in the bladder, along with oxygen ("fire air") that is collected as a gas in the bladder.

ure 4 is another bladder apparatus. Its use can be illustrated as follows. Place chalk ($CaCO_3$) in the bottom of the bladder and tightly tie a knot at point *B*, sealing off the bottom. Add acid solution into the main part of the bladder *AA*, squeeze out residual air and tie this section off at C. Then fasten the opening *D* tightly around an inverted bottle stoppered tightly with a cork. When ready, the knot at *B* can be untied, that at *D* untied and the cork (still inside the bladder) removed and gaseous products (fixed air in this case) collected. Scheele employed a number of different reactions to obtain oxygen, including heating saltpetre, red calx of mercury (HgO), and heating "black manganese" (MnO_2) with sulfuric acid or phosphoric acid. Scheele's studies of respiration of oxygen included bees (*Fig. 5*).

Scheele made a number of other extremely valuable contributions to chemistry.[1] He added oil of vitriol to fluorspar (CaF_2) and distilled it. He found extensive corrosion of the retort containing the reaction mixture; all the luting used to seal the apparatus was now brittle and friable, and a white deposit was formed on all interior surfaces of the apparatus. Scheele had formed hydrofluoric acid (HF), which attacked the glass vessel (silicon dioxide) and formed gaseous SiF_4. When he duplicated the experiment but added water to the receiver, he was surprised to find a layer of gelatinous silica on top of the water.

Scheele was, of course, well aware of the source of "urinous phosphorus," as it had been originally been reported by Henning Brand a century earlier (see the earlier essay in this book, p. 122). His friend Gahn had discovered calcium phosphate in bone and horn around 1769 or 1770.[1] Scheele treated burned hartshorn with nitric acid, precipitated gypsum with sulfuric acid, concentrated the filtrate and distilled the resulting phosphoric acid over charcoal, and collected the resulting phosphorus.[1]

Scheele discovered chlorine when he treated "black manganese" with hydrochloric acid and isolated a suffocating, greenish-yellow gas.[1] The bleaching properties of chlorine were readily apparent. Scheele's discovery that plumbago (graphite) burned completely to yield fixed air allowed him to conclude that "black lead" was, in reality, pure carbon, just like diamond.[1]

In February 1775, Scheele was still an apothecary's assistant (*studiosus pharmaciae*), but his fame was such that he was elected as a member of the Royal Academy of Sciences of Sweden.[1] At this point, he decided to finally achieve the stature of a full apothecary. Herman Pohl, who owned the pharmacy privilege that permitted him to own an apothecary shop in Köping, died in 1775, leaving his widow, Sara Margaretha Sonneman Pohl, searching for a buyer.[2] Scheele reached an agreement with her that was almost subverted by another buyer. However, his widespread and justified fame forced Fru Pohl to obey public sentiment and stick by her initial agreement.[2] Although not much seems to be known about their relationship, Roald Hoffmann and Carl Djerassi employ some literary license to speculate sensitively about this relationship in their play *Oxygen*.[7] A letter that Scheele sent to Lavoisier in October 1774, which was never answered,[8] plays a vital role in the play. We do know that on his deathbed, Scheele married Fru Pohl so that she could inherit the pharmacy business from him.[2]

---

1. J.R. Partington, *A History of Chemistry*, Macmillan and Co. Ltd., London, 1962, Vol. 3, pp. 205–234.
2. C.C. Gillispie, *Dictionary of Scientific Biography*, Charles Scribners Sons, New York, Vol. XII , 1975, pp. 143–150.
3. A. Greenberg, *A Chemical History Tour*, John Wiley and Sons, New York, 2000, pp. 153–156.
4. C.W. Scheele, *Chemical Observations and Experiments on Air and Fire*, London, printed for J. Johnson, 1780.
5. C.W. Scheele, *Chemische Abhandlung von der Luft und Feuer, nebst einer Vorbericht von Torbern Bergman*, verlegt von Magnus Swederus, Uppsala und Leipzig, 1777. This is one of the rarest, most desired books in the field of rare chemistry book collecting. I am grateful to The Roy G. Neville Historical Chemical Library (California) for providing this image.
6. Greenberg, op. cit., pp. 135–137.
7. C. Djerassi and R. Hoffmann, *Oxygen*, VCH-Wiley, Weinheim, 2001.
8. J.-P. Poirier, *Lavoisier—Chemist, Biologist, Economist*, R. Balinski (transl.), University of Pennsylvania Press, Philadelphia, 1996, pp. 76–83.

LAUGHING GAS OR SIMPLY "SEMIPHLOGISTICATED NITROUS AIR" (?)

Pneumatic chemistry, the collection and handling of gases ("airs"), began in the 1720s thanks to the work of Stephen Hales and later improvements by William Brownrigg.[1,2] Thus, prior to the 1770s the only gases obtained in pure form and characterized to some degree were fixed air (carbon dioxide) and inflammable air (hydrogen); both were termed "factitious airs" using the nomenclature of Henry Cavendish.[3] In the early 1770s, improving the techniques of pneumatic chemistry, Dr. Joseph Priestley isolated and characterized a series of new "airs" that included, among others, oxygen ("dephlogisticated air," O_2), nitrogen ("phogisticated air," N_2), nitric oxide ("nitrous air," NO), and nitrous oxide ("dephlogisticated nitrous air," N_2O).[4] Unbeknownst to Priestley, the Swedish apothecary Carl Wilhelm Scheele was performing a parallel series of experiments and had isolated and characterized oxygen ("fire air") in 1772 (at least two years earlier than Priestley) and possibly even in 1771.[5] Both Priestley and Scheele were firmly anchored in phlogiston theory. Figure 111 depicts some of Priestley's pneumatic apparatus including his pneumatic tub (*Fig. 1*).

Some of the confusions engendered by the phlogiston theory can be briefly illustrated by the nomenclature described above. For example, "factitious airs" referred to gases derived from solids. Heating chalk (calcium carbonate, $CaCO_3$—both "chalk" and "calcium" are related to the Latin *calx or calcis*) produces "fixed air" (carbon dioxide, CO_2). Figure 111 (see *Fig. 11*)[6] depicts Priestley's apparatus for collecting factitious airs from solids heated in a gun barrel placed partly in a hearth. The gas is collected in a tube that is filled with mercury and inverted into a dish also containing mercury (*Fig. 11* in Figure 111). Oxygen can be obtained by heating solid calxes including mercuric oxide (HgO) or manganese dioxide (MnO_2). Carbon dioxide and oxygen are, thus, both "factitious airs." So far, so good. However, adding zinc to sulfuric acid ("oil of vitriol") produces another "factitious air"—"inflammable air" (hydrogen, H_2)—"clearly" obtained from the solid metal. Indeed, Cavendish thought that inflammable air *was* phlogiston when he isolated it in 1766. In fact, the hydrogen comes from the aqueous acid solution since water is a compound of hydrogen and oxygen. But this understanding was achieved only in the early 1780s.

Priestley found that when the "nitrous air" (NO), obtained by treating copper with *aqua fortis* (nitric acid, HNO_3), was stored over iron filings—a new gas was produced in which a candle burned more intensely than in either common air or his "nitrous air." Clearly, this new "air" wanted phlogiston just as oxygen ("dephlogisticated air") did and hence was also dephlogisticated. Since it was superior to "nitrous air" in supporting combustion, Priestley logically named it "dephlogisticated nitrous air."

Now recall the device of considering "phlogisticated" to mean "minus oxygen" and "dephlogisticated" to mean "plus oxygen."[7] We now know that the new gas was nitrous oxide (N_2O)—laughing gas—a story for another day.[8] In fact, it has *less* oxygen relative to "nitrous air" and, on the basis of oxygen content, should have been called "phlogisticated nitrous air" (or even worse, "semiphlogisticated nitrous air"). However, the degree of dephlogistication ("hunger" for

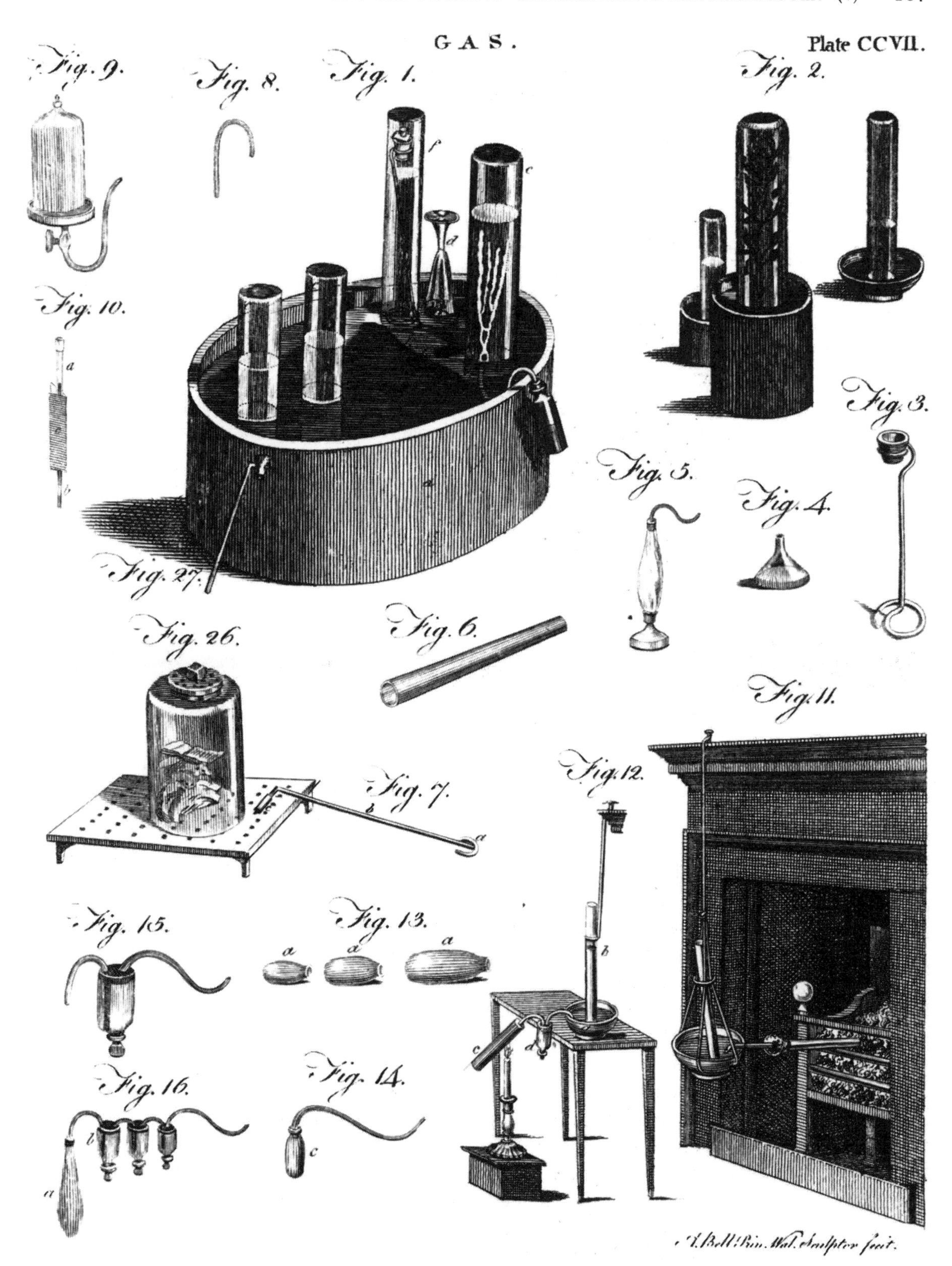

FIGURE 111. ▪ Collage of Joseph Priestley's pneumatic apparatus. A gun barrel is used as reaction chamber and placed in a fireplace. The resulting gases, if water-soluble, are collected over mercury (*Fig. 11*).

the essence of fire) did not have a straightforward relationship to oxygen content. Indeed, chlorine, discovered by Scheele,[5] could ignite hydrogen. Even Lavoisier believed that chlorine must contain oxygen—but it does not. In fact, the observed chemistry of nitrous oxide reflected its structure and reactivity, factors that could not possibly be understood in the late eighteenth century, rather than oxygen content. The confusion was understandable.

---

1. J.R. Partington, *A History of Chemistry*, Vol. 3, Macmillan & Co., Ltd, London, 1962, pp. 112–127.
2. A. Greenberg, *A Chemical History Tour*, John Wiley & Sons, New York, 2000, pp. 122–130.
3. Partington, op. cit., pp. 302–319.
4. Partington, op. cit., pp. 237–268.
5. Partington, op. cit., pp. 205–229.
6. This is an early nineteenth-century engraving depicting Priestley's pneumatic apparatus. The collection of carbon dioxide from the gun barrel over a dish of mercury is to be found in J. Priestley, *Experiments and Observations on Different Kinds of Air, and Other Branches of Natural Philosophy*, Vol. III, Thomas Pearson, Birmingham, 1790, Plate II, as well as earlier editions.
7. R. Hoffmann and V. Torrence, *Chemistry Imagined—Reflections on Science*, Smithsonian Institution Press, Washington, DC, 1993, pp. 82–85.
8. Greenberg, op. cit., pp. 163–165.

EULOGY FOR EUDIOMETRY

Eudiometers, at their simplest, were nothing more than inverted graduated tubes for measuring volumes of gases collected over water (or above mercury if the gases were water-soluble; see *Fig. 11* in Figure 111). The real power of eudiometry derived from the ability to measure volume changes when different gases reacted. Joseph Priestley employed very simple eudiometers for his studies of factitious airs such as "nitrous air" (NO or nitric oxide, isolated in 1772) and "dephlogisticated air" (O_2 or oxygen, isolated by him in 1774). Figure 112 depicts such a eudiometer (see g)[1] and glassware for handling gases in a eudiometry experiment. The three vessels labeled *f* in Figure 112 are typically one-, two-, and 4 ounce measures for gases, initially filled with water, but inverted in a pneumatic trough and filled with the gas in question. For example, one measure of a gas, "common air," could be transferred to a "reaction chamber"—a glass jar filled with water and inverted in a pneumatic trough. When an equal measure of "nitrous air" is added, a reddish gas forms immediately then disappears. The reaction would be allowed to continue for about two minutes and the gaseous contents transferred into the eudiometer. For "common air," the typical result was that 1.36 measures, of the 2.00 mixed, remained in the eudiometer.[2] Eudiometry also indicated that one volume of pure "dephlogisticated air" reacted completely with two volumes of "nitrous air." Some 35 years later (combining the discoveries of Dalton, Gay-Lussac, and Cannizzaro), chemists would understand the gaseous reaction to be

$$2\,NO + O_2 \rightarrow 2\,NO_2 \text{ (reddish gas)}$$

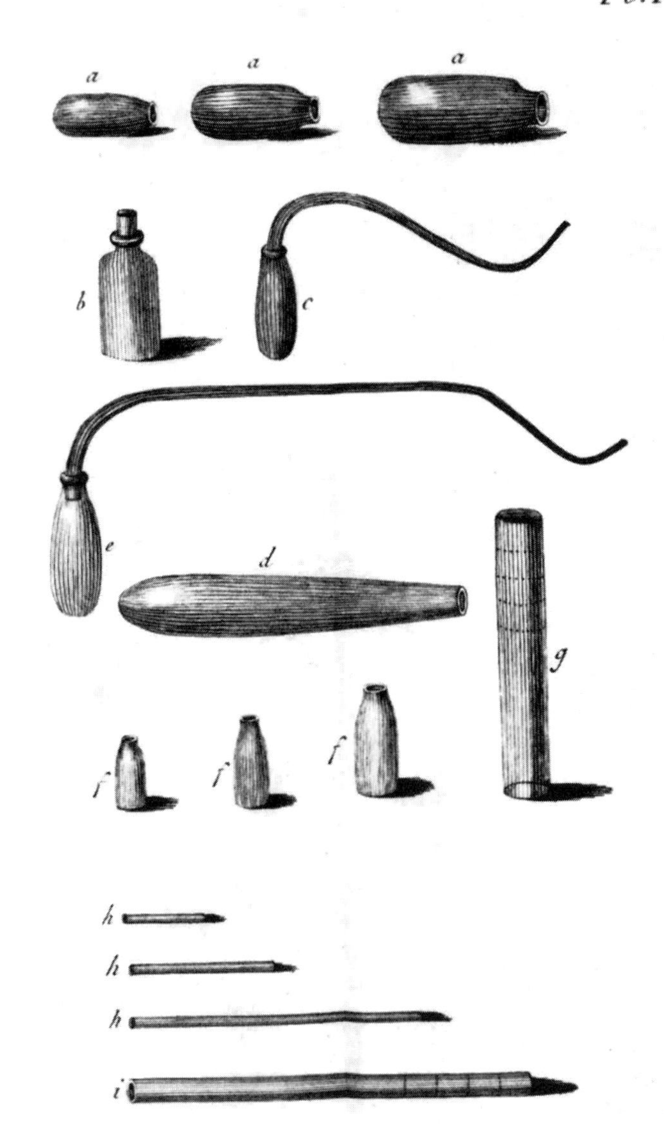

FIGURE 112. ▪ Drawing *g* depicts the simplest possible eudiometer (a graduated tube sealed at one end). Measured volumes of gas are introduced and changes in final volume measured. Priestley's earliest eudiometers simply measured the "goodness" (oxygen content) of air by reaction with freshly generated "nitrous air" (actually nitric oxide, NO). (From Priestley, *Experiments and Observations on Different Kinds of Air,* 1790 edition.)

We know also that nitrogen dioxide dissolves in water to form nitric acid:

$$3\ NO_2 + H_2O \rightarrow 2\ HNO_3 + NO$$

The NO regenerated would recycle in the reaction vessel, and eventually virtually all of it would react with oxygen and become nitric acid. Actually, these complications, combined with the fact that NO was considerably more water-soluble than oxygen, led to lots of inconsistencies depending on which gas was added first, the extent of shaking of the reaction vessel, and how long the gases were allowed to react. Eventually, Cavendish introduced consistency into the technique.[2]

It is now easy to understand the eudiometry results for mixing of "common air" and "nitrous air." We know that air is roughly 20% (by volume) oxygen. Thus, from 1.0 measure of "common air," an 0.2 measure of oxygen will disappear upon reaction with an 0.4 measure (from the original 1.0) of "nitrous air." The remaining volume of gas in the eudiometer is 1.4 measures—the total volume lost is 0.6 measure. In reality, 1.36 measures would typically remain (see above); take the volume lost (0.64 measure), divided by 3, and the result is a bit over 0.21. Thus, roughly 21% of "common air" is oxygen. If a sample of "common air" had been reduced to 10% oxygen (through calcination of a metal or introduction of a mouse into the container), mixing of 1.0 measure each of "common air" and "nitrous air" would leave a volume of 1.7 measures in the eudiometer. The "goodness" of this sample was only 1.7, compared to the "goodness" of common air (1.36).[2] Priestley was delighted that he no longer had to kill mice to test the "goodness of air" (see *Fig. 26* in Figure 111).[2,3] He and others tested the hypothesis that city air had less oxygen than country air. By this measure at least, these two "types" of "common air" had equal "goodness." However, Priestley (and Scheele independently) discovered that the air dissolved in water had slightly more oxygen than that found in "common air" (since nitrogen is a bit less water-soluble than oxygen).[2]

Throughout the latter part of the eighteenth and the early part of the nineteenth centuries, a variety of modifications were devised in which the eudiometer served as both volumetric cylinder *and* reaction chamber.[4] Sulfur powder or red phosphorus could be suspended in a dish surrounded by the air in question and situated above the water in the eudiometer. A powerful magnifying glass would ignite the solids, ultimately forming water-soluble acids, and diminishing the air volume accordingly. Although hydrogen ("inflammable air") and oxygen had been ignited in different admixtures starting in the 1770s—two volumes of hydrogen and one volume of oxygen gave the loudest "pop"—the eudiometer of Volta (see eudiometer *b* with electrodes *f* in *16* at the bottom center of Figure 113[5]) efficiently and accurately employed an electric spark to ignite the two gases and measure the remaining volume.

If we think of the eudiometer in its original incarnation as an apparatus employed to measure oxygen in air, it has long been as extinct as the dodo. Unlike the dodo, we can bring the eudiometer back to life for student demonstration purposes. In their PBS (Public Broadcasting Service) televison series, Philip and Phylis Morrison did just that.[6] They assembled a Volta-style eudiometer and added two "thimblefuls" each of hydrogen and oxygen gas. When they ignited the mixture with an electric spark, the gases expanded [the "slow-mo" (slow-motion) camera indicated no loss of gas bubbles from the bottom of the eudiometer] and left behind one "thimbleful" of gas. An afternoon of repetitions did not change the result. In their words: "The residue was simply the amount of oxygen that could not be taken up into water, always one volume of the total of four. Again, we learned H_2O. Something deep within water appears to know simple arithmetic."[6]

FIGURE 1 ■ For full description, see page xvii.

FIGURE 3 ■ For full description, see page xvii.

FIGURE 7 ■ For full description, see page xvii.

Liber de arte Distil
landi de Compositis.
Das büch der waren kunst zü distillieren die
Composita vñ simplicia/vnd dz Büch thesaurus pauperû/Ein schatz d' armë ge-
nät Micariû/die brösamlin gefallen võ dë büchern d' Artzny/vnd durch Experimët
võ mir Jheronimo brüschwick vff geclubt vñ geoffenbart zü trost denë die es begerë.

FIGURE 32 ■ For full description, see page xvii.

FIGURE 92 ■ For full description, see page xvii.

FIGURE 117 ■ For full description, see page xvii.

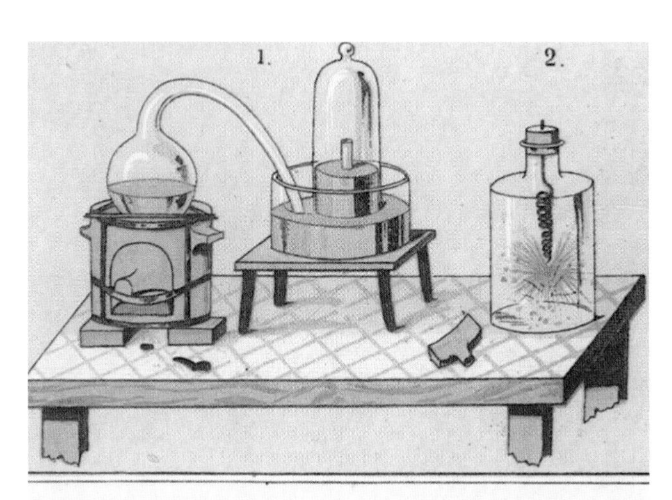

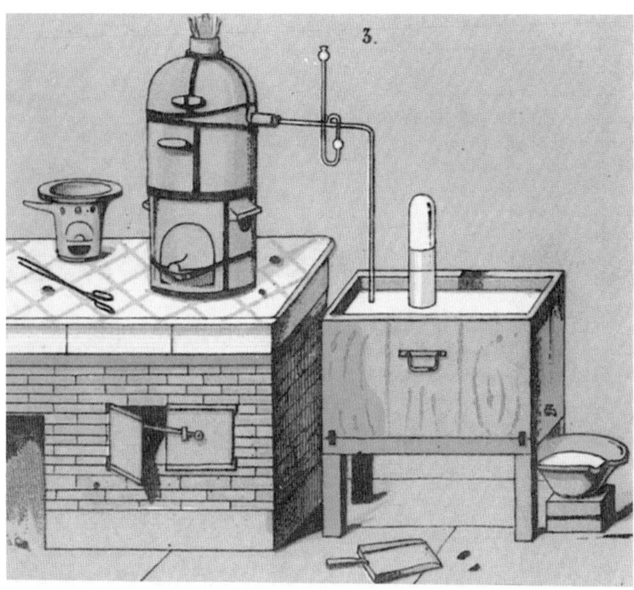

FIGURE 139 ■ For full description, see page xvii.

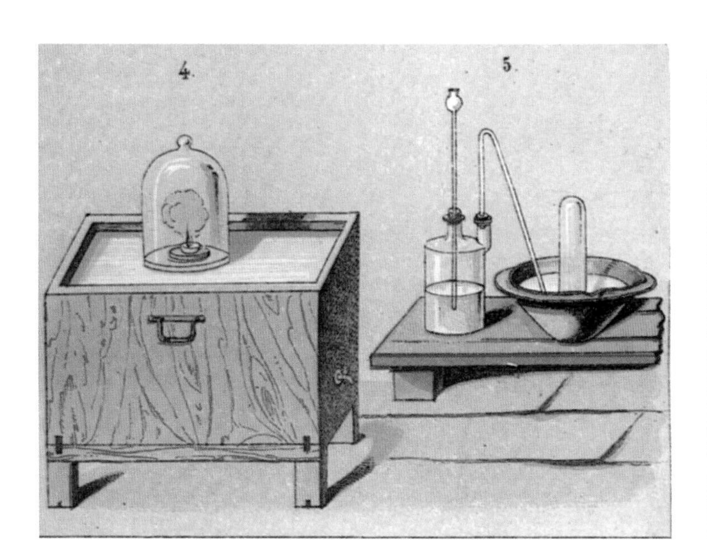

FIGURE 140 ■ For full description, see page xvii.

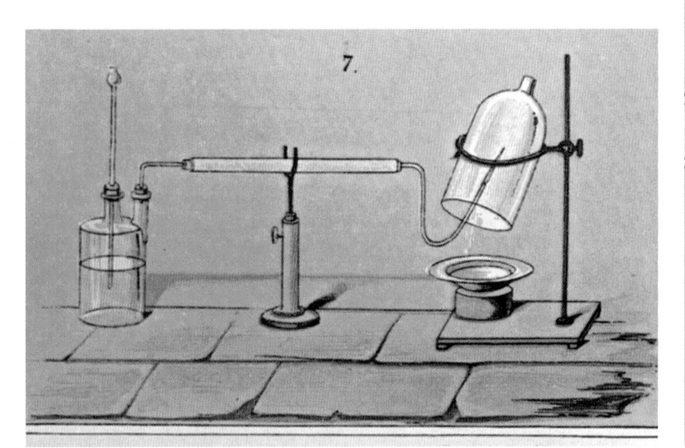

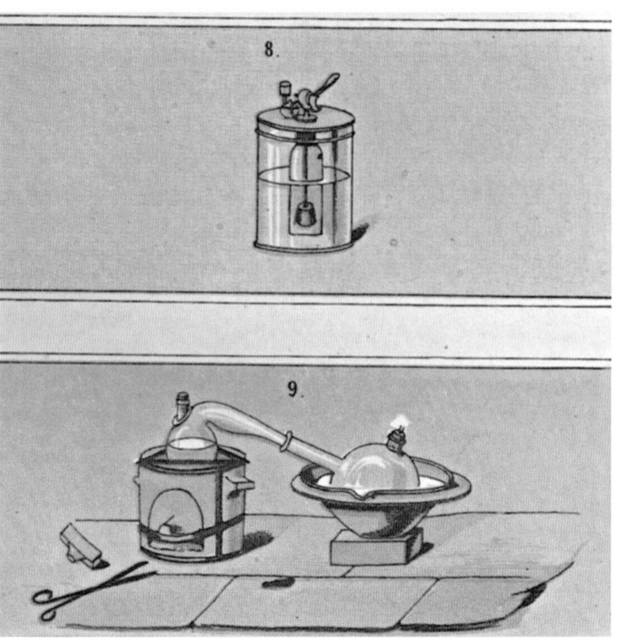

FIGURE 141 ■ For full description, see page xvii.

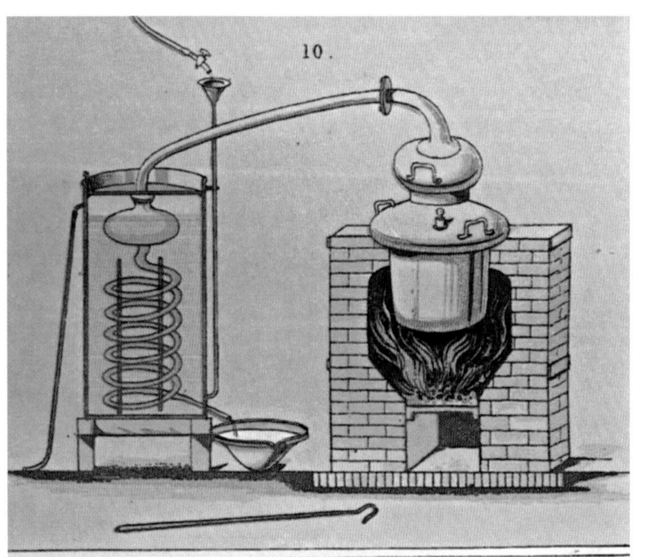

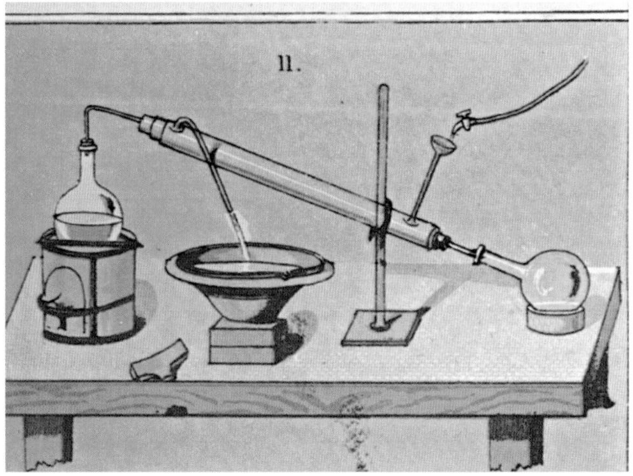

FIGURE 142 ■ For full description, see page xvii.

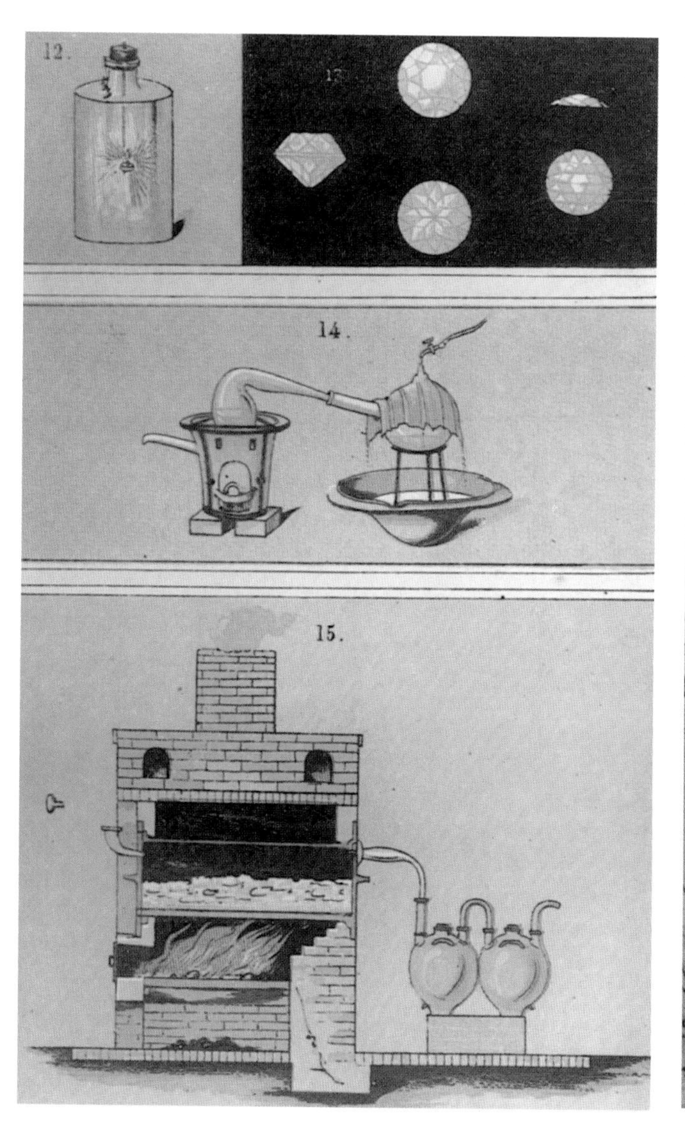

FIGURE 143 ■ For full description, see page xvii.

FIGURE 144 ■ For full description, see page xvii.

FIGURE 156 ■ For full description, see page xvii.

FIGURE 164 ■ For full description, see page xvii.

FIGURE 176 ■ For full description, see page xvii.

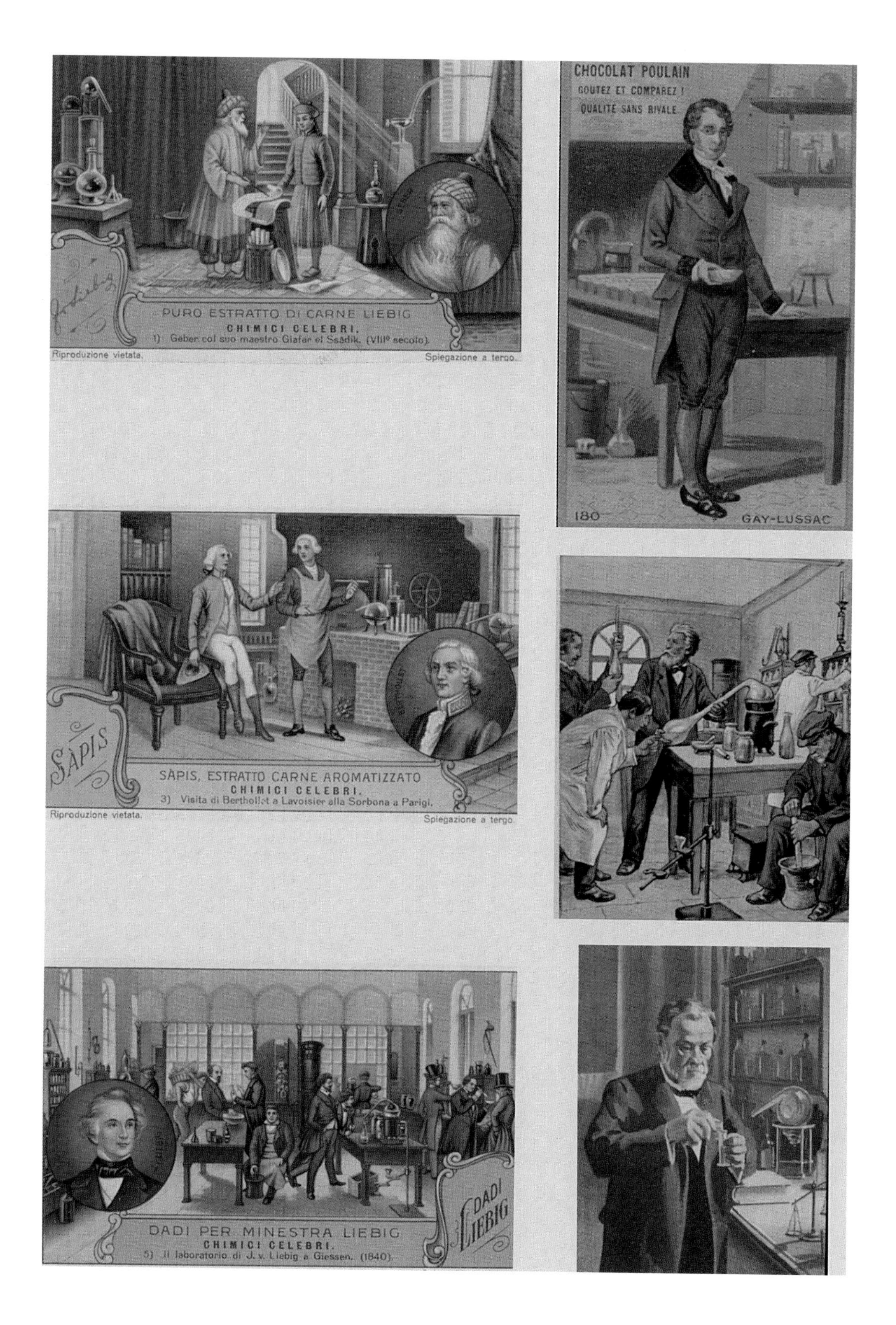

FIGURE 177 ■ For full description, see page xvii.

FIGURE 178 ■ For full description, see page xvii.

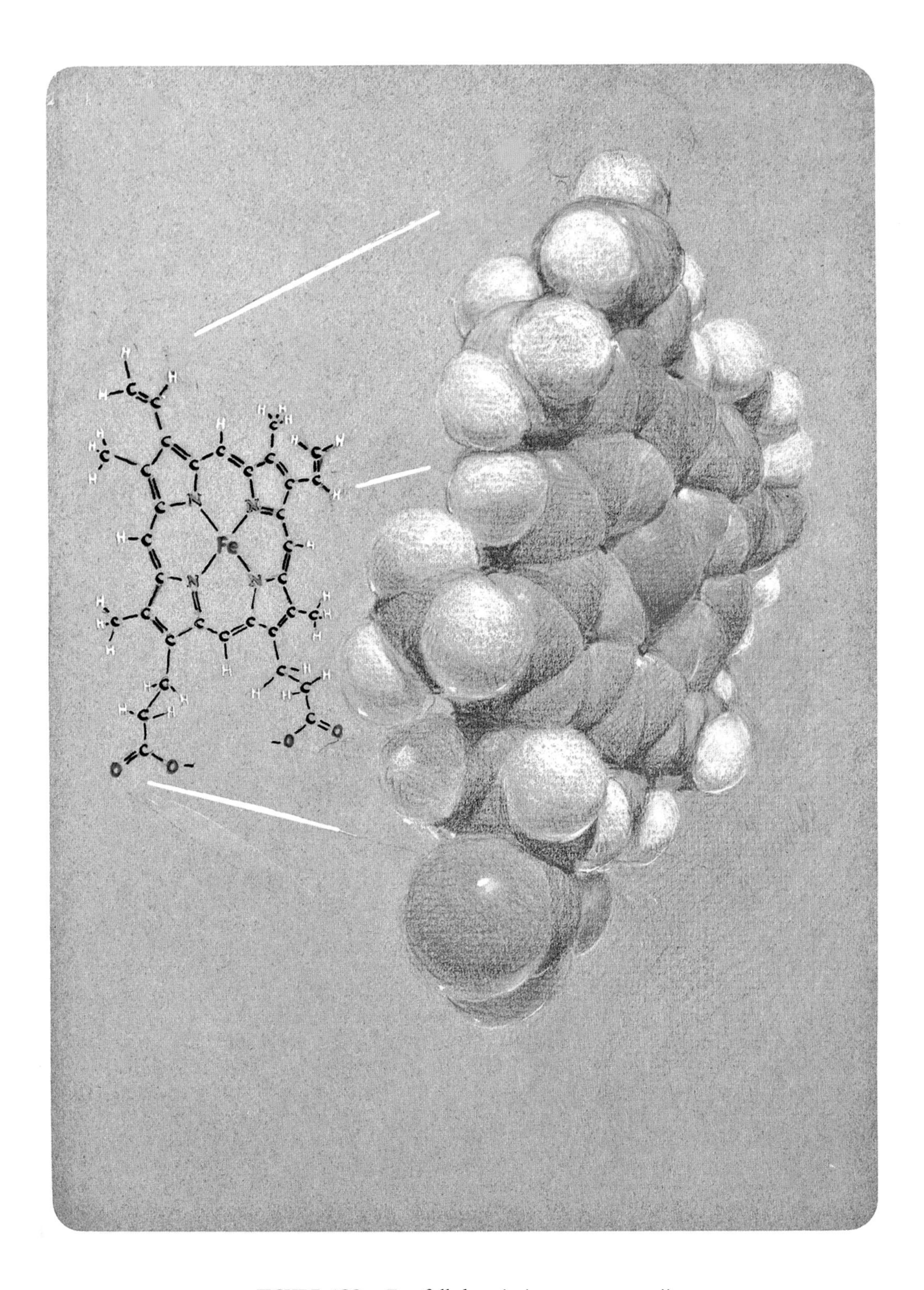

FIGURE 182 ■ For full description, see page xvii.

FIGURE 187 ■ For full description, see page xvii.

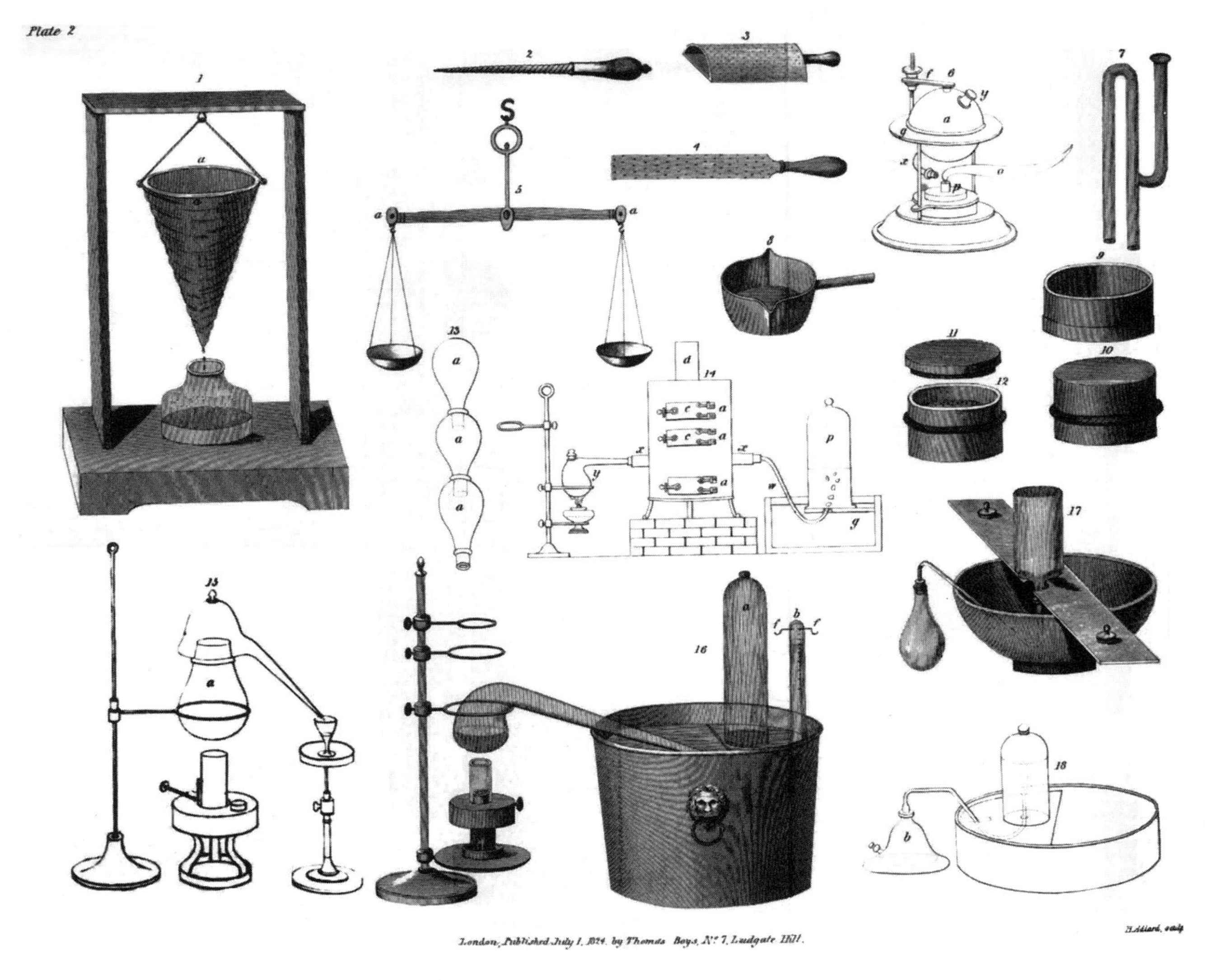

FIGURE 113. ■ Alessandro Volta's eudiometer (see bottom center figure, *16*—tube standing on the right in the tub). Combination of hydrogen and oxygen in this eudiometer could be ignited and the change in volume measured. (From Accum's 1824 *Explanatory Dictionary of the Apparatus and Instruments . . . .*)

1. J. Priestley, *Experiments and Observations on Different Kinds of Air, and Other Branches of Natural Philosophy*, Thomas Pearson, Birmingham, 1790, Vol. 1, pp. 20–30; Vol. 3, Plate IV.
2. J.R. Partington, *A History of Chemistry*, Vol. 3, Macmillan and Co, Ltd., London, 1962, pp. 252–263; pp. 321–328.
3. A. Greenberg, *A Chemical History Tour*, John Wiley & Sons, New York, 2000, pp. 137–139.
4. [F. Accum] (i.e., "A Practical Chemist"), *Explanatory Dictionary of the Apparatus and Instruments Employed in the Various Operations of Philosophical and Experimental Chemistry with Seventeen Quarto Copper-Plates*, Thomas Boys, London, 1824, pp. 100–110, which describes 10 eudiometers (Priestley, Pepy, Scheele, De Marti, Humbolt, Hope, Seguin, Bertholet, Davy, Volta).
5. Accum, op. cit., Plate 2), figure *16*.
6. P. Morrison and P. Morrison, *The Ring of Truth—an Inquiry into How We Know What We Know*, Random House, New York, 1987, pp. 191–193.

WATER WILL NOT "FLOAT" PHLOGISTON

We have imbibed, sweated, and excreted water since time immemorial—so it might be nice to know what it's made of. Water boils, freezes, and is recovered unchanged from salts and other "earths." It is absolutely elemental to our very existence, as is air. One of the four Aristotelian elements, water can be transmuted to "air" by adding heat, to "earth" by removing wetness, and it "neutralizes" its contrary element—fire. Its status as an element survived Robert Boyle's scathing criticism of the ancients in his 1661 classic *The Sceptical Chymist*. As late as 1747, Ambrose Godfrey, Boyle's very capable assistant, reported the chemical conversion of water to earth,[1] an experimental conclusion once and for all time refuted in 1770 by Antoine Lavoisier.[2] So, when and how did we learn the true nature of water, or how to get "From H to eau," as Philip Ball so wittily phrases it?[3]

In 1766, Cavendish reported releasing a "flammable air" (hydrogen) that he equated to phlogiston, the very essence of fire, from metals through reaction with aqueous acids. Eight years later, Priestley reported isolating "dephlogisticated air" (oxygen). It must have been something of a disappointment when the two were mixed and didn't immediately consume each other[4] to give . . . to give *what?* . . . "dephlogisticated phlogiston"?—effectively nothing? Priestley first ignited this mixture in 1775 in a $1\frac{1}{2}$-inch bottle with a $\frac{1}{4}$ inch opening and later "amused himself by carrying these corked or stoppered bottles about and exploding them."[5] Formation of water, in the combustion of hydrogen in common air, was first noted in 1776/77 by Pierre Joseph Macquer, who observed dew collected on a cold porcelain dish above the flame. Similar observations were made in England by John Warltire, who burned hydrogen and oxygen in various combinations of volumes; the best proportion was 2 : 1.[5] But seeing residual water appear in chemical reaction vessels was commonplace. One explanation was that water had to be squeezed out of air before it could become fully phlogisticated. Listen carefully, dear reader; phlogiston theory was creaking loudly now and beginning to crumble. In fact, gases occupy huge volumes compared to liquids—

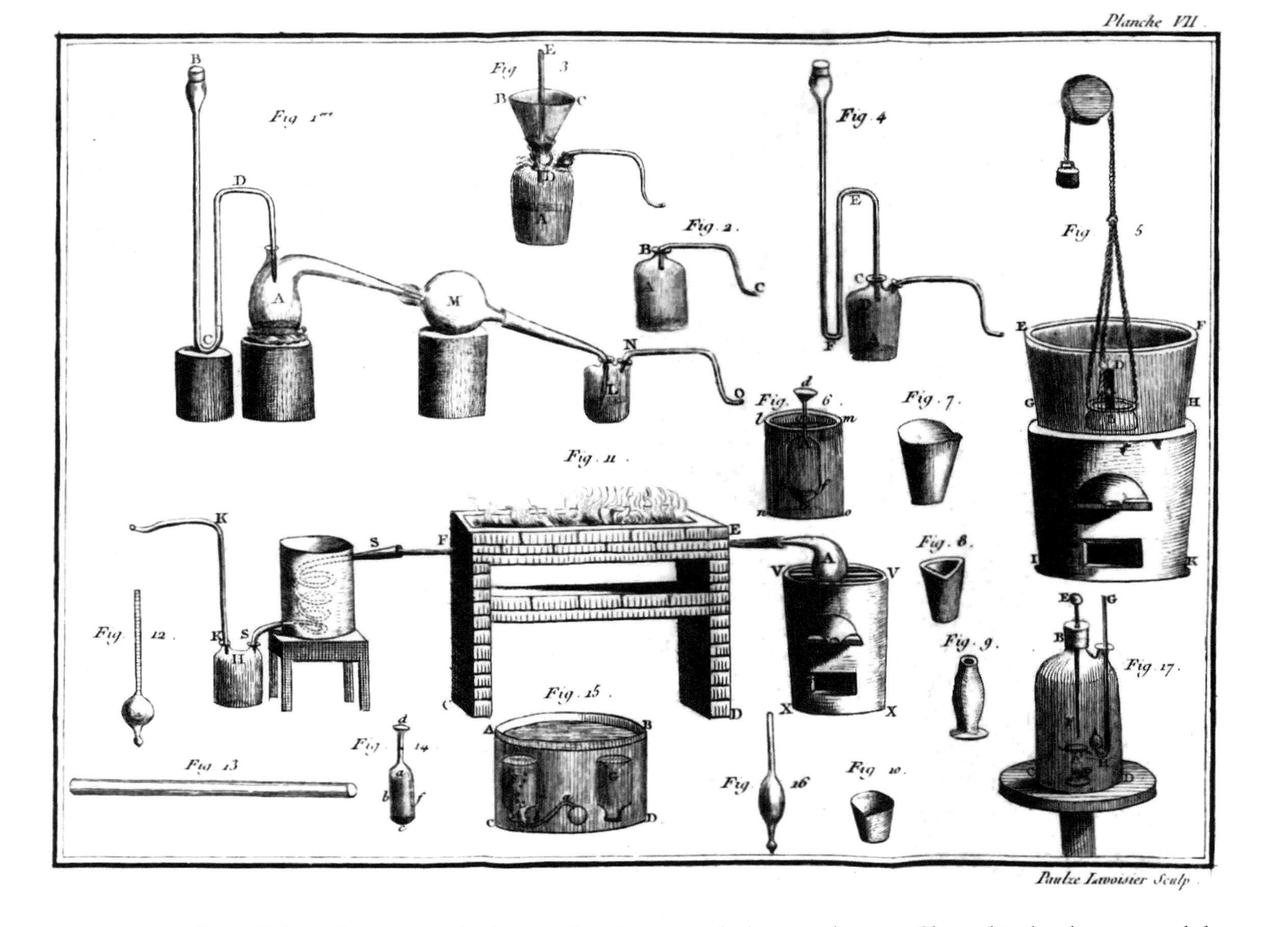

FIGURE 114. ▪ *Figure 11* shows the apparatus for *decomposition* of water into hydrogen and oxygen. Charcoal is placed in an annealed glass tube in the furnace. Over the fire, charcoal is oxidized to CO_2 that is collected while hydrogen is collected separately. Alternatively, iron plates could be placed in a tube and heated red hot, forming iron oxide and releasing hydrogen in the presence of steam. (From Lavoisier, *Traité Elémentaire de Chimie*, *seconde édition*, 1793.)

one liter of the two gases properly mixed will yield about 10 drops of water—rather easy to dismiss.

Suffice it to say, we will not solve "The Water Controversy"[6] here. The key figures are Henry Cavendish, James Watt, and Antoine Lavoisier. Volta's eudiometer (see Figure 113) stimulated Priestley, Cavendish, and others to use this technique. Although James Watt (1736–1819), inventor in 1765 of a vastly improved steam engine, made a strong case that he first recognized that water is a compound rather than an element, it was Cavendish who is accorded the discovery. He was the first to experimentally combine the exact proportions of hydrogen and oxygen and find their mass conserved in liquid water. However, anchored in phlogiston theory, Cavendish's interpretation was that "flammable air" (hydrogen) was really water plus phlogiston (Φ) while dephlogisticated air (oxygen) was really water lacking phlogiston:[5,7]

$$\begin{array}{cccc} \text{Hydrogen} & + \quad \text{oxygen} & \rightarrow & \text{water} \\ (\text{Water} + \Phi) & (\text{water} - \Phi) & \rightarrow & \text{water} \end{array}$$

It would seem logical that if hydrogen were pure phlogiston and oxygen were "dephlogisticated air," then their "marriage" might produce, if anything, "phlogisticated air" or "azote" (nitrogen). Instead, water is found—it has body and considerable density. It must have been there all along, right?

It was Lavoisier who finally provided absolutely conclusive evidence for water's composition. Water played an important role in his career. His first paper, published at the age of 22, established one of Lavoisier's *leitmotifs*.[8] He calcined (heated) gypsum (calcium sulfate), collected and weighed the water of crystallization, and then reconstituted the original substance by adding the water collected to the anhydrous salt. As noted above, he disproved the claim by others that water was converted to earth upon heating. His claim for primacy in the discovery of the composition of water bears striking resemblance to his claim for the discovery of oxygen; the historical record indicates in each case, that he (1) did *not* make the initial discovery, (2) withheld some information from rivals, but (3) completely and correctly explained the breakthroughs with his antiphlogistic theory of oxidation.[9,10]

In Figure 114 we see Lavoisier's apparatus for *decomposing water* (*Fig. 11*)[11] in which an annealed glass tube (*EF*, surrounded by clay mixed with crushed stoneware and reinforced with an iron bar) is placed into furnace *EFCD*; connected at one end to retort A, which produces steam; and at the other to condenser *SS*, which drips unreacted water into *H*. Gas from the tube leaves through *KK* and is appropriately purified and collected. The first experiment found that the quantity of steam lost from A, and passing through the empty red-hot tube, precisely equaled the quantity of water collected in *H*. In the second experiment, 28 grams of charcoal were placed in the tube—these were gone following prolonged exposure to steam in the red-hot tube and produced 100 grams of carbonic acid gas (carbon dioxide), and 13.7 grams of "a very light gas . . . that takes fire" (hydrogen) with a loss of 85.7 grams of water. It was known that 100 grams of carbonic acid gas contained 72 grams of oxygen and 28 grams of carbon (*remember, no atoms or formulas yet*). Thus, 85.7 grams of water decomposed into 13.7 grams of hydrogen and 72 grams of oxygen—the *Ferme Générale*, in which

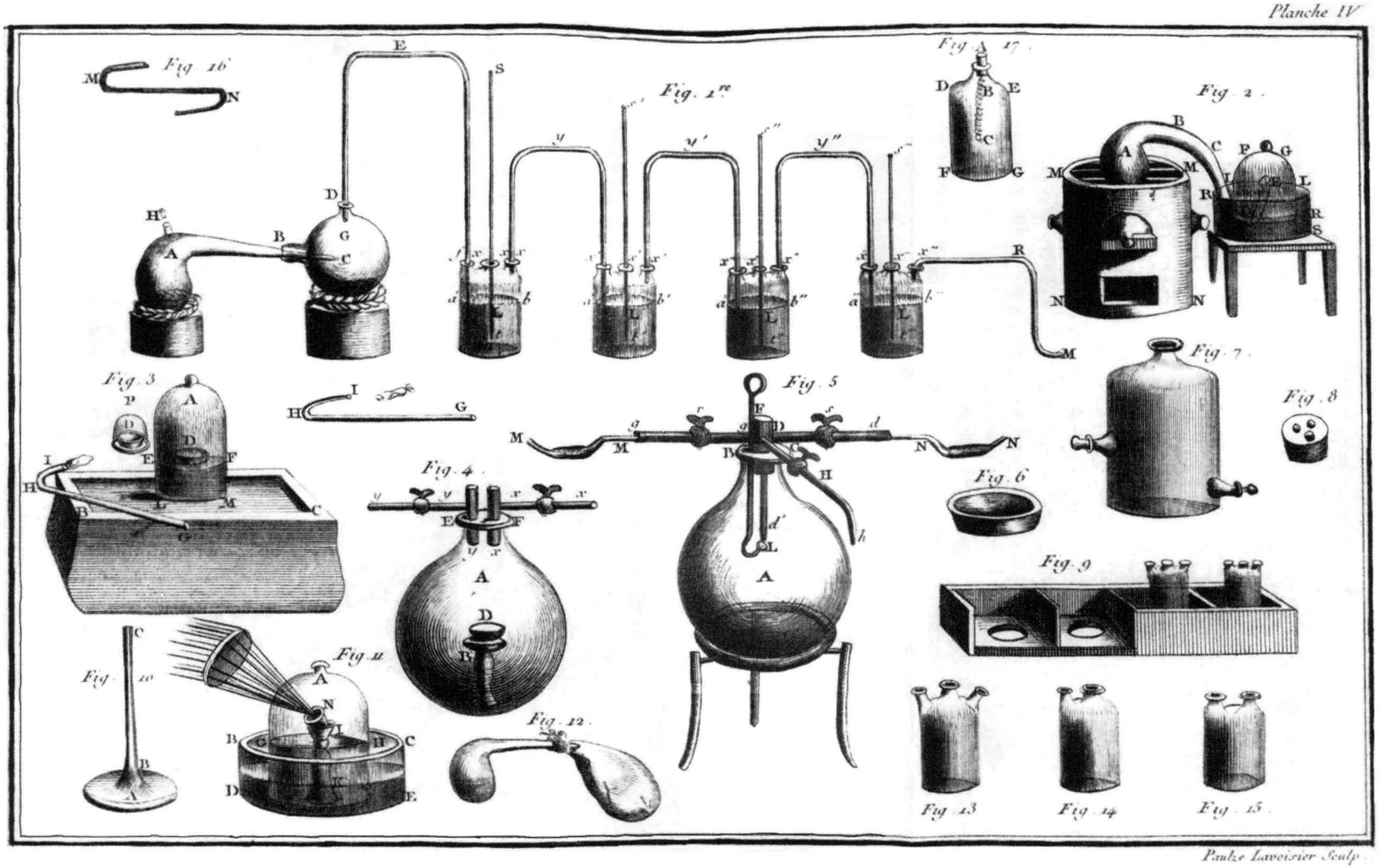

FIGURE 115. ■ *Fig.* 5 shows the apparatus for *synthesis* of water from hydrogen and oxygen. Once the correct amounts are added, a spark is set off at L and quantitative reaction occurs. (From Lavoisier, *Traité Elémentaire de Chimie*, *seconde édition*, 1793.)

Lavoisier was a shareholder, would have been well satisfied with this neat accounting. In the third experiment, 274 grams of soft iron, in thin plates rolled up into spirals, are exposed to steam in the red-hot tube. No carbonic acid gas is found this time. Instead, the iron is now a black oxide (the same as produced by combustion of iron in oxygen) and its weight augmented by 85 grams of oxygen. The amount of hydrogen collected was 15 grams; the amount of water lost, 100 grams. From these two very different experiments, the same result—water is 85% oxygen and 15% hydrogen.

In Figure 115 we see Lavoisier's apparatus for *synthesizing water* (*Fig.* 5).[11] The 30-pint glass balloon *A* conducts pure oxygen from the left (through drying tube *MM* filled with powdered calcium chloride or similar) and pure hydrogen from the right (through drying tube *NN*). An electric spark will be supplied in the vicinity *Ld*. On the basis of the results of water decomposition above, 85 grams of oxygen and 15 grams of hydrogen are added slowly with periodic sparking. *Et voilá*—100 grams of liquid water!

Phlogiston had "met its Waterloo." Water's decomposition into the elements hydrogen and oxygen and its reconstitution from these elements sounded a death knell for phlogiston theory. The same result was achieved by electrolysis of water in 1789.[12] The light gaseous "essence of fire" released from iron by aqueous acids was phlogiston, according to Cavendish in 1766. How ironic to learn that "flammable air" comes from the aqueous acid, not the metal, and that it would be Cavendish who effectively discovers it! But confidentially, friends, at the true end of the second millennium, we still do not fully understand why hydrogen and oxygen don't immediately consume each other when mixed and form water without the aid of a spark.[4]

---

1. Godfrey (Ambrose and John), *A Curious Research into the Element Water; Containing Many Noble and Useful Experiments on that Fluid Body*, T. Gardener, London, 1747.
2. J.R. Partington, *A History of Chemistry*, Vol. 3, Macmillan & Co, Ltd., London, 1962, pp. 379–381.
3. P. Ball, *Life's Matrix: A Biography of Water*, Farrar, Straus and Giroux, New York, 2000, pp. 141–147.
4. A very high level of quantum chemical calculations seemingly clarifies some, but not all, of the mystery of hydrogen's very slow reaction with oxygen (see M. Filatov, W. Reckien, S.D. Peyerimhoff, and S. Shaik, *Journal of Physical Chemistry* A, Vol. 104, p. 12014 (2000). I thank Professor Joel F. Liebman for making me aware of this article.
5. Partington, op. cit., pp. 325–338.
6. Partington, op. cit., pp. 344–362.
7. W.H. Brock, *The Norton History of Chemistry*, W.W. Norton & Co., New York, 1993, pp. 109–111.
8. J.-P. Poirier, *Lavoisier—Chemist, Biologist, Economist* (transl. R. Balinski), University of Pennsylvania Press, Philadelphia, 1993, pp. 13–16.
9. Partington, op. cit., pp. 402–410; 436–453.
10. J.-P. Poirier, op. cit., pp. 76–83.
11. A. Lavoisier, *Elements of Chemistry in a New Systematic Order, Containing All the Modern Discoveries*, second edition (transl. R. Kerr), London, 1793, pp. 135–149. See Plates VII and IV, respectively, for the apparatus for decomposition and synthesis of water.
12. Partington, op. cit., p. 457.

BEN FRANKLIN—*DIPLOMATE EXTRAORDINAIRE*

Here is a problem worthy of Benjamin Franklin, America's greatest diplomat: How do you retain the friendships of two close friends who disagree absolutely and fundamentally on chemical theory? Joseph has been your friend since the 1760s, took your advice and published *The History and Present State of Electricity.*[1] At considerable risk, this English clergyman steadfastly supported the independence of your beloved America. You met Antoine in 1772. He is brilliant, dashing, and wealthy, and his fellow countrymen venerate you (Figure 116).[2] His beautiful, gifted, and wealthy wife Marie Anne speaks to you in French, English, or Latin, as you wish, and has painted your portrait in oil (Figure 117).[3] The Lavoisiers host "smashing" parties and stimulating salons. In contrast, Joseph and Mary Priestley live an ascetic life and preside over sober teas. Joseph published the first paper describing the discovery of what he calls "dephlogisticated air." But he is firmly anchored in the century-old phlogiston theory. Antoine calls Joseph's air "oxigene" and believes that this theory is as dead as the dodo,[4] whose extinction roughly coincided with the birth of phlogiston. So what is a great diplomat to do? The answer? Read both schools of thought in depth and leave no record whatsoever about where you stood on the greatest chemical controversy of the Enlightenment.[5]

Neutral as he was on phlogiston, Franklin nonetheless made some highly original and insightful chemical speculations. One of the most fascinating is his statement of the conservation of matter in a 1752 letter[6] composed when Lavoisier was but nine years old:[5,6]

> The action of fire only *separates* the particles of matter, it does not *annihilate* them. Water, by heat raised in vapour, returns to the earth in rain; and if we collect all the particles of burning matter that go off in smoke, perhaps they might, with the ashes, weigh as much as the body before it was fired: And if we could put them in the same position with regard to each other, the mass would be the same as before, and might be burnt over again.

Although the law of conservation matter is strongly and quite properly associated with Lavoisier, its chemical consequences were stated explicitly at least a century earlier and, indeed, the concept dates back to antiquity.[7] Nonetheless, Franklin's views on this matter are not widely known and his statement even suggests a specific experiment to verify the law. Franklin also reports witnessing the flammability of swamp gas (methane), in New Jersey, no less, over a decade before it was isolated by Alessandro Volta:[5]

> choose a shallow place, where the bottom could be reached by a walking-stick, and was muddy: the mud was first to be stirred with the stick, and when a number of bubbles began to arise from it, the candle was applied. The flame was so sudden and so strong, that it catched his ruffle and spoiled it, as I saw.

Dudley Herschbach notes that Franklin's observations over the course of six decades "convinced him of the danger of lead poisoning."[8] According to Her-

FIGURE 116. ■ Portrait of Benjamin Franklin in the *Ouevres De M. Franklin*, Paris, 1773. The poem is translated in footnote 2 of this essay; see p. 182. (This book belonged to Dr. Werner Heisenberg, courtesy of his son Professor Jochen Heisenberg.)

schbach, "he attributed a sudden attack of 'Dry Bellyach' that had beset a family to drinking rainwater collected from a leaded roof. He noted that trees planted around the house years before had grown tall enough to shed leaves on the roof, thereby creating acid that corroded the lead and 'furnished the water . . . with baneful particles'."[8]

Franklin and Lavoisier shared an interest in gunpowder, specifically the major (75%) component: saltpetre. Around 1775, as the American Revolution began to heat up, the English forbade shipments of gunpowder from Europe.[9] The Royal Navy was a convincing argument. In the same year the Continental Congress authorized publication of the pamphlet *Several methods of making salt-petre; recommended to the inhabitants of the United Colonies, by their representatives in Congress*. Sections were authored by Franklin and Dr. Benjamin Rush, who subsequently became signers of the Declaration of Independence.[9] Barn stalls and

FIGURE 117. ▪ Madame Lavoisier was instructed in painting by the famous artist Jacques Louis David. This is a photo of the oil portrait she painted of her close friend Benjamin Franklin. See color plates. (Courtesy of a relative of Benjamin Franklin.)

household chamber pots became vital sources of saltpetre. The French were more than eager to sell saltpetre and gunpowder to the American Colonies in order to weaken their long-time foes the English. Franklin dealt very successfully with his *bon ami* Lavoisier, who was an official in the *Régie de Poudres*, the government organization governing the production and quality of gunpowder. Figure 118 is from a report principally authored by Lavoisier on the fabrication of saltpetre.[10] It shows a factory in which earths and ashes rich in manure are washed, the water evaporated, and the remaining liquors cooled to allowed crystallization of saltpetre. In the lower left is a figure of a hydrometer (Lavoisier terms it an "aerometer") used to measure densities of the remaining liquors to assess when they are ready to crystallize.

Ballooning had its origins in France. The Montgolfier brothers gave the first public demonstration of a hot-air balloon on June 5, 1783. On August 27, 1783, J.-A.-C. Charles (of $V = kT$ fame)[11] made the first ascent in a balloon filled with "inflammable air" (hydrogen). Franklin was enthusiastic about ballooning and was said to quip that Montgolfier was the father of the balloon and

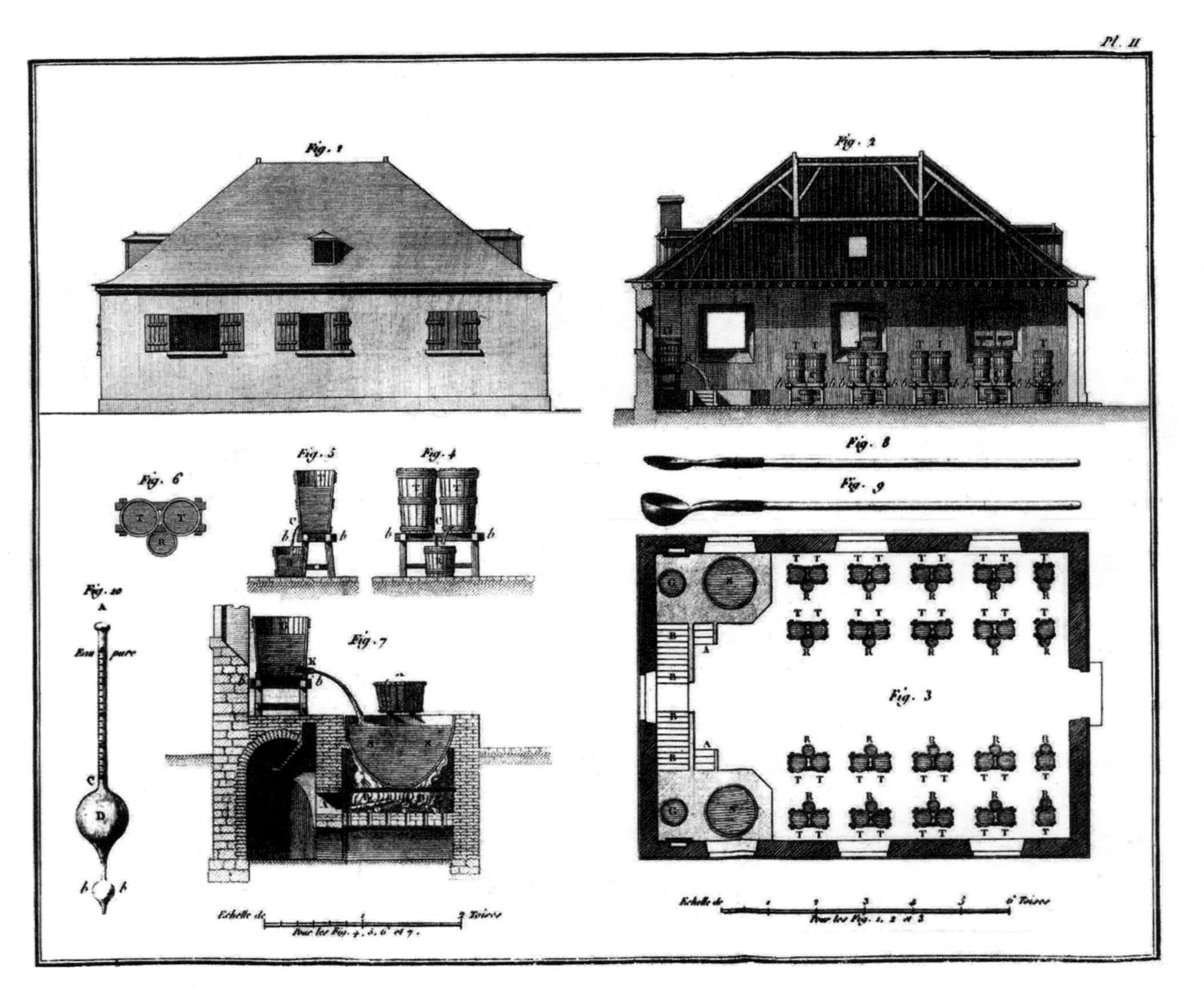

FIGURE 118. ■ A saltpetre factory described in a 1779 report (the figure is from the 1794 reprint) authored by Lavoisier and others on saltpetre manufacture.

Charles its wet nurse.[9] Both Franklin and Lavoisier served on committees that evaluated and exchanged information on ballooning. Strikingly, Lavoisier and Franklin also served on a committee to investigate the phenomenon of "human magnetism" pioneered by Franz Anton Mesmer. Although Franklin corresponded with Mesmer, both he and Lavoisier were skeptical of the claimed phenomenon, and the Committee's 1784 report was negative. We now know that Mesmer had, during his investigations, discovered hypnosis and the power of suggestion. Here is Franklin concluding an otherwise skeptical letter about Mesmer's work:[9]

> There are in every great rich city a number of persons who are never in health, because they are fond of medicines and always taking them, and hurt their constitutions. If these people can be persuaded to forbear their

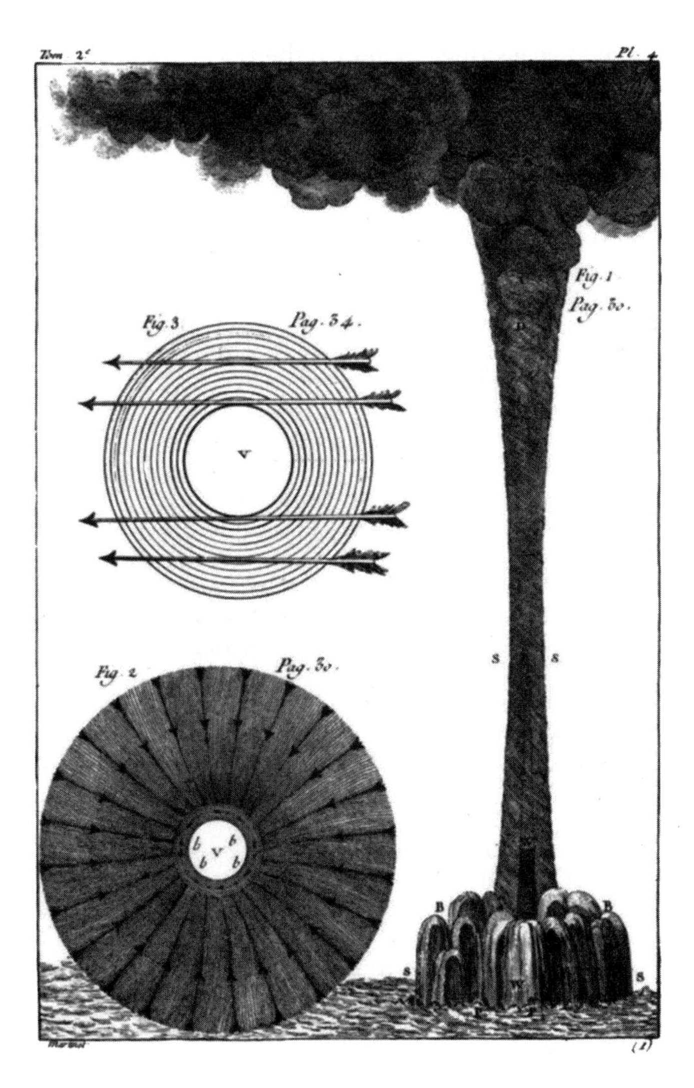

FIGURE 119. ■ Franklin was one of the earliest earth scientists. This figure shows a water spout and explains its origin. Franklin explained the effects of the Gulf Stream on weather in the northeastern United States. (From *Ouevres De M. Franklin*, courtesy of Professor Jochen Heisenberg.)

> drugs in expectation of being cured by only the physicians finger or an iron rod pointing at them, they may possibly find good effects tho' they mistake the cause.

Franklin is, of course, best known for his studies of electricity, and Herschbach compares their revolutionary importance with those of Newton or Watson and Crick.[12] Franklin believed electricity to be a fluid—*excess* corresponding to "positive" electricity and *deficit* to "negative" electricity. Indeed, his invention of the lightning rod around 1772 protected gunpowder storehouses and Franklin happily used his "thunderhouse demo" to show the efficacy of this invention.[12] However, his wide-ranging interests also led him to explain why the winds of a "noreaster" actually come from the southwest and, similarly, how the Gulf Stream affects climate in the northeastern United States. Figure 119 depicts Franklin's explanation of a waterspout in the ocean.[2] Clearly, he can be considered one of the fathers of earth science. Franklin invented the glass harmonica and *may* have composed a string quartet (in the key of F—what else?)—a slightly mischievous opus since the left hands of the surprised musicians remain idle throughout.[12]

---

1. J.R. Partington, *A History of Chemistry*, Vol. 3, Macmillan & Co., Ltd, London, 1962, pp. 237–256.
2. M. Barbeu Dubourg (transl.), *Oevres de M. Franklin*, Chez Quillau, Paris, 1773. I am grateful to Professor Jochen Heisenberg for providing for review the copy belonging to his father, Dr. Werner Heisenberg. The poem under the frontispiece portrait of Franklin was translated by my colleague Professor Jean Benoit as follows:

   He has conquered Heaven's fire
   He has helped the arts to blossom in wild climates
   America places him at the head of the sages
   Greece would have placed him amongst their gods

   It is abundantly clear that the French lionized Franklin. The "wild climates" referred to is a French Enlightenment view of the cultural milieu (or lack thereof) in the New World.
3. I am grateful to a relative of Benjamin Franklin for supplying a photograph of the oil-on-canvas portrait of Franklin by Mme. Lavoisier that is in his possession.
4. *The New Encyclopedia Britannica*, Encyclopedia Britannica, Inc. Chicago, 1986, Vol. 4, p. 148.
5. D.I. Duveen and H.S. Klickstein, *Annals of Science*, Vol. 11, No. 2, pp. 103–128 (1955).
6. I am grateful to Professor Dudley Herschbach for making me aware of this letter and its importance.
7. Partington, op. cit., pp. 377–378.
8. D. Herschbach, *Environmental Encyclopedia*. I thank Professor Herschbach for making me aware of this aspect of Franklin's work.
9. D.I. Duveen and H.S. Klickstein, *Annals of Science*, Vol. 11, No. 4, pp. 271–308 (1955); Vol. 13, No. 1, pp. 30–46 (1957).
10. [A. Lavoisier et al.], *Instruction sur l'Establissement des Nitrières, et sur la Fabrication de Salpêtre*, Cuchet, Paris (1794) (original edition 1779).
11. Charles' law for ideal gases—the volume of a gas is directly proportional to its absolute temperature.
12. D. Herschbach, *Harvard Magazine*, Cambridge, UK, Nov.–Dec. 1995, pp. 36–46.

OKAY, I NOW KNOW WHAT "OXIDATION" MEANS, BUT WHAT IS "REDUCTION"?

Thanks to Lavoisier's work toward the end of the eighteenth century, we understand that metals add oxygen to form oxides and that combustion of organic matter adds oxygen to both carbon and hydrogen. Thus, propane forms carbon dioxide and water as it lights our grills. Oxidation of nonmetals such as nitrogen, phosphorus, and sulfur form oxides that behave as acids while metal oxides behave as bases. The oxidation concept was extended in the nineteenth century—the lower-valence form of copper could be oxidized to the higher-valence form (even if no oxygen were involved). For example, yellow cuprous oxide (Cu_2O, which contains 11.1% oxygen) adds oxygen to form black cupric oxide (CuO, which contains 20.1% oxygen), and cuprous chloride (CuCl) may be oxidized to cupric chloride ($CuCl_2$) in the absence of oxygen. Indeed, that means that chlorine is also an oxidizing agent, as is iodine.

Now it so happens that the "reduction" concept, the opposite of "oxidation," is centuries older than its "contrary." What is the origin of the older term? The first instinct is to posit that "reduction" might refer to the fact that conversion of a calx, such as CuO, to the metal is accompanied by weight reduction.[1] However, despite the fact that *some* seventeenth-century chemists (Rey, Boyle, LeFevre, etc.) first reported that metals are lighter than their calxes, the "reduction" concept was effectively in use much earlier.[2]

Happily, perusal of a fat, old dictionary provides the needed insight. Although a popular two-pound dictionary,[3] offers "to lessen in any way" as the first of 15 definitions of the word "reduce," an older weightier tome, a 20-pound dictionary,[4] offers its primary definition as "to bring back; to lead to a former place or condition; restore" in full agreement with the Latin root *reducere*, "to lead back." And so there amiable reader is, I believe, the crucial insight. The ancient artisans assumed the pure metal to be the reference state and the action of reversion to the metal (from its calx, for example, by heating with charcoal) was understood to be "reduction" (reversion to its pristine state).[2] And this is also interesting since for most metals (but certainly not gold), the true former (earthly) state is not the metal but a salt, often a sulfide that must be roasted to obtain the metal. John Read[5] noted that the process of "putrefaction" (Fourth Key of Basil Valentine; see Figure 4) involves blackening of the initial alchemical mixture accompanied by the "death" of metals baser than gold. This is actually the roasting and oxidation of metals and their sulfide ores. The reverse is their "resurrection" (Eighth Key of Basil Valentine; Figure 5), the restoration of the original metals, and reunion with their souls through the chemical process we call "reduction."

It has also long been known that heating a calx in a stream of hydrogen gas causes reduction of the calx to the lighter pure metal. (Oh! Did I mention that water is also formed?) So hydrogen gas is a reducing agent. Reduction by hydrogen of unsaturated fats, however, produces saturated fats, which weigh *more* than the corresponding unsaturated fats. The reduced, saturated fat contains more calories than does the unsaturated fat so . . . oh, never mind! In any case, the modern definition of "reduction," which unifies all of these diverse operations, is "a process in which a substance gains one or more electrons" (oxidation is the "contrary").[6]

Well, darn it, as a chemist I know what I mean when I say "reduction." To paraphrase Popeye The Sailor—"I yam what I yam and that's all [*spit*] what I yam!"[7]

---

1. C. Cobb and H. Goldwhite, *Creations of Fire*, Plenum Press, New York, 1995, p. 8.
2. J.R. Partington, *A History of Chemistry*, Vol. 2, Macmillan & Co. Ltd., London, 1961, p. 19.
3. *College Edition—Webster's New World Dictionary of the American Language*, The World Publishing Company, Cleveland and New York, 1964, p. 1219.
4. *Webster's New Twentieth Century Dictionary of the English Language Unabridged*, second edition, The World Publishing Company, Cleveland and New York, 1956, p. 1514. This is also the primary definition in the Oxford English Dictionary.
5. J. Read, *From Alchemy to Chemistry*, Dover Publications, Inc., New York, 1995, p. 33.
6. T.L. Brown, H.E. LeMay, Jr., and B.E. Bursten, *Chemistry—the Central Science*, Prentice Hall, Upper Saddle River, NJ, 1997, pp. G–11, G–13. Pauling notes a failed attempt by Professor E.C. Franklin of Stanford University to remove this confusion by coining the words "de-electronation" (for oxidation) and "electronation" (for reduction) [see L. Pauling, *General Chemistry*, privately printed (Edwards Brothers, Inc. Lithographers—Ann Arbor), Pasadena, 1944, p. 65].
7. Popeye's neologisms and puns ("vitalicky"; "I know what rough is, but what's roughined?") have outlasted those of Professor Franklin (see note 6 above).

THE MAN IN THE RUBBER SUIT

Antoine Lavoisier "buried" phlogiston theory and, in so doing, explained the basis of combustion and calcinations such as the rusting of iron. However, it is less widely appreciated that it was Lavoisier who first demonstrated that metabolism is simply a very slow combustion process. Where this metabolism actually occurred, heart, lungs, or both places, remained a mystery to him.

It was apparent to John Mayow as early as 1674 that respiration removed something from atmospheric air and the remaining depleted air could not support life or combustion.[1] Mayow's observations were strengthened a century later by the work of Scheele, Priestley, and Lavoisier, who isolated, manipulated, and characterized the "airs" they studied. When a mouse was confined in a bell jar containing atmospheric air, the air soon became "mephitic" or "deadly" because its full complement of "vital air" (oxygen) was depleted.[2] If the "mephitic air" (~99% nitrogen) were recharged with oxygen in the correct 4 : 1 proportion, a mouse would live equally long as it would in atmospheric air. The isolation and characterization of the mouse's exhalation gas, "fixed air" (carbon dioxide), was only first reported in 1756 by Joseph Black.[3] In 1777, Lavoisier concluded that animal metabolism combines carbon and oxygen to produce carbon dioxide, just as carbon unites with oxygen during combustion.[4,5]

The ice calorimeter, designed by Pierre Simon de Laplace, was first employed by Lavoisier during winter, 1782/83.[6–8] Heat from the reaction vessel was measured by the quantity of ice in the surrounding metal jacket that melted and was collected as water. Lavoisier and Laplace measured the heat given off by many chemical processes, including the combustion of charcoal. They also measured the heat produced by a living guinea pig.

By burning charcoal and measuring the "fixed air" produced, Laplace and Lavoisier equated formation of 1 ounce of fixed air to melting 26.692 ounce of ice.[6] Over a 10-hour period the amount of fixed air collected from a guinea pig's exhalations (224 grains, where 576 grains = 1 ounce) equated, on this basis, to melting 10.14 ounces of ice. The actual heat given off by the guinea pig over ten hours was greater, melting 13 ounces of ice. Although there were errors in the determination of the heat of combustion of charcoal, the main discrepancy was that the guinea pig "burned" not only carbon but also the hydrogen in its "fuel" to form water, thus releasing additional heat.[9] However, Lavoisier did not understand the true nature of water in early 1783 or that it was a product of respiration.

The discovery that water was not an element, but a compound of hydrogen and oxygen was absolutely critical to Lavoisier's growing understanding of respiration. In 1774, he ignited "flammable air" (hydrogen), isolated eight years earlier by Cavendish, in the presence of "vital air" and tried to collect the resulting "air" over water.[10] Naturally, the small quantity of water vapor generated went unnoticed. Although credit for the discovery of water's composition remains somewhat controversial, most chemical historians attribute it to Cavendish in 1783.[11] However, it was Lavoisier who determined the precise composition both in its decomposition into the elements and its synthesis from the elements (see Figures 114 and 115) and he reported these findings in early 1784. Water was not an element but a compound and a combustion product of compounds containing hydrogen. Thus, he realized that the water exhaled and sweated by animals was likely to be a product of respiration. The remaining problem for the "master of the chemical balance" was that a complete accounting of masses, input versus output, had not yet been demonstrated. Unfortunately, no human being had yet been hermetically sealed in a closed flask for precise studies of mass balance.

However, in 1790 Lavoisier and his assistant Armand Seguin conducted studies in which Seguin was sealed completely in a suit made of elastic-rubber-coated taffeta.[5,12] He breathed pure oxygen through a tube glued airtight around his lips and exhaled through this tube. The scene was depicted in a drawing by Madame Lavoisier (Figure 120)[13]—Seguin is seated at the left, while Lavoisier is standing center right and providing oxygen and Madame Lavoisier is taking notes. The disappearance of oxygen was carefully measured and the exhaled carbon dioxide and water vapor collected and measured. The amount of oxygen inhaled closely matched the quantity exhaled in the forms of carbon dioxide and water. In order to fully account for the mass balance of water, the mass of the man in the rubber suit was measured carefully before and after experimental periods. Perspiration, and other "effluvia," trapped in the suit were measured to an amazing accuracy of 18 grains in 135 pounds[5]—an expensive balance, indeed! When it came to Lavoisier's apparatus, money was no object. Figure 121[13] depicts the same experiment but Seguin exerting himself by peddling a treadle. The uptake of oxygen was considerably greater. On November 17, 1790, Seguin and Lavoisier presented a memoir that stated in part: [5]

FIGURE 120. ▪ Lavoisier was the first to demonstrate that respiration is in fact combustion. This is a drawing by Madame Lavoisier (depicted at right) of her husband conducting respiration experiments on his assistant Armand Seguin, completely enveloped in a rubber suit. Seguin survived and eventually became extremely wealthy as an army contractor. (Courtesy Professor Marco Beretta.)

FIGURE 121. ▪ Another drawing of Lavoisier's respiration studies on his assistant Seguin—the man in the rubber suit. Madame Lavoisier depicts herself drawing this scene. (Courtesy Professor Marco Beretta.)

> Respiration is only a slow combustion of carbon and hydrogen, similar in every way to that which takes place in a lamp or lighted candle and, from this viewpoint, breathing animals are actual combustible bodies that are burning and wasting away.

---

1. J.R. Partington, *A History of Chemistry*, Macmillan and Co. Ltd., London, 1961, 1961, Vol. 2, pp. 577–614.
2. Partington, op. cit., 1962, Vol. 3, pp. 205–234; 237–297.
3. Partington 1962, op. cit., pp. 130–143.
4. Partington 1962, op. cit., pp. 471–479.
5. J.-P. Poirier, *Lavoisier—Chemist, Biologist, Economist* (transl. R. Balinski), University of Pennsylvania Press, Philadelphia, 1996, pp. 300–309.
6. Partington 1962, op. cit., pp. 426–434.
7. Poirier, op. cit., pp. 135–140.
8. A. Greenberg, *A Chemical History Tour*, John Wiley and Sons, 2000, pp. 150–152.
9. Modern calorimetric data indicate that combustion of carbon (graphite) sufficient to produce exactly 1 ounce of carbon dioxide would melt 26.86 ounces of ice. If sufficient glucose ($C_6H_{12}O_6$) were burned to collect the same 1 ounce of CO_2, one might naively have expected in 1783 that 26.86 ounces of ice would be melted. However, we know that formation of 0.41 ounces of H_2O would accompany the 1 ounce of CO_2 formed in glucose combustion. The extra heat from formation of water added to the heat from formation of carbon dioxide would melt 31.89 ounces of ice.
10. Poirier, op. cit., pp. 140–144.
11. Partington 1962, op. cit., pp. 325–338.
12. Partington 1962, op. cit., pp. 471–479.
13. I am grateful to Professor Marco Beretta for supplying these images.

"POOR OLD MARAT"? I THINK NOT!

Jean-Paul Marat is considered today to have been a minor scientist and was so judged by the *Académie des Sciences* over two centuries ago. He remains, however, famous and infamous as an impassioned and uncompromising "Friend of the People"—a major actor in the triumphs, excesses, and tragedies of the French Revolution. Although Marat himself was murdered on July 13, 1793, some 10 months before the execution of Lavoisier, he certainly helped to inflame passions and create the atmosphere that led the brilliant aristocrat to the guillotine on May 8, 1794.

Born in the Swiss canton of Neuchâtel in 1743, Marat left home in 1759, spent 6 years in France and 11 in England and Scotland, writing philosophical tracts that gained some international notice.[1–4] One of these, *The Chains of Slavery* (1774), was said to have first advanced his idea of the "aristocratic" plot.[3] Marat started to attend medical classes in France around 1760, then moved to England and practiced medicine beginning in 1765. Following 10 years of successful practice, he was awarded the honorary degree of Doctor of Medicine at the College of St. Andrews in Scotland in 1775.[5] Although Samuel Johnson was critical of St. Andrews' practice of selling degrees (he said that the college would "grow richer by degrees"), the two medical faculty members recommending him were highly regarded.[5] Marat then resettled in France, proceeded to ingratiate himself with the aristocracy, and was appointed physician to the personal guards of Comte D'Artois, the youngest brother of King Louis XVI. He charged his aristocratic clients[4,6] almost 1 louis (24 livres) per consultation, or roughly $1000. today, and was considered a very accomplished physician.

Beginning in 1778, Marat began a series of scientific investigations of the "imponderables" light, heat, fire and electricity and ". . . began to lay siege to the Academy of Science."[4] Figure 122 is from Marat's 1780 book[7] in which he explained the nature of heat and fire. He identified a *fluide igné*[8] or fiery fluid that, in some ways, anticipated Lavoisier's *calorique* (caloric). He posited that when a hot object contacts a cold object, *fluide igné* is passed from the warmer object to the colder until the contents are equal. Marat viewed heat and fire as two closely related effects having the same origin. Heat is produced when energy input is only moderate and fire when energy input is high.[8] This physical interpretation of a continuum from heat to fire neglected the dramatic chemistry of fire. According to Marat, all substances must contain *fluide igné*, or they could not reach the temperature of their surroundings. It was the movement of this fluid, not its mere presence, that generated heat and fire.[8] Among a list of substances commonly recognized as flammable (carbon, camphor, naphtha, essential oils, alcohol, phosphorus), which Marat describes as "highly impregnated with *fluide igné*," he also lists nonflammable "fixed alkali" (or sodium carbonate, Na_2CO_3 today).[8] The confusion probably derives from release of "fixed air" (carbon dioxide) from heating this salt just as "fixed air" is "released" when carbon-containing substances are burned. Marat explained the fact that a flame will soon be extinguished in an enclosure as follows: "the air, violently expanded by the flame and unable to escape, dramatically compresses and smothers it."[4]

The shadowy figures in Figure 122 were said by Marat to be images of actual *fluide igné* escaping from a lighted candle (*Fig. 1*), a burning piece of charcoal (*Fig. 2*), and a red-hot cannonball (*Fig. 3*). These images were obtained using Marat's "solar microscope," a dark room with a tiny hole allowing entrance of a very thin beam of light that passes through the lens of a microscope and onto a screen. The small visible cone of a candle flame, for example, is imaged, as is the surrounding column of *fluide igné*. America's ambassador to France, Benjamin Franklin, attended one of Marat's demonstrations. He exposed his bald pate to the solar microscope, and it was duly observed that "we see it surrounded by undulating vapors that all end in a spiral. They look like those flames through which painters symbolize Genius."[4]

Marat's attempt for recognition by the French Academy of Science was rejected in May 1779. In June 1780 Lavoisier called the Academy's attention to a paper by Marat that implied that the Academy had endorsed his *fluide igné* findings.[4] The Academy rebutted this assertion, and Marat now included the Academy and Lavoisier in particular on a lifelong enemies list. It should be noted that the wealthy and brilliant Lavoisier had become a member of the Academy at the age of 25. His rejection may have been the beginning of what Gottschalk refers to as Marat's "martyr complex."[9]

French society underwent enormous change throughout the eighteenth century with considerable stresses and cracks developing in the social fabric.[10] The authority of the French monarch remained absolute. In the hands of an inspirational king, Louis XIV—"The Sun King," the French remained reasonably quiescent. However, the next two rulers of the *Ancien Régime*, Louis XV and

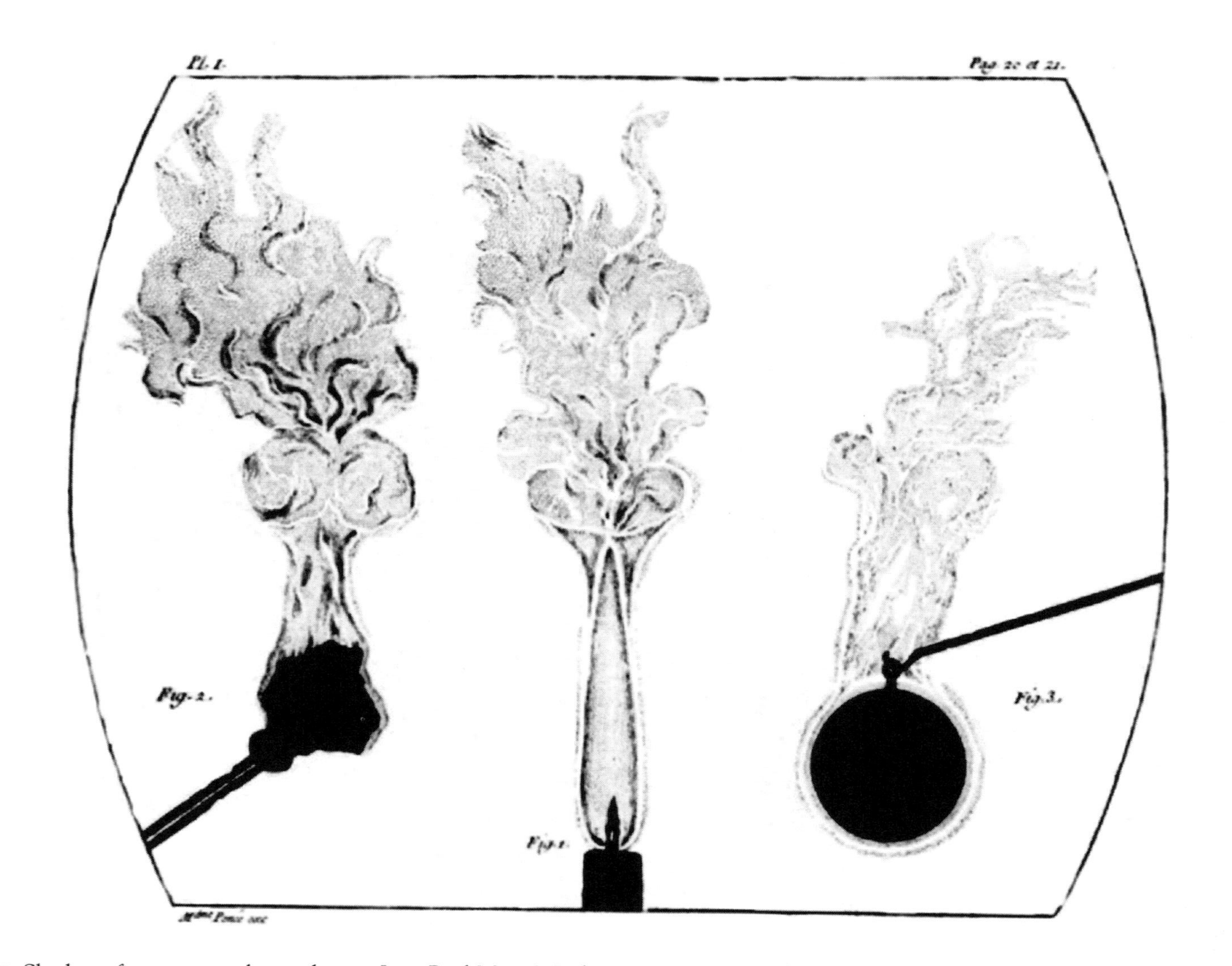

FIGURE 122. ▪ Shadowy figures were obtained using Jean Paul Marat's "solar microscope" in a dark room. Marat postulated a *fluide igné* that could escape from heated or burning materials (he makes no real distinction between physical and chemical processes). Here he claims to have observed this imponderable substance escaping from (a) a lighted candle, (b) a burning piece of charcoal, and (c) a red-hot cannonball. Lavoisier played a major role in denying Marat's application for membership in the Academy of Sciences and thus earned his undying hatred, (From Marat's 1780 *Recherches Physiques sur le Feu.*)

Louis XVI, were relatively ineffectual. The revolution began stirring largely in urban areas during 1788, but its most dramatic expression, the storming of the Bastille, occurred on July 14, 1789. Not long afterward, Marat, whose political activism dramatically increased during the 1780s, began to edit a newspaper, *Ami du Peuple* ("Friend of the People"), to raise and incite revolutionary fervor.[1–4] Marat initially supported a constitutional monarchy but quickly embraced the views of the rabid antimonarchists who forcibly brought the royal family from Versailles to Paris in October 1789. However, the monarchy remained intact, and Marat's criticism of the king's Finance Minister caused the revolutionary to flee to England briefly during 1790.

Poirier uses "Cultural Revolution"[11] to describe France's revolt against the intellectual authority of knowledge similar, by implication, to that seen in China during the 1960s. The Academy's 1784 criticism of mesmerism, in which both Lavoisier and Franklin played lead roles, was now attacked as elitist.[12] Two excerpts from Marat's 1791 pamphlet *Les Charlatans Modernes* illustrate the vulnerability of the academicians:[13]

> At the head of them all would have to come Lavoisier, the putative father of all the discoveries that have made such a splash. Because he has no ideas of his own, he makes do with those of others.

Recall that this was after Lavoisier had totally revolutionized chemistry. Marat's description of the Academy was no less demagogic:[14]

> A collection of vain men, very proud to meet twice a week to chatter idly about fleurs-de-lys; they are like automatons accustomed to following certain formulas and applying them blindly. . . . What a pleasure it is to see the mathematicians yawn, cough, spit, and snigger when a chemistry paper is being read, and the chemists snigger, spit, cough, and yawn when a mathematics paper is presented.

The French Revolution became increasingly radicalized and bloodthirsty during the next few years.[10] The more conservative of the revolutionary factions, the Girondins, supported the constitutional monarchy. However, it became ever clearer that the king would never abandon the arrogant and uncompromising aristocracy. A brief war with neighboring Prussia was instigated by the king, who had tried unsuccessfully to flee France, in June 1791. The hope was that a foreign war would quell the civil war. The plot backfired, and a "second revolution" overthrew the monarchy in August 1792. The National Convention, comprising relatively conservative Girondins and more radical Montagnards, was established to develop a new constitution. The king was convicted in December and executed in January 1793. Queen Marie Antoinette was also imprisoned and ultimately guillotined in October 1793.

By that time, the Montagnards had defeated the Girondins.[10] Indeed, the most radical factions of the Montagnards defeated their own bourgeoisie factions. Marat was, with Robespierre and Danton, among these most radical factions. Extended periods of hiding in cellars and sewers apparently contributed to a painful and disfiguring skin condition that Marat relieved with frequent baths. On July 13, 1793, Charlotte Corday, regarded now as an innocent, possibly brainwashed, tool of the Girondins, gained entry to Marat's home and stabbed him to death in his bathtub. The "Death of Marat" was memorialized by artist Jacques Louis David in a dramatic painting (Figure 123). David himself seems to have been a fashionable radical who had no problem "tacking with the prevailing wind." He had been Madame Lavoisier's art teacher and also collected a fee, roughly equivalent to $300,000 today, for painting the Lavoisiers' portrait in 1789.[15] Nevertheless, David attacked the academicians during the revolution, and was commissioned to paint the Marat portrait. David dressed Marat's body in Roman style for the funeral.[16] The "Law of Suspects," which began the Reign of Terror in October 1793, as a response to Marat's murder, devoured Marie Antoinette and half a year later Lavoisier and his father-in-law in its ravenous maw. Ultimately, popular revulsion with this bloodbath caused some moderation in the later years of the decade, but France did not truly stabilize until Napolean Bonaparte imposed a military dictatorship in late 1799.

Historical viewpoints change with the times. The nineteenth century was unkind to Marat. However, in 1964 the German playwright Peter Weiss published a fascinating play that is often referred to by the abbreviated title "Marat/Sade."[17] Marat is portrayed more sympathetically in the play within this play. It is 1808 and the Marquis de Sade, a fallen nobleman and writer, is interred in the Asylum of Charenton. He is directing a play that dramatizes the murder of Marat. A Chorus intones "Poor Old Marat" during sections of the play (singer Judy Collins popularized the lyrics during the 1960s). The play dramatizes the murder of Marat. Charlotte Corday is depicted as an automaton—a sort of

FIGURE 123. ▪ Marat was trained as a physician and had a very profitable practice. He became radicalized throughout the 1780s and was a formidable member of the most radical groups after 1789. Extended periods of hiding in the sewers of Paris may have provoked a painful skin disease that he treated with countless baths. It was in the bath that he was stabbed to death by one Charlotte Corday on July 13, 1793. The scene (shown here in black and white) was painted by Jacques Louis David and is exhibited in the Musée Royaux des Beaux-Arts in Brussels. The events leading up to and including Marat's assassination were dramatized in 1964 by playwright Peter Weiss.

doomsday machine for Marat. Lavoisier makes a brief cameo appearance in the play. The play's critical juxtaposition is between the committed but fanatical Marat and the Marquis who has lived a life both intellectual and debauched testing the boundaries of human nature. Sade has sympathy for the common people and their revolution, but he is cynical, world-weary and deeply frightened of fanatic crusaders like Marat. In the play, there is this exchange between them:[18]

Sade: I don't believe in idealists
who charge down blind alleys
I don't believe in any of the sacrifices
that have been made for any cause
I believe only in myself

Marat: [*turning violently* to SADE]
I believe only in that thing which you betray
We've overthrown our wealthy rabble of rulers
disarmed many of them though
many escaped
But now those rulers have been replaced by others
who used to carry torches and banners with us
and now long for the good old days
It becomes clear
that the Revolution was fought
for merchants and shopkeepers
the bourgeoisie
a new victorious class
and underneath them
ourselves
who always lose the lottery

Historically, Marat and Sade never actually conversed.[17]

---

1. L.R. Gottschalk, *Jean Paul Marat—a Study in Radicalism*, Benjamin Blom, New York, 1927.
2. C.D. Conner, *Jean Paul Marat—Scientist and Revolutionary*, Humanity Books, Amherst, 1998.
3. *The New Encyclopedia Britannica*, Encyclopedia Britannica, Inc., Chicago, 1986, Vol. 7, pp. 813–814.
4. J.-P. Poirier, *Lavoisier—Chemist, Biologist, Economist*, University of Pennsylvania Press, Philadelphia, 1993, pp. 110–112.
5. Gottschalk, op. cit., pp. 4–5. See also Conner's spirited defense of Marat's medical training—Conner, op. cit., pp. 33–34.
6. Poirier, op. cit., p. 428.
7. [J.P.] Marat, *Recherches Physiques sur le Feu*, chez C. Ant. Jombert, Paris, 1780
8. Marat, op. cit., pp. 17–21.
9. Gottschalk, op. cit., pp. 1–31.
10. *The New Encyclopedia Britannica*, op. cit., Vol. 19, pp. 483–502.
11. Poirier, op. cit., pp. 328–333.
12. Poirier, op. cit., p. 159.
13. Poirier, op. cit., p. 196.
14. Poirier, op. cit., p. 329.
15. Poirier, op. cit., p. 1.
16. Poirier, op. cit., p. 330.
17. P. Weiss, *The Persecution and Assassination of Jean-Paul Marat as Performed by the Inmates of the*

Asylum of Charenton Under the Direction of the Marquis De Sade, English version By Geoffrey Skelton, Atheneum, New York, 1965.
18. Weiss, op. cit., pp. 41–42. Copyright Suhrkamp Verlag Frankfurt am Main 1964. Permission to reprint English version (ISBN 0-7145-0361-4) courtesy Marian Boyars Publishers, London (UK) 1965.

POOR OLD LAMARCK

It is sad that the only thing we learn in school about Jean Baptiste Lamarck (1744–1829)[1] is that he explained the long limbs and necks of giraffes by noting that they must continuously stretch and extend themselves, thus strengthening and slightly elongating their necks and legs during their lifetimes, *and* that these acquired improvements are inherited by their offspring. Successive generations would continue to "improve" in this manner—we might now say "evolve." This explanation was offered almost 60 years before the publication of *The Origin of Species* by Charles Darwin in 1859, which presented evolution as an observed fact and offered natural selection as its mechanism. It would be another 6 years before Gregor Mendel would present his observations on hybrid peas, a further 35 years before the significance would be fathomed, and another 50 years before Watson and Crick would explain it. Nonetheless, we remember Lamarck for his wrong theory just as Brooklyn Dodger pitcher Ralph Branca is forever remembered for the homerun ball he threw to Bobby Thompson on October 3, 1951 that allowed the New York Giants to snatch the pennant from its rightful owners. Branca, who proudly wore number 13 on his uniform, had a very respectable lifetime record (88 wins; 68 losses;[2] "You could look it up"—C. Stengel[3]), but he will always be remembered for that gray, infamous Manhattan afternoon. Lamarck could just as well have worn number 13. He was one of 11 children born into the "semi-impoverished lesser nobility of Northern France."[1] He married three (possibly four) times—his three known wives dying early of illness, his total of eight children including one deaf son, one insane son, and all but one of his children were consigned to lives of poverty. In order to pay for his funeral in 1829, his family had to sell his books and scientific collection at public auction and appeal to the *Académie des Sciences* for funds.[1]

In fact, Lamarck made numerous important contributions to science in the late eighteenth and early nineteenth centuries. He coined the terms "biology" and "invertebrate" and developed a taxonomy system that was, in some respects, easier to use than that of Linnaeus.[1] He was widely recognized as one of the leading experts on invertebrate biology and was one of the first paleontologists to relate fossils to living creatures and to try to account for the differences between, say, fossil clams and their living relatives. This naturally led him to try to explain the sources of these differences. While he never used the term "evolution," he was certainly a protoevolutionist.[1] Lamarck had a unified view of Nature that was almost mystical in nature. Only living organisms could make "organic matter." These organisms could change (read "evolve") through the generations by strengthening and improving themselves—a very alchemical idea—strengthening of human characteristics leading to human perfection—the removal of impu-

FIGURE 124. ■ Jean Baptiste Lamarck was an important biologist who, unfortunately, is widely remembered for his incorrect theory of acquired traits. His chemistry, however, was very outmoded, and in his one chemistry text Lamarck tried to describe the repulsion between corpuscles of matter as they absorbed heat that caused them to expand and repel. (From Lamarck's 1794 *Recherches Sur Les Causes Des Principaux Faits Physiques*.)

rities gradually perfecting and transmuting base metals into gold. When organisms died, the decomposing organic matter would become mineral matter. Indeed, he resisted the growing reductionism (he termed them "small facts") in science. Moreover, Lamarck, as a professor and curator at the *Muséum National d'Histoire Naturelle*, arranged the invertebrates according to systematic classification rather than the random "cabinet of curiosities" typical of earlier such museums. We see Lamarck's pioneering methods in the halls of dinosaurs and mammals in modern museums and, indeed, in the evolutionary organization within these halls.

Unfortunately, Lamarck's resistance to the "small ideas" developing in chemistry fixed him firmly in the pre-Boyleian seventeenth century. He retained his belief in the four elements and was particularly fascinated by the different forms assumed by the element fire. Lamarck believed that "matter of fire" and "matter of electricity" were essentially one and the same. This is not so surprising. If one is a great distance from a large fire, it is still possible to see the sky light up as the fire intensifies or as a new fuel source inflames. The appearance is not very different from the diffuse appearance of lightning obscured by clouds. Lamarck was well aware that Benjamin Franklin had demonstrated the electrical nature of lightning. Moreover, it was also apparent by the end of the eighteenth century that both fire and electricity caused chemical change.

Figure 124 is from Lamarck's first and most important chemical work,[1] which was published in 1794.[4] The top part of the figure shows two cork balls suspended over a hook by a silk thread. The light cork balls have been electrified through friction, and they separate (see *Fig. B* at the top). The reason offered by Lamarck is that electrification causes each sphere to be surrounded by superlight "matter of electricity." The pressure of the air pushes in on all sides equally (the thread causes a bit of distortion), thus separating the two electrified spheres. If the charged cork balls were forced to touch (*Fig. A*), the regions of electric matter between them would overlap and the overall shape of the electric matter would be oval (not the optimal spherical) with a small gap at the top. Air pressure would seep in and force these balls apart (*Fig. B*). In the bottom of Figure 124 we see a vase full of water placed over a fire. Again, air pressure keeps the flame concentrated under the vase, and the easiest path for the fire, according to Lamarck, is through the vessel and into the water. As more fire is absorbed, the water molecules are surrounded by ever larger coats of superlight "matter of fire"—the water gets warmer (and less dense). Eventually, the low density and high energy of these particles, forced by the downward pressure of the atmosphere, cause them to vaporize and carry away "matter of fire" with them. (Come to think of it fellow teachers—what an interesting way to represent the latent heat in molecules of water vapor!)

And so, we wish we could say some nice things about Lamarck's chemistry. However, his early contributions to biology and its museum display for the public were of great value and do honor to his memory.

---

1. C.C. Gillispie (Editor-in-Chief), *Dictionary of Scientific Biography*, Vol. VII, Charles Scribner's Sons, New York, 1973, pp. 584–593. His full name, for the record, was Jean Baptiste Pierre Antoine de Moncet de Lamarck.
2. D.S. Neff and R.M. Cohen, *The Sports Encyclopedia: Baseball*, St. Martin's Press, New York, 1989.

3. P. Dickson, *Baseball's Greatest Quotations*, HarperCollins Publishers, New York, 1991, p. 427.
4. J.B. Lamarck, *Recherches sur les Causes des Principaux Faits Physiques*, Tome Premier, Chez Maradan, 1794, pp. 198–204.

MON CHER PHLOGISTON, "YOU'RE SPEAKING LIKE AN ASS!"

The jubilant revolutionaries who overthrew the monarchy in August 1792 were convinced that France had been born anew. On 8 *frimaire* An II (Year 2), that is, November 28, 1793, Antoine Laurent Lavoisier and his father-in-law Jacques Paulze, reported for internment to the Port-Libre prison.[1] Lavoisier's situation had become increasingly perilous as the revolution radicalized and closed in all around him, stripping him of offices, colleagues, and the ability to travel, and after some days of hiding in Paris, his very freedom. He would lose his life in late spring 1794.

But in 1788 Lavoisier was at the height of his prestige and authority. As one of 40 wealthy partners in the *Ferme Générale*, he was a shareholder in the company empowered to collect taxes on all imports. This included the very salt that sustained people's lives. The General Farm also exercised some control of the flow of this revenue into the Royal Treasury. It therefore wielded considerable influence on France's fiscal policy. Lavoisier himself had, as a partner in the Farm, become a member of the board of directors of the Discount Bank, the "banker's bank," which advanced money to the Royal Treasury and supplied the gold and silver for minting into coins.[2] A world-class economist, he would soon become its president. Poirier nicely states this situation—a "private company was controlling the loans made to the government by a private bank."[2]

During most of 1788 Lavoisier wrote his masterpiece *Traité Elémentaire de Chimie* and it was published in early 1789 (see Figure 125).[3,4] The project had started as an attempt to provide an accessible introduction to chemistry, evolved to update the 1787 *Méthode de nomenclature* chimique [3,4] and became the most important treatise in the history of chemistry. It included the first modern list of chemical elements (33 in number, including the "imponderables" light and caloric).[3–5] Lavoisier was very much concerned with public education[6] and pedagogy, and this is reflected in his textbook. As late as September 22, 1793, he advocated to the Convention the education of a technologically proficient populace. However, in the prevailing climate of the "Cultural Revolution,"[7] Robespierre and the Jacobins wanted a more ideological education. This debate came to an abrupt halt with the Reign of Terror in October 1793.[6]

The year 1788 was a triumphant one for the Lavoisiers, even as the winds of revolution were stirring. Madame Lavoisier wrote to Jean Henri Hassenfratz, Director of the Arsenal, seeking suggestions for celebrating the victory of their chemical revolution. He suggested a portrait of the Lavoisiers and an allegorical play in which oxygen defeats phlogiston.[8,9] The portrait of the Lavoisiers was completed in 1789 by Jacques Louis David, for a fee of 7000 livres ($280,000 in current money)[9] and now sits in the Metropolitan Museum of Art. The only tangible evidence of a brief satirical play or masque is found in a letter by a Dr. von E** published in Crell's journal *Chemische Annalen*.[8,9]

TRAITÉ
ÉLÉMENTAIRE
DE CHIMIE,
PRÉSENTÉ DANS UN ORDRE NOUVEAU
ET D'APRÈS LES DÉCOUVERTES MODERNES;
Avec Figures :

Par M. LAVOISIER, de l'Académie des Sciences, de la Société Royale de Médecine, des Sociétés d'Agriculture de Paris & d'Orléans, de la Société Royale de Londres, de l'Institut de Bologne, de la Société Helvétique de Basle, de celles de Philadelphie, Harlem, Manchester, Padoue, &c.

TOME PREMIER.

A PARIS,
Chez CUCHET, Libraire, rue & hôtel Serpente.

M. DCC. LXXXIX.

Sous le Privilège de l'Académie des Sciences & de la Société Royale de Médecine.

FIGURE 125. ▪ Title page from the first edition (1789) of Lavoisier's 1789 masterpiece, *Traité Elémentaire De Chimie*, the first modern textbook of chemistry.

Another, earlier masque is, however, creatively imagined in the play *Oxygen*, authored by two distinguished modern chemists, Carl Djerassi and Roald Hoffmann.[10] The year 2001 marks the hundredth anniversary of the Nobel Prize. In modern Stockholm the 2001 chemistry Nobel committee is informed, in secret, that they will also choose the first "retro-Nobel Prize." The committee quickly reaches consensus that the discovery of oxygen and its role in chemistry and respiration merits the first "retro–Nobel." Should it go to the Swede Carl Wilhelm Scheele who first isolated "fire air" (oxygen) in 1771 (or 1772) but did not publish the work until 1777? Should it go to Joseph Priestley, who independently discovered "dephlogisticated air" (oxygen) in 1774 and promptly published his discovery in the same year? Both Scheele and Priestley erroneously believed that their "air" drew phlogiston from burning or rusting substances. Or should it go to Antoine Laurent Lavoisier, whose intellectual synthesis fully explained the role of oxygen in combustion, calcination, and respiration? *Oxygen* actually begins in the Stockholm of 1777, where we meet Marie Anne Paulze Lavoisier, Mary Priestley, and Sara Margaretha Pohl, assistant and companion to Scheele, in a sauna. Their husbands have been summoned to Sweden to perform experiments before King Gustav III, who will render *The Judgement of Stockholm*. This play-within-a-play might somehow help the 2001 committee to settle its (or at least the play-goers') predicament. On the evening before the royal command chemistry demonstrations, the Lavoisiers perform their brief masque before King Gutav III and the assembled company. The Priestleys, Scheele, and Fru Pohl become increasingly uncomfortable and, finally, quite upset upon its conclusion. Antoine plays the vanquished "Phlogiston" and Marie Anne the victorious "Oxygène" in this play-within-a-play-within-a-play. Leading up to their masque's conclusion we have Madame Oxygène addressing Monsieur Phlogistique:[11]

> *Mon cher* monsieur, you're speaking like an ass!
> You know there's no such thing—negative mass!
> A revolution is about to dawn
> In chemistry, as Oxygen is born.
> Phlogiston is a notion of the past,
> Disproved and set aside, indeed, surpassed.

Madame Lavoisier was surely one of the most fascinating figures in the history of chemistry.[12] She plays the central role in *Oxygen*, and a mysterious note long hidden in her *nécessaire*[12] solves, in the play at least, a chemical riddle more than two centuries old. As to who wins the first "retro-Nobel"—that is for you, gentle reader, to guess—but first read the play.

Widespread resentment of Lavoisier, which accompanied admiration of this awesome polymath, predated the French Revolution. As a member of the Academy of Sciences, his investigation, along with Franklin and others, that dismissed mesmerism as bad science, was resented by a populace that wanted to believe in it.[13] Another longstanding grievance was the Farm's "watering of tobacco" prior to distribution to distributors.[14] More serious, however, was Lavoisier's role in the General Farm's collection of taxes. Imagine a powerful holding company comprised of 40 individuals, the purpose of which was to zealously collect taxes for the Royal Treasury but not before they had taken a very tidy profit. The salt tax was widely despised—salt was a staple for preserving meat—and indeed, is a very substance of life.[15,16] The salt tax was one of Lavoisier's specific responsibilities for the General Farm. Another was the tax upon imports into Paris. Lavoisier, using the genius in accountancy that he applied to chemistry, had realized late in the 1770s that only four-fifths of the goods needed to supply the inhabitants of Paris were actually reported and taxed.[17] The remaining fifth was being smuggled with a loss to the Royal Treasury (and, incidentally, the General Farm). His solution, a wall with toll gates surrounding Paris, was accepted in 1787 and built at a cost of 30 million livres ($1.2 billion).[17] Once again, try to imagine a private company owned by 40 of the wealthiest individuals in the United States, walling in New York City and building palatial toll gates at taxpayer expense for use by the Internal Revenue Service. One of the accusations leveled against Lavoisier years later was that the wall built around Paris confined the city's air to the detriment of its citizens' health.

But the ground was beginning to shake and soon Lavoisier would experience a foreshadowing of his own fate. As Director of the Gunpowder Administration, he had authority over gunpowder shipments from the Arsenal. Not long after the storming of the Bastille on July 14, 1789, mysterious transfers of gunpowder from the Arsenal were observed by citizens who concluded that a Royalist counterattack was in the offing.[18] Lavoisier was seized and held briefly in custody—some members of the crowd gathered along his route of transport demanded summary execution. However, he explained the shipments in detail, was declared innocent, and released. On March 20, 1791, the National Assembly abolished the General Farm.[19] In the aftermath, Lavoisier's business affairs were found to be aboveboard. The learned academies were abolished in August, 1793.[20] Academicians who had not abandoned their "elitist" views and wholeheartedly joined the people were now in danger. The "Reign of Terror" fully radicalized the Revolution, and Lavoisier, Paulze, and 26 other members of the *Ferme Générale* were guillotined over the course of 35 minutes on May 8, 1794.[21]

---

1. J.-P. Poirier, *Lavoisier—Chemist, Biologist, Economist* (R. Balinski, transl.), University of Pennsylvania Press, Philadelphia, 1993, pp. 346–369.
2. Poirier, op. cit., pp. 220–221.
3. Poirier, op. cit., pp. 192–197.
4. J.R. Partington, *A History of Chemistry*, Macmillan & Co. Ltd., London, Vol. 3, 1962, pp. 484–487.
5. A. Greenberg, *A Chemical History Tour*, John Wiley & Sons, New York, 2000, pp. 143–146.
6. Poirier, op. cit., pp. 336–345.
7. Poirier, op. cit., pp. 328–335.
8. The author is grateful to Dr. Jean-Pierre Poirier for supplying a copy of Hassenfratz's letter.
9. Poirier, op. cit., pp. 1–3.
10. C. Djerassi and R. Hoffmann, Oxygen, Wiley-VCH, Weinheim, 2001. I am grateful for permission from Professor Djerassi and Professor Hoffmann to use this section from their play and for their helpful comments.
11. Djerassi, op. cit., pp. 42–45.
12. R. Hoffmann, *American Scientist*, Vol. 90, No. 1 (Jan.–Feb. 2002), pp. 22–24.
13. Poirier, op. cit., pp. 154–159.
14. Poirier, op. cit., pp. 23–28, 115–116, 166–170.
15. Poirier, op. cit., p. 120.
16. P. Laszlo, *Salt: Grain of Life*, Columbia University Press, New York, 2001.
17. Poirier, op. cit., pp. 170–173.
18. Poirier, op. cit., pp. 241–245.
19. Poirier, op. cit., pp. 272–273.
20. Poirier, op. cit., pp. 333–335.
21. Poirier, op. cit., pp. 381–382.

REQUIEM FOR A LIGHTWEIGHT

Although Phlogiston Theory was vanquished during the 1780s, it is worthwhile to summarize some of the definitions and arguments for and against phlogiston.[1] These will be limited almost entirely to coverage in the present book and is not meant to be a thorough treatment.

1. *What Were the Origins of Phlogiston Theory?* Diverse cultures had ancient beliefs in dualities (male–female; yin–yang; Sol–Luna; sulfur–mercury). This was modified by Paracelsus and others to the *tria prima:* sophic sufur, sophic mercury, and sophic salt, which constitute matter in various proportions. Becher (in the seventeenth century) recognized three "earths." One of these, *terra pinguis* or "fatty earth," was said to be present in combustible and metallic substances. It is analogous to sophic sulfur. Becher's theory was modified by Stahl (early eighteenth century) who coined the term "phlogiston" (Φ) to replace *terra pinguis*.

2. *What Is the Nature of Phlogiston (Φ)?* Phlogiston (Φ) is frequently defined as "the essence of fire." Sometimes Φ is simply identified as the fire released from a burning substance. Phosphorescence of a substance was considered to be a visual manifestation of the Φ stored in that substance. White phosphorus is loaded with Φ since it phosphoresces and can also ignite spontaneously. Phlogiston was usually considered to be an imponderable (superlight or even lacking mass) substance (often a fluid). However, fire did not always accompany release of Φ and, thus, might be merely one manifestation of its release.

3. *What Chemical Phenomena Did Phlogiston (Φ) Explain?* Most powerfully, it was a unifying theory for combustion of matter *and* for formation of calxes (often what we call oxides). This was by no means obvious. It is important to note that until the mid-eighteenth century, gases arising from combustion, such as carbon dioxide, were simply seen as "airs" and not collected.

$$\text{Charcoal (contains } \Phi) + \text{heat} \rightarrow \text{ash} + \Phi$$

$$\text{Copper (contains } \Phi) + \text{heat} \rightarrow \text{copper calx} + \Phi$$

$$\text{Charcoal (contains } \Phi) + \text{copper calx} + \text{heat} \rightarrow \text{ash} + \text{copper (contains } \Phi)$$

Cavendish collected the "flammable air" derived from "dissolving" metals in aqueous acids. What remained upon evaporation of the solution was the calx. The "flammable air" he collected had only 7% of the density of "atmospherical air." It appeared that the superlight, superflammable gas "obviously" released from the metal might well be Φ itself.

$$\text{Copper (contains } \Phi) + \text{sulfuric acid} \rightarrow \text{copper calx} + \Phi\text{? ("flammable air")}$$

Other "airs" could also remove Φ from metals:

$$\text{Copper (contains } \Phi) + \text{nitric acid} \rightarrow \text{copper calx} + \text{"nitrous air" (contains } \Phi)$$

What Lavoisier called *oxygen* today was termed "dephlogisticated air" by Priestley. It comprised one fifth of the atmosphere and had great affinity for Φ. "Nitrous air" and "flammable air" each carry the same amount of Φ since one volume of each will lose all of its Φ to one-half volume of "dephlogisticated air." Atmospheric air that absorbs Φ is "wounded" and when fully phlogisticated is "deadly" or "mephitic." What remains is "mephitic" or "phlogisticated" air, or nitrogen, which had earlier been named "azote" ("without life").

Food contains Φ; fatty food is particularly rich in Φ

4. *What Were Phlogiston's Failures?* Most notably, *increases* in weight upon *loss* of Φ:

Copper (contains Φ) + heat → copper calx + Φ
(63.5 grams) (79.5 grams)

This had been noted since the sixteenth century. If the law of conservation of matter holds, then Φ has negative mass (–16.0 grams in this case). When gases were collected, starting in the mid-eighteenth century, the results of exhaustive combustion of charcoal would be

Charcoal (contains Φ) + heat → ash + "fixed air" + Φ
(60 grams) ($\ll$1 gram) (ca 220 grams)

The mass of the gas generated was quite consequential and inconsistent with loss of Φ unless it had negative mass (~–160 grams in the present case). Also of immense significance was the problem of the composition of water. Water generally was unnoticed (or remarked upon) as a product of combustion. Combustion of "flammable air" (Φ?) by combination with "dephlogisticated air" might be expected to give "dephlogisticated phlogiston" (simply *nothing*?) or perhaps simply "air" devoid of Φ—possibly nitrogen? Instead, the product was water. Water could likewise be split chemically to give hydrogen ('flammable air") and oxygen ("dephlogisticated air"). It was now obvious that the Φ derived from "dissolving" copper in sulfuric acid came from the acid solution (in the form of "flammable air" or hydrogen) rather than from the metal itself. Similarly, the Φ derived from "dissolving" copper in nitric acid also came from the acid solution (in the form of "nitrous air" or nitric oxide, NO) rather than from the metal itself. There were other very fundamental questions, including "Where did Φ go once it was lost?" Why did the volume of "atmospherical air" decrease by one-fifth when Φ was gained? Did the "dephlogisticated air" lose its "elasticity" due to "injury"? There were countless other, more subtle, quantitative problems. Here is one—if 63.5 grams of copper is "dissolved" in sulfuric acid, all of its Φ is released in one volume of"flammable air" with total conversion of the metal to its calx. If 63.5 grams of copper is "dissolved" in nitric acid, all of its Φ is released in two-thirds volume of "nitrous air" with total conversion of the metal to its calx. In the first case, all of the Φ will be removed from the 1 volume of "flammable air" by one-half volume of "dephlogisticated air." In the second case, all of the Φ will be removed from the two-thirds volume of "nitrous air" by one-third volume of the very same "dephlogisticated air." The numbers simply do not add up.

5. *Consolidation.* Combustion and calx formation are both examples of chemical combinations with the oxygen in the air. That is why calxes are heavier than their metals and why the products of combustion weigh more than the combustibles (when oxygen is neglected). A useful strategy is to employ Roald Hoffmann's suggestion of considering Φ as "minus oxygen."[2] Thus, oxygen is lost from the atmosphere in combustion rather than the atmosphere gaining Φ. Metals gain oxygen, rather than losing Φ, when they form calxes. Substances are oxidized by oxygen, which is, of course, an oxidizing

agent. Thus, gain of Φ corresponds to reduction; Φ would be a reducing agent. Try it. It's fun.

---

1. My father, Murray Greenberg, suggested this essay.
2. R. Hoffmann and V. Torrence, *Chemistry Imagined—Reflections on Science*, Smithsonian Institution Press, Washington, DC, 1993, pp. 82–85.

SECTION VI
A YOUNG COUNTRY AND A YOUNG THEORY

"IT IS A PITY SO FEW CHEMISTS ARE DYERS, AND SO FEW DYERS CHEMISTS"

A daring entrepreneurial, yet practical, spirit imbued the citizens of the nascent United States of America, and it was exemplified by the precocious young physician John Penington. A student of Dr. Benjamin Rush at the University of Pennsylvania and a contemporary of Dr. Caspar Wistar, Penington completed his medical education at Edinburgh, where he wrote the following to his former teacher in 1790:[1]

> Alas, dear sir, I despair of meeting a Rush, or a Wistar, here, it is not the character of the professors at Edinburgh, to take the youthful inquirer by the hand & accompany him in the road of true knowledge.—Pride and reserve prevail among the professors, idleness & dissipation in the generality of the students: and for want of *proper company*, I have hitherto retreated to books and a solitary walk: in short I find nothing here likely to corrupt my patriotism.

In 1789, at the age of 20, Penington formed, in Philadelphia, the first chemical society in America[2–4] (possibly the world's first[3]) and authored the first American chemistry book—*Chemical and Economical Essays* (see Figure 126),[5,6] a work that had favorably impressed Thomas Jefferson.[7] His chemical society lasted briefly and was succeeded by The Chemical Society of Philadelphia, which was founded in 1792 [Dr. James Woodhouse (M.D.)—first president].[2,3] In 1793, he was one of six co-signatories to an endorsement for the Hopkins process for making potash (KOH) and pearl ash (K_2CO_3), for which the first U.S. patent was granted.[1]

Chemical and Economical Essays is a spirited book, and its flavor may be sampled from his picturesque comparison between the "practical chemist" and the "mere theorist" that retains a bit of resonance even today:[8]

> Chemists themselves belong to two great and distinct classes, which, it is a pity are not connected; in the one class we may rank those who perform a great number of operations by heat and mixture, without ever knowing the secondary causes of the effects produced; these are called practical chemists, such as are dyers, who cannot account for, or conceive why, alum, for instance, should be of use in their art; or why galls and copperas should produce a black dye; such also are tanners, who cannot explain the action of oak bark upon the hides; such likewise are many apothecaries, who can make aqua fortis. &c. &c. but know nothing of the rationale of the process; the other class is the mere theorist, who is well acquainted with the "effects of heat and mixture" upon all bodies, and can account for them all, but never soils his fingers with a piece of charcoal, or has had occasion to break a crucible; such a chemist can inform us admirably how the changes of colour in dying are pro-

CHEMICAL

AND

ECONOMICAL ESSAYS,

DESIGNED TO ILLUSTRATE

THE CONNECTION BETWEEN THE THEORY AND PRACTICE OF CHEMISTRY, AND THE APPLICATION OF THAT SCIENCE TO SOME OF THE ARTS AND MANUFACTURES OF THE UNITED STATES OF AMERICA.

"IT IS A PITY SO FEW CHEMISTS ARE DYERS, AND SO FEW DYERS CHEMISTS."

BY JOHN PENINGTON.

PHILADELPHIA:

PRINTED BY JOSEPH JAMES.

M, DCC, XC.

FIGURE 126. ▪ Title page of the first chemistry textbook, as opposed to a syllabus, pamphlet, translation, or reprinting of a foreign text, published in the United States. Its precocious author, John Penington of Philadelphia, published the book when he was 21 years old. The prior year he formed the first chemical society in the United States. Trained as a physician under Benjamin Rush at the University of Pennsylvania, Penington died at the age of 25, during the yellow-fever epidemic of 1793 as he struggled to save lives.

duced, but would be unable to produce them himself; he can account for the action of oak bark upon animal substances, without ever having smelt the odour of a tan-yard; he could explain the theory and process of making aqua fortis; and perhaps were he to attempt to make it, he would be two hours kindling a fire in his furnace, break his distillery apparatus, and suffocate himself with the fumes.

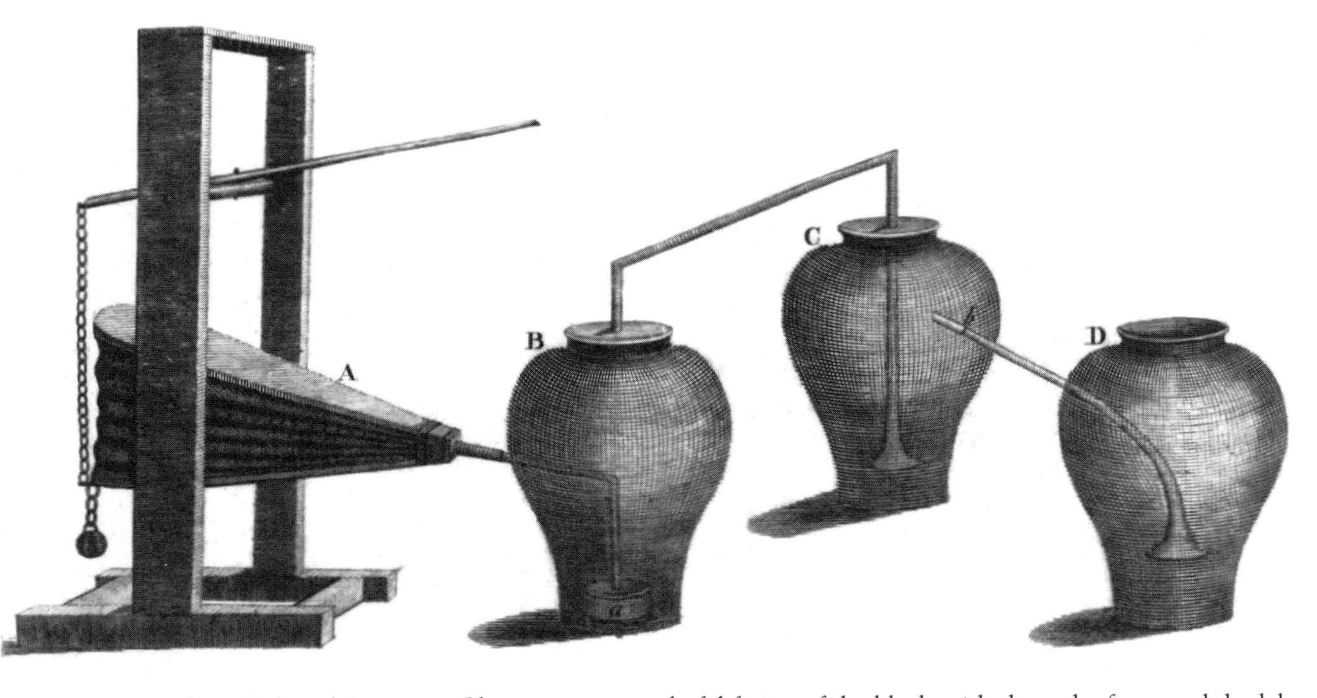

FIGURE 127. ■ Post-Colonial American Chemistry—a wonderful fusion of the blacksmith shop, the farm, and the laboratory. Early American (1790) apparatus for producing sulfuric acid—three 30-gallon crocks joined by lead pipe and connected to a bellows. (From Penington, *Chemical and Economical Essays*, see Figure 126.)

Figure 127, taken from Penington's book,[9] depicts an apparatus that is a rather striking hybrid of chemistry vessels (actually stoneware—30-gallon oil jars connected by leaden pipes) and bellows typical of a smith's shop. The purpose is the production of sulfuric acid. More than 10 years earlier, Lavoisier burned sulfur under oxygen in a closed vessel by using a powerful magnifying lens and collecting the resulting sulfuric acid. However, since small amounts of gas occupy huge volumes, very little sulfuric acid can be produced in practical vessels in this manner. Alternatively, saltpetre (KNO_3) can be used as a highly condensed source of oxygen, but it is very expensive. Penington's hybrid apparatus is a kind of flow reactor rather than batch reactor since it introduces fresh reagent oxygen continuously into the reaction. Penington notes the extreme exertions needed to push the bellows and suggests a modification in which a small iron still continuously passes steam into vessel *B* (he misstates it as C) in order to provide a continuous source of pressure to aid the bellows.[9]

Dr. Penington, who devised a method of heat-preserving milk (prior to pasteurization), began his medical practice in Philadelphia in 1792. He was stricken in the yellow-fever epidemic of 1793 that claimed the lives of one-fifth of the city's population and "continued to attend patients until he also succumbed."[6] He was 25 years old.

---

1. W. Myles, in *Chymia*, No. 4 (H.M. Leicester, ed.), University of Pennsylvania Press, Philadelphia, 1953, pp. 76–77.
2. W. Myles, In *Chymia*, No. 3 (H.M. Leicester, ed.), University of Pennsylvania Press, Philadelphia, 1950, pp. 95–113.
3. E.F. Smith, *Chemistry in America*, D. Appleton & Co., New York and London, 1914, pp. 12–43.
4. The Chemical Society of Philadelphia disbanded between 1805 and 1810 and was succeeded in 1811 by the Columbian Chemical Society, also formed in Philadelphia (see notes 2 and 3 above).
5. J. Penington, *Chemical and Economical Essays*, Joseph James, Philadelphia, 1790.
6. W.D. Williams and W.D. Myles, *Bulletin of History of Chemistry*, Vol. 8, p. 18, 1990. Earlier American chemistry publications, by Rush and the blind lecturer Henry Moyes, were small specialized pamphlets, not books. I am grateful for correspondence concerning Penington and his book with Professor William D. Williams, Harding University, Searcy, Arkansas. Harding University houses the combined Americana chemical book collections of Dr. Williams and historian Dr. Wyndham D. Myles and has been officially designated an historic chemical site.
7. A. Greenberg, *A Chemical History Tour*, John Wiley & Sons, New York, 2000, pp. 189–191.
8. Penington, op. cit., pp. 4–5.
9. Penington, op. cit., pp. 146–150.

TWO EARLY VISIONS: OXIDATION WITHOUT OXYGEN *AND* WOMEN AS STRONG SCIENTISTS

The Chemical Society of Philadelphia began in 1792 and was succeeded by the Columbian Chemical Society in 1811.[1] There are few traces of the earlier society, but copies of the annual address delivered by Thomas P. Smith on April 11, 1798 (Figure 128)[2] are known.[3] Smith, only 21 or 22 years old, was a member of the Society's Nitre Committee.[1] The committee placed announcements in news-

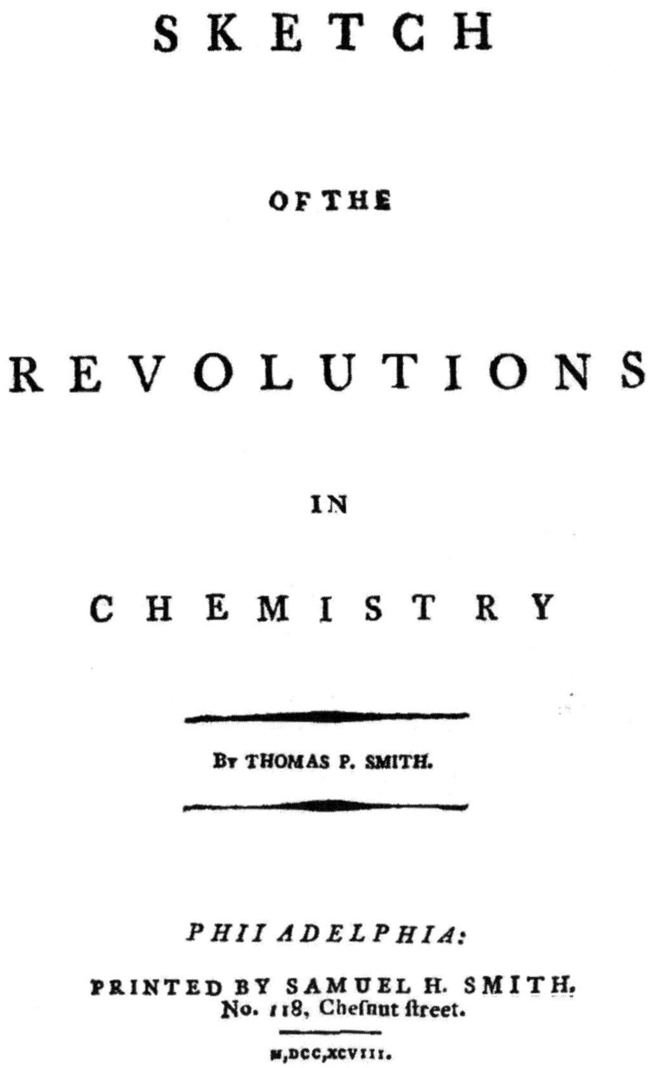

SKETCH

OF THE

REVOLUTIONS

IN

CHEMISTRY

By THOMAS P. SMITH.

PHILADELPHIA:

PRINTED BY SAMUEL H. SMITH,
No. 118, Chefnut ftreet.

M,DCC,XCVIII.

FIGURE 128. ▪ Title page from Thomas P. Smith's lecture before The Chemical Society of Philadelphia on April 11, 1798. The precocious Mr. Smith dared to imagine a world of scientific contributions by female chemists and stretched the notion of oxidation to include (correctly, as it soon turned out) oxidation by gases other than oxygen.

papers asking citizens to provide any information they had on niter, a component of gunpowder, by mail ("post paid", mind you) to Mr. Smith at No. 19 North Fifth Street or to the other four committee members, including Society President Dr. James Woodhouse (No. 13 Cherry Street).[1]

An amusing aspect of young Mr. Smith's oration was that he politely ignored the Society's expectation that "it shall contain all the discoveries made in the science of chemistry during the preceding year."[2] Instead, he gave a quite excellent brief history of chemical revolutions through the end of the eighteenth century. But we shall dwell briefly on two conjectures near the end of his presentation.

For the first one, Smith briefly summarizes Lavoisier's theory of combustion:

1. Combustion is never *known* to take place without the presence of oxigene.
2. In every *known* combustion there is an absorption of oxigene.
3. There is an augmentation of weight in the products of combustion equal to the weight of the oxigene absorbed.
4. In all combustion there is a disengagement of light and heat.

He then poses a simple, but imaginative, question: "Should we conclude because those substances which burn the readiest in oxigene will not burn in any other gas, that no substances are to be found that will?"

Actually, another such gas was already known—but misunderstood. Scheele had isolated chlorine gas in 1773, by dissolving pyrolusite (MnO_2) with cold acid of salt (HCl or muriatic acid).[4] Scheele had learned that pyrolusite was "dephlogisticated" (a good oxidizing agent). He logically considered chlorine to be dephlogisticated acid of salt.[4] It was later found to support the glow of a taper better than air, explode with hydrogen when kindled by a glowing taper, and burn phosphorus, ammonia, bismuth, antimony, powdered zinc, and other active metals.[5] However, Lavoisier argued that all acids contain oxygen (oxygen means "acid former") and the French school called chlorine (Cl_2) *oxygenated muriatic acid*. Note the perfect mirror-image consistency between Scheele's and Lavoisier's nomenclature for chlorine: "oxygenated" = "dephlogisticated." Since sulfuric and similar acids released oxygen from pyrosulite, the idea that MnO_2 released its oxygen to muriatic acid was not far-fetched. Lavoisier's view remained dominant, if questioned, until Humphry Davy proved some 30 years later that chlorine contained no oxygen and was, therefore, a pure element—a view finally accepted by Berzelius in the 1820s.[6] So young Smith was correct—there is combustion without oxygen. And in another hundred years the "Tyrannosaurus Rex"[7] or "Tasmanian Devil"[8] of elements, fluorine gas, would be isolated and found to support vigorous, even explosive, spontaneous combustion. Indeed, its affinity for hydrogen is much stronger than oxygen's and it will liberate oxygen gas from water with much heat:

$$2\ H_2O\ (liq) + F_2\ (gas) \rightarrow 4\ HF\ (aq\ soln) + O_2\ (gas) + heat$$

Smith nears the end of his brief history of chemistry as follows:[2,9]

> I shall now present you with the last and most pleasing revolution that has occurred in chemistry. Hitherto we have beheld this science entirely in the hands of men; we are now about to behold women assert their just, though too long neglected claims, of being participators in the pleasures arising from a knowledge of chemistry. Already have Madame Dacier and Mrs. Macauly established their right to criticism and history. Mrs. Fulhame has now laid such bold claims to chemistry that we can no longer deny the sex the privilege of participating in this science also.[10] What may we not expect from such an accession of talents? How swiftly will the horizon of knowledge recede before our united labours! And what unbounded pleasure may we not anticipate in treading the paths of science with such companions?

(A vision of Marie and Pierre Curie 100 years into the future?)

Smith died, from an accidental gun mishap, in 1802 on an ocean voyage to Europe where he was to continue his chemical and mineralogical studies.[1,3] He was, like Dr. John Penington, only 25 when he perished.

---

1. W. Miles, in *Chymia*, Vol. 3, University of Pennsylvania Press, Philadelphia, 1950, pp. 95–113.
2. T.P. Smith, *Annual Oration Delivered before the Chemical Society of Philadelphia, April 11, 1798—A Sketch of the Revolutions in Chemistry*, Samuel H. Smith, Philadelphia, 1798.
3. E.F. Smith, *Chemistry in America*, D. Appleton and Co., New York and London, 1914, pp. 12–47 (reprints in full Thomas P. Smith's oration).
4. J.R. Partington, *A History of Chemistry*, Vol. 3, Macmillan & Co, Ltd., London, 1962, pp. 212–214.
5. Partington, op. cit., pp. 540–542, 572.
6. W.H. Brock, *The Norton History of Chemistry*, W.W. Norton & Co., New York and London, 1993, p. 154.
7. G. Rayner-Canham, *Descriptive Inorganic Chemistry*, Freeman, New York, 1996, pp. 349–352.
8. A. Greenberg, *A Chemical History Tour*, John Wiley & Sons, New York, 2000, p. 239.
9. Smith is referring to Anne Dacier (1654–1720), a renowned European classicist, editor, and translator and Mary Ludwig Hays McCauley (1754–1832), who joined her husband Hays at the Battle of Monmouth (New Jersey) on June 28, 1778. Mrs. McCauley hauled pitchers of well water back and forth for the soldiers, and when Hays collapsed from the heat, she took his place at the cannon for the remainder of the battle (see *The New Encyclopedia Britannica*, Encyclopedia Britannica, Inc., Chicago, 1986, Vol. 3, pp. 841–842; Vol. 7, p. 611. "Molly Pitcher" has been honored by a battle monument and, profoundly, by the naming of the food court/gasoline station complex at Exit 7 of the New Jersey Turnpike.
10. Smith's footnote here is as follows: *Mrs. Fulhame has lately written an ingenious piece entitled "An Essay on Combustion, with a view to a new art of dyeing and painting, wherein the phlogistic and anti-phlogistic hypotheses are proved erroneous". Since the delivery of this oration she has been elected a corresponding member of this Society.* (see also Greenberg, op. cit., pp. 156–159 and K.J. Laidler, The World of Physical Chemistry, Oxford University Press, Oxford, 1993, pp. 250–252; 277–278. Laidler calls Mrs. Fulhame a "forgotten genius").

EXCLUSIVE! FIRST PRINTED PICTURES OF DALTON'S MOLECULES

> there are things which exist with solid and everlasting body, which we show to be the seed of things and their first-beginnings, out of which the whole sum of things now stands created.[1]
>
> There is then a void, mere space untouchable and empty. For if there were not, by no means could things move; for that which is the office of body, to offend and hinder, would at every moment be present to all things; nothing, therefore, could advance, since nothing could give the example of yielding place.[2]
>
> But as it is, because the fastenings of the first-elements are variously put together, and their substance is everlasting . . . No single thing then passes back to nothing, but all by dissolution pass back into the first-bodies of matter,[3]

Thus, does the Latin poet Lucretius speak to us, from a distance of 2000 years, in *De rerum natura* (*The Nature of Things*) "justifying"

1. That atoms are the ultimate and indestructible "seeds" of matter
2. The existence, indeed requirement, of empty space (void or vacuum)
3. The law of conservation of matter

Before Lucretius is awarded a share of the "*retro*–Nobel Prizes"[4] in chemistry, physics, *and* literature, we must admit that these were purely philosophical premises—no scientific hypotheses were tested experimentally. Lucretius' epic poem summarized views of earlier Greek philosophers including Democritus, Leucippus, and Epicurus (see Robert Boyle's ironic comment, pp. 109–110).

An early near-scientific theory of corpuscles or atoms was published by Daniel Sennert (1574–1637), Professor of Medicine at the University of Wittenberg as early as 1618.[5] The French Philosopher René Descartes (1596–1650) fathered analytical geometry, but his contributions to physics and chemistry were not significant.[6] He believed in atomlike ultimate particles that packed together such that the universe contained no voids ("Nature abhors a vacuum"). All motion in the universe had to be coordinated in a form of "cosmic gridlock." In contrast were the views of Pierre Gassendi (1592–1655),[7] a classicist who studied Epicurus and adopted the Epicurean concept of atoms and voids rather than the Cartesian continuum. Gassendi found a firm scientific argument for the true existence of void and vacuum in Torricelli's barometer, invented in 1643.

Robert Boyle and Isaac Newton, both strong adherents of alchemy, advanced a corpuscular theory of matter. It is worth remembering that, since they believed that lead could be transmuted to gold, there could be no "uniquely" gold or lead corpuscles. Their views were influenced by Gassendi.[8]

Boyle's law (1662) had demonstrated that if a gas expanded, for example, to eight times its volume, its pressure would decrease by a factor of 8. The density was also found to decrease by 8. Now, one might imagine that "thinning" of a "cartesian fluid" could decrease its density by expanding its continuum of "atoms." However, it is even harder to rationalize the reduction in pressure using this model. Instead, a model imagining the gas to be composed of individual "corpuscles," separated in space, might explain such behaviors in terms of Newtonian physics. The rarified gas would have, on average, even greater space between corpuscles. In 1687, Newton attempted to explain Boyle's law by postulating repulsion between hard corpuscles in the gaseous fluid.[9] Relating the repulsion to centrifugal force, he predicted that it would be inversely proportional to the distance between the centers of the atoms. Thus, reduced pressure in the gas was the result of reduced repulsion between corpuscles now further separated. Just as Newton did not attempt to explain the nature of the gravitational force that drew objects to the earth, he also did not attempt to understand the source of this mysterious repulsive force between atoms.[9]

John Dalton's earliest atomic theory originated in 1801 and was purely physical in nature.[10] Its basis was Boyle's law and his own law of partial pressures (see next essay). But his truly fundamental breakthrough, which occurred in 1803, was to produce the modern paradigm that ties together everything we know today about chemistry. Dalton's atomic law was the culmination of the chemical revolution that had occurred during the preceding three decades.[11,12]

Lucretius' poem suggests that the Law of Conservation of Matter has been assumed for at least two millennia. It was certainly a fundamental scientific assumption during the scientific revolution. However, it was Lavoisier who propounded the view that, unless all material mass could be accounted for in a chemical reaction, one could not even try to understand it. Critical, too, were Richter's establishment of tables of equivalents and his concept of stoichiometry and Proust's law of definite composition, that successfully survived his debate with Berthollet.[13] Dalton's notebook entry on September 6, 1803 (his thirty-seventh birthday) includes the first symbolic drawings and relative weights of his atoms.[11]

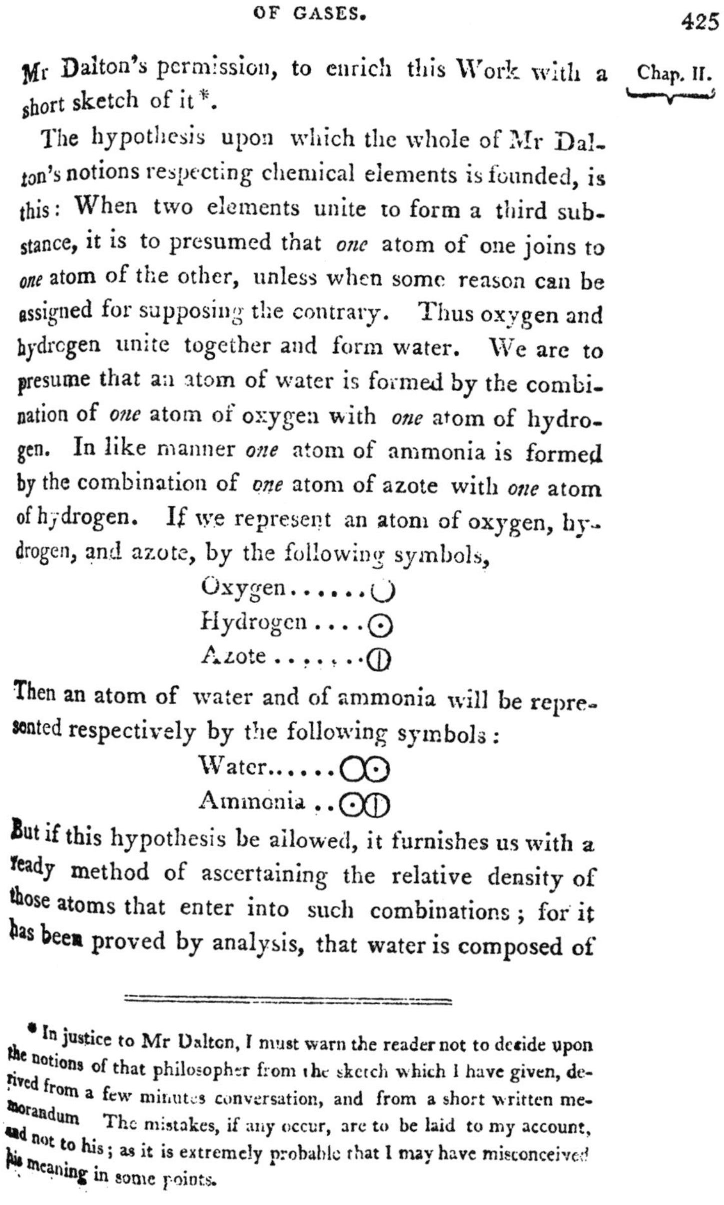

OF GASES. 425

Chap. II.

Mr Dalton's permission, to enrich this Work with a short sketch of it*.

The hypothesis upon which the whole of Mr Dalton's notions respecting chemical elements is founded, is this: When two elements unite to form a third substance, it is to presumed that *one* atom of one joins to *one* atom of the other, unless when some reason can be assigned for supposing the contrary. Thus oxygen and hydrogen unite together and form water. We are to presume that an atom of water is formed by the combination of *one* atom of oxygen with *one* atom of hydrogen. In like manner *one* atom of ammonia is formed by the combination of *one* atom of azote with *one* atom of hydrogen. If we represent an atom of oxygen, hydrogen, and azote, by the following symbols,

Oxygen......○
Hydrogen....⊙
Azote.......⦶

Then an atom of water and of ammonia will be represented respectively by the following symbols:

Water......○⊙
Ammonia..⊙⦶

But if this hypothesis be allowed, it furnishes us with a ready method of ascertaining the relative density of those atoms that enter into such combinations; for it has been proved by analysis, that water is composed of

* In justice to Mr Dalton, I must warn the reader not to decide upon the notions of that philosopher from the sketch which I have given, derived from a few minutes conversation, and from a short written memorandum. The mistakes, if any occur, are to be laid to my account, and not to his; as it is extremely probable that I may have misconceived his meaning in some points.

FIGURE 129. ■ Although John Dalton developed a physical atomic theory in 1801 and extended it to chemistry in 1803, he did not publish his theory until 1808. However, Thomas Thomson at the University of Edinburgh was an early advocate of atomic theory and, with the Dalton's permission, published its first printed discussion in 1807 (see *A System of Chemistry*, 3rd ed., London, 1807).

Thomas Thomson[14] received his M.D. degree at Edinburgh in 1799, where he was inspired by Joseph Black. Starting in 1800 he lectured on chemistry at Edinburgh and published the first edition of his comprehensive *A System of Chemistry* in 1802. Thomson visited Dalton in 1804 and enthusiastically adopted his atomic theory. Interestingly, the first published statement of Dalton's theory appeared in the third edition (1807) of Thomson's five-volume chemical treatise.[15] Dalton's *Chemical Philosophy* was published the next year.[11,12] It is thrilling to read Thomson's polite and tentative remarks and view the first printed pictures of atoms as they appeared in his book (Figure 129):[14]

> We have no direct means of ascertaining the density of the atoms of bodies; but Mr. Dalton, to whose uncommon ingenuity and sagacity the philosophic world is no stranger, has lately contrived an hypothesis which, if it prove correct, will furnish us with a very simple method of ascertaining that density with great precision.

The Quaker Dalton postulated a principle of "greatest simplicity" and thus assumed, for example, that water was comprised of one atom each of hydrogen and oxygen and ammonia comprised of one atom each of nitrogen and hydrogen (see Figure 129). This led to values of atomic weights in 1803 that we would now see as anomalies[11] (e.g., "*azot*" or nitrogen = 4.2; oxygen = 5.5; where hydrogen assumed = 1.0). Dalton was, however, also aware that "carbonic acid" contained twice the weight of oxygen per weight of carbon than did the newly discovered "carbonic oxide."[16] Similar findings were extant for oxides of nitrogen.[11] Thus, this law of multiple proportions (e.g., CO_2 vs. CO), developed by Dalton, was a clear corollary of his atomic theory.

---

1. C. Bailey (transl.), *Lucretius on the Nature of Things*, Oxford University Press, Oxford, 1910, p. 43.
2. Bailey, op. cit., p. 38.
3. Bailey, op. cit., p. 35.
4. The idea of a "*retro*–Nobel Prize" forms the premise for the inventive play *Oxygen* by Carl Djerassi and Roald Hoffmann; see C. Djerassi and R. Hoffmann, *Oxygen*, Wiley-VCH, Weinheim, 2001.
5. J.R. Partington, *A History of Chemistry*, Macmillan & Co. Ltd., London, 1961,Vol. 2, pp. 271–276.
6. Partington, op. cit., pp. 430–441.
7. Partington, op. cit., pp. 458–467.
8. Partington, op. cit., pp. 502–507.
9. Partington, op. cit., pp. 474–477.
10. Partington, *A History of Chemistry*, Macmillan & Co., Ltd., London, 1962, Vol. 3, pp. 765–782.
11. Partington (1962), op. cit., pp. 782–786.
12. A. Greenberg, *A Chemical History Tour*, John Wiley & Sons, New York, 2000, pp. 170–175.
13. Greenberg, op. cit., pp. 168–170.
14. Partington (1962), op. cit., pp. 716–721.
15. T. Thomson, *A System of Chemistry in Five Volumes*, third edition, Bell & Bradfute, and E. Balfour, London, 1807, Vol. III, pp. 424–431.
16. Partington (1962), op. cit., pp. 271–276.

ATMOSPHERIC WATER MOLECULES AND THE MORNING DEW

John Dalton[1] recorded atmospheric measurements throughout his long scientific life and at 6 A.M. on July 27, 1844 he made his final diary entry—"little rain this day"—in a feeble hand just before he died.[2] This lifelong interest led him to try to understand the occurrence of water vapor in the air (why doesn't it simply condense?). Another puzzler was why air was completely homogeneous—why didn't the denser oxygen component settle out from the more abundant, lighter nitrogen gas? Could it be that the two gases formed a weak 1 : 4 compound susceptible to displacement of nitrogen by more reactive substances, such as metals or hydrogen, having higher affinities for oxygen?

A decade before he enunciated his atomic theory, Dalton had found that the amount (pressure) of water vapor in air or introduced into a vacuum depended solely on the temperature of the liquid water in equilibrium.[3] (Dalton also developed the concept of the dewpoint).[3] This suggested that water vapor did not form a chemical compound with air (else, why would it enter an evacuated vessel?). It also suggested that water's vapor pressure and very existence were completely independent of other gases in the air. In 1801, his experimental studies permitted a more general statement of what we now call Dalton's law of partial pressures:[4]

> When two elastic fluids, denoted by A and B, are mixed together, there is no mutual repulsion amongst their particles; that is, the particles of A do not repel those of B, as they do one another. Consequently, the pressure or whole weight upon any one particle arises solely from those of its own kind.

In modern terms, we could write this as

$$P_{\text{total}} = P_{\text{oxygen}} + P_{\text{nitrogen}} + P_{\text{water vapor}}$$

This is an interesting, if not very straightforward notion. Why should "particles" of "A" (e.g. nitrogen) repel other "A" particles but not "particles" of "B," such as oxygen, toward which they remain indifferent?

Dalton's first atomic theory was a "physical one." From his 1801 presentation[5] we see his depiction of the four atmospheric gases (water, oxygen, nitrogen, and carbonic acid). Separately, each gas repels like "atoms" (top of Figure 130), but mixed "atoms" of different gases do not repel or attract (bottom of Figure 130). Dalton, modest Quaker that he was, nonetheless compared his theory to Newton's law of universal gravitation.[5] This comparison was *not* immodest. A few years later, Dalton would realize that his theory explained chemistry as well as physics.

Dalton's explanation mixes Lavoisier's caloric theory with Newton's mechanical repulsion theory and then adds a dash of his own special ingredient. First, it is important to recall that when combined in a metal oxide or calx, the element oxygen is in its "fixed" state. Thus, according to Lavoisier, "oxygen gas" is actually "oxygenated caloric" since heat was required in order to free the ele-

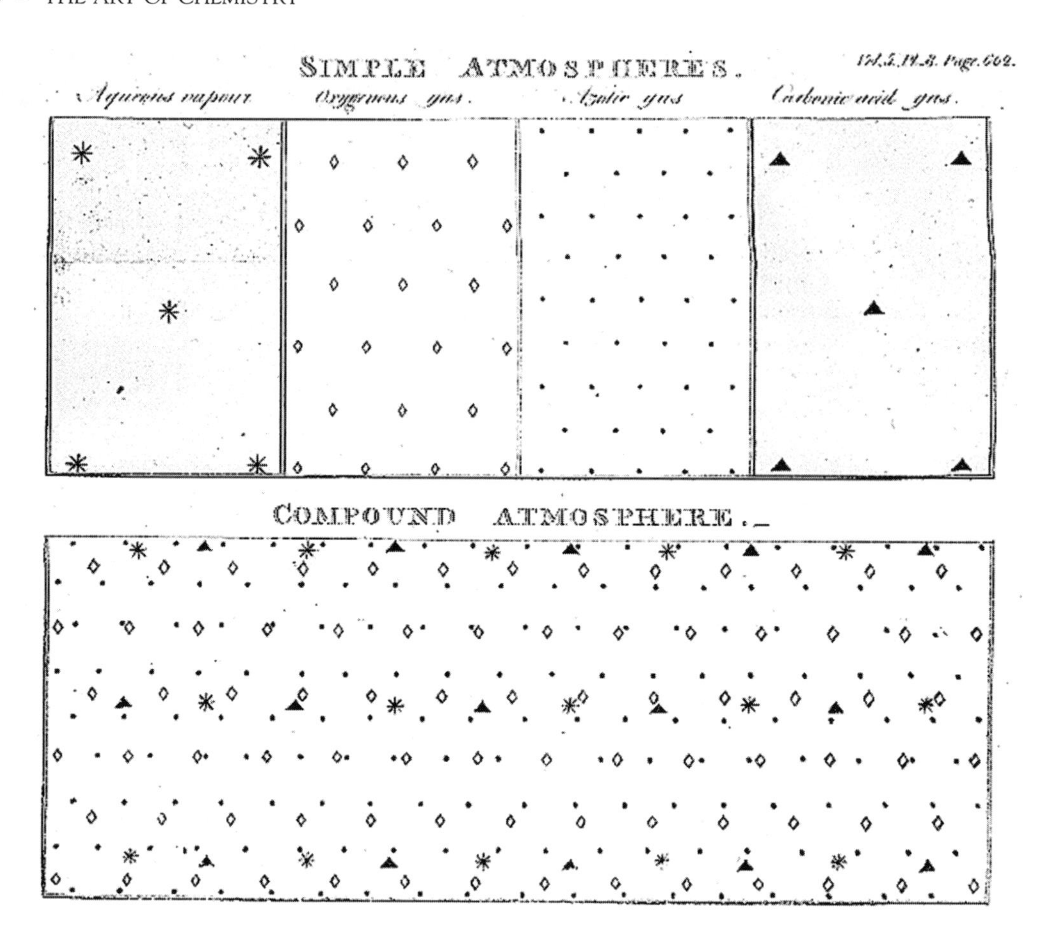

FIGURE 130. ■ Graphical description of Dalton's law of practical pressures printed in the 1802 *Memoirs of the Literary and Philosophical Society of Manchester*. According to Dalton's theory, the corpuscles of a single atmospheric gas (e.g., water, oxygen, nitrogen, carbon dioxide) repel each other but not corpuscles of other gases. Thus, the gases may be mixed ("superimposed: as in the bottom of this figure) without any interaction; hence, the atmosphere is freely mixed, not layered. See Figure 131, which depicts Dalton's attempt at explaining these phenomena.

ment from its calx. Similarly, boiling water adds caloric to form vapor. Note, in an earlier essay, how Lamarck's contemporary picture (Figure 124) of evaporation of water places "jackets" of caloric around corpuscles of water so as to increase the space between them causing these corpuscles to move into the gas phase.

Dalton's clever "take" on this problem of repulsion amongst like molecules was diagrammed in Part II of his *Chemical Philosophy* (Figure 131).[6] At the top part of this figure are represented "jackets" of caloric surrounding gaseous molecules. (*Note;* Hydrogen gas is thought to be monoatomic.) The lower part of the figure demonstrates why like molecules, such as *azote* (nitrogen), repel each other while different types of gas molecules are mutually indifferent. Since at a given temperature, the sizes and "jackets" of caloric surrounding all azote gas molecules must be equal in dimension, lines of force are totally aligned and repulsion occurs.[7] The same is true for repulsion between hydrogen molecules. However, such perfect alignment does not occur between hydrogen and *azote*, and they have totally independent existences and make their own independent contributions to

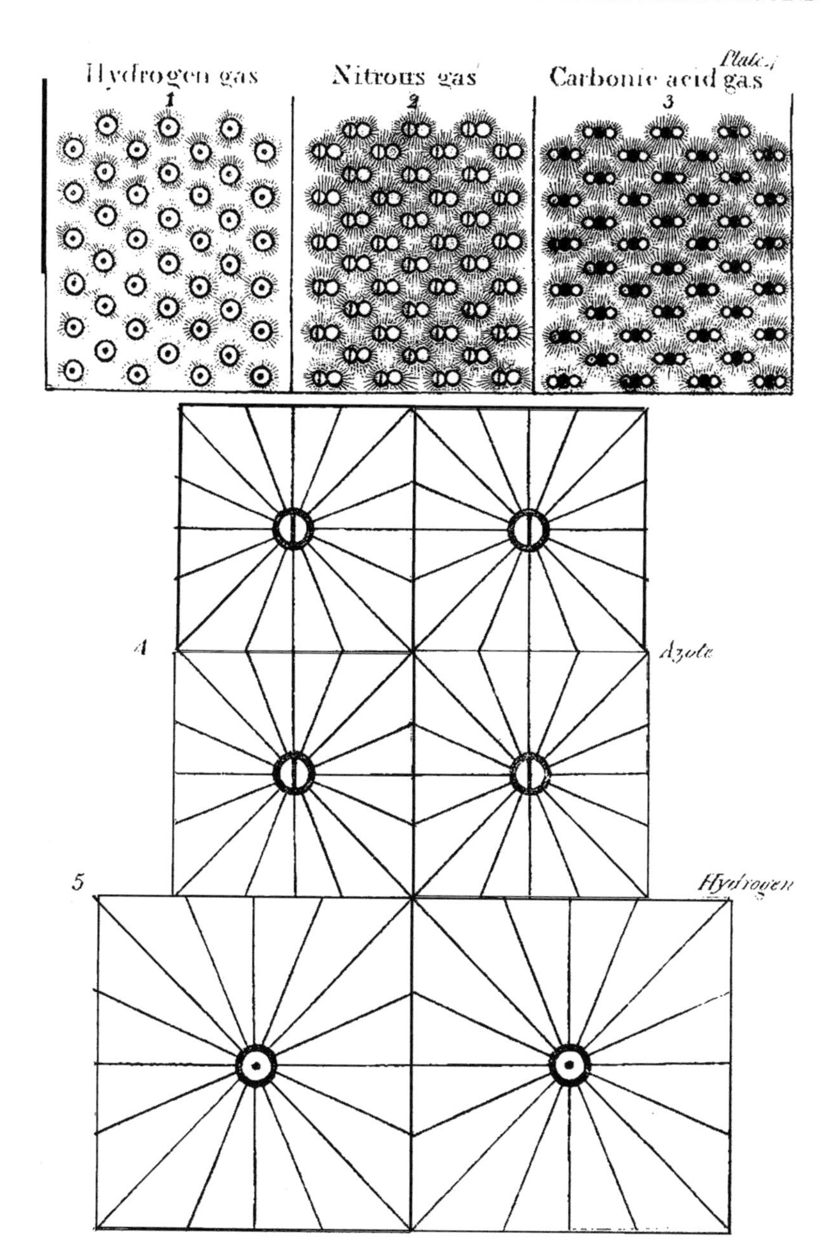

FIGURE 131. ■ Figures of jackets of caloric and "lines of force" postulated by Dalton to explain why like molecules of gas must repel each other so that gases of different densities remain mixed rather than layered in the atmosphere (from Dalton, *A New System of Chemical Philosophy*, Part I, Manchester, 1808).

the total pressure of a mixture. It was vital for Dalton that like "atoms" repelled and different "atoms" did not, since this prevented separation and layering of bulk gases according to gas densities.

Not only is air uniformly mixed at sea level, rather than consisting completely or mostly of the denser gas, oxygen, but the same mixture persists high into the atmosphere. This was known at the time Dalton was formulating his atomic theory. In 1804, Joseph Louis Gay-Lussac piloted a balloon some 23,000

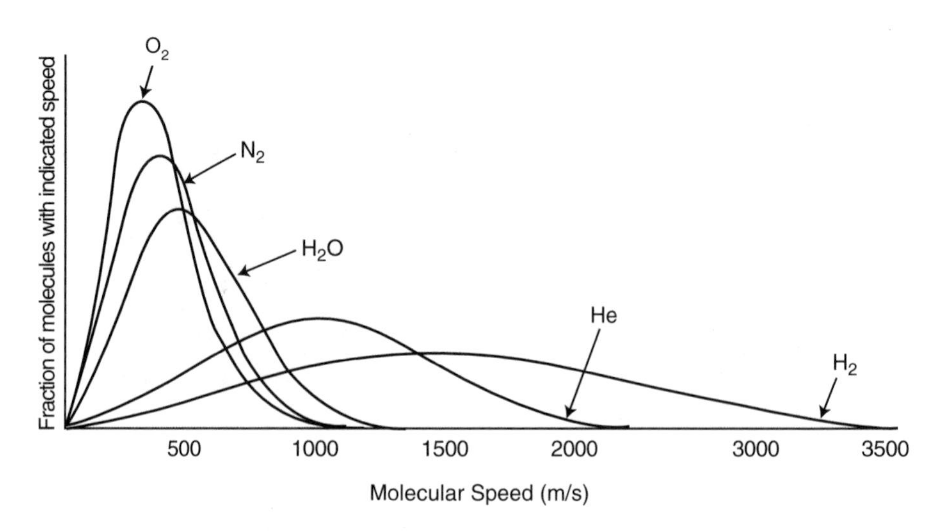

FIGURE 132. ■ The distribution of velocities of gas molecules at 25°C. Note how much greater the velocities are for ultralight hydrogen molecules and helium atoms. These substances thus escape the earth's atmosphere, unlike the heavier gases. (Adapted from Brown et al., *Chemistry—the Central Science*).

feet above Paris, and collected two air samples, which were shown to have the same composition as air at sea level.[8]

We now know that the troposphere, stratosphere and mesosphere extending up to about 60–70 miles above the earth have essentially the same uniform chemical composition.[9] Oxygen, nitrogen, water vapor, and the other atmospheric gases move at velocities[10] very far below the velocity needed to escape the earth's gravitational pull [11 kilometers per second (km/s) or 22,000 miles per hour (mph)].[11] The fact that these gaseous molecules have negligibly weak attractions for each other, the natural tendency to maximize disorder (increase entropy) and constant mixing by winds driven by the earth's rotation guarantees total mixing of the atmospheric gases. Figure 132 depicts the velocity distributions of simple gaseous molecules.[10] The two lightest gases, hydrogen and helium, show drastically different distributions relative to the other gases in this figure. Even though their average velocities are well below the earth's escape velocity, the most rapidly moving "outlier molecules" will escape into space and over time these gases are lost from the earth's atmosphere. Thus, although hydrogen is the most abundant element in the universe, only the minutest traces are found in the earth's atmosphere as the result of continuing high-energy processes that split water molecules. Similarly, the minute traces of helium in the atmosphere are due to fresh outgassing of radioactive materials that emit *alpha* particles (helium nuclei).

---

1. J.R. Partington, *A History of Chemistry*, Macmillan & Co. Ltd., London, 1962, Vol. 3, pp. 755–822.
2. Partington, op. cit., p. 760.
3. Partington, op. cit., pp. 762–765.
4. Partington, op. cit., pp. 765–767.
5. J. Dalton, *Memoirs of the Literary and Philosophical Society of Manchester*, Vol. V, Part II, Codell and Davies, London, 1802, pp. 535–602.

6. J. Dalton, *A New System of Chemical Philosophy*, Part I, Manchester, 1808; Part II, Manchester, 1810 (see Plate 7).
7. Partington, op. cit., pp. 778–781.
8. J.R. Partington, *A History of Chemistry*, Macmillan & Co. Ltd., London, 1964, Vol. 4, p. 78.
9. *The New Encyclopedia Britannica*, Encyclopedia Britannica, Inc., Chicago, 1986, Vol. 14, p. 311.
10. T.L. Brown, H.E. LeMay, Jr., and B.E. Bursten, *Chemistry—the Central Science*, seventh edition, Prentice-Hall, Upper Saddle River, NJ, 1997, pp. 364–368.
11. S. Mitton (ed.), *The Cambridge Encyclopedia of Astronomy*, Crown Publishers Inc., New York, 1977, pp. 193–195.

'TIS A BONNIE CHYMISTRIE WE BRRRING YE

Philadelphia is the birthplace of chemistry in America. But Philadelphia's chemical roots extend back to the Scottish Enlightenment and the University of Edinburgh in particular. David Hume and Adam Smith spent a considerable part of their lives in Edinburgh, and historian Jan Golinski notes its "stimulating local environment" and quotes Tobias Smollett, who called it a "hotbed of genius."[1,2] The first Chair of Chemistry in America was awarded in 1769 to Dr. Benjamin Rush at the University of Pennsylvania (see Figure 133 for the title page to his 1770 syllabus of lectures).[3,4] Rush was a co-signer with Benjamin Franklin of the Declaration of Independence. He obtained his M.D. degree at the University of Edinburgh and studied chemistry with Dr. Joseph Black.[5] Black's most important discovery of course was the isolation and characterization of "fixed air" (carbon dioxide), published in 1756. However, his other great contribution was his influence on students who attended his lucid and up-to-date lectures. Black's lecture notes were published posthumously by his student John Robison in 1803.[6] Samuel Latham Mitchill, John Maclean, and Benjamin Silliman, Sr., the first professors of chemistry at Columbia (1792), Princeton (1795), and Yale (1802), all studied with Black at Edinburgh.[7,8] Prior to Rush's appointment, the first person to teach chemistry as part of the course at the medical college of the University of Pennsylvania was John Morgan, who, of course, learned his chemistry from Black at Edinburgh.[9]

It is fair to say that other colonies besides Pennsylvania showed an early interest in chemistry. Indeed, Jefferson of Virginia and Adams of Massachusetts weighed in with rather strongly held opinions.[10] James Madison, the future author of the Constitution and fourth president of the United States, included chemistry in his natural philosophy lectures at the College of William and Mary. He even published a letter on his chemical experiments on the "Sweet Springs" in the *Transactions of the Philosophical Society*.[11]

Dr. Black studied chemistry at the University of Glasgow in the laboratory of William Cullen (1710–1790).[12] He moved to Edinburgh with Cullen and presented the dissertation for his M.D. degree in 1754.[5] Black succeeded Cullen as professor of chemistry at Glasgow in 1756 and at Edinburgh in 1766. Cullen had learned his chemistry by reading the works of Boerhaave in English translation. Cullen's original works included studies on bleaching and salt manufacture. His

12 OLD CHEMISTRIES

SYLLABUS

Of a COURSE

OF

LECTURES

ON

CHEMISTRY.

PHILADELPHIA: Printed 1770.

FIGURE 133. ▪ Dr. Benjamin Rush was educated by Dr. Joseph Black at the University of Edinburgh. In 1769 he was awarded the first Chair in Chemistry in America, at the University of Pennsylvania. Here is the title page for his course syllabus. (From Smith, *Chemistry in America*). Dr. Rush was a signer of the American Declaration of Independence and worked with co-signer Benjamin Franklin to perfect saltpetre.

essay[13] "Of the Cold produced by evaporating Fluids, and of some other means of producing Cold" extended Boerhaave's studies of thermometry:

> when a thermometer had been immersed in spirit of wine, tho' the spirit was exactly of the temperature of the surrounding air, or somewhat colder; yet upon taking the thermometer out of the spirit, and suspending it in the air,

the mercury in the thermometer, which was of Fahrenheit's construction, always sunk two or three degrees.

Philadelphia was the home of the first chemical society, formed in 1789, by John Penington, who studied chemistry in Edinburgh.[9] Two years later, James Woodhouse, a student of Rush, and therefore "chemical grandson" to Black, founded the Chemical Society of Philadelphia.[9] In 1794, Rush tried to convince Joseph Priestley, persecuted for his political beliefs, to leave England and resettle in Philadelphia, home of Priestley's longtime friend Benjamin Franklin, since deceased.[14] Priestley eventually chose the more bucolic Northumberland. In summary, is it not appropriate that two of the most important resources for chemical historians in America, the Edgar Fahs Smith collection at the University of Pennsylvania and the Chemical Heritage Foundation, reside in the City of Brotherly Love?

---

1. J. Golinski, *Science as Public Culture—Chemistry and Enlightenment in Britain, 1760–1820*, Cambridge University Press, Cambridge, UK, pp. 13–25.
2. A. Herman, *How the Scots Invented the Modern World: The True Story of How Western Europe's Poorest Nation Created Our World and Everything in It*, Crown Publishers, New York, 2001.
3. E.F. Smith, *Old Chemistries*, McGraw-Hill, New York, 1927, pp. 11–14.
4. E.F. Smith, *Chemistry in America*, D. Appleton and Co., New York, 1914.
5. J.R. Partington, *A History of Chemistry*, Macmillan & Co. Ltd., London, 1962, Vol. 3, pp. 130–143.
6. J. Black, *Lectures on the Elements of Chemistry, delivered in the University of Edinburgh, by the late Joseph Black, M.D. . . . now published from his Manuscripts, by John Robison, LL.D.*, Edinburgh, Mundell & Sons for Longman & Rees, 1803. The American edition was published in 1807. My copy of this three-volume set was purchased at an auction of books de-accessed by the Franklin Institute in Philadelphia in 1986. The signature of its original owner, Adam Seybert, a student of Dr. Rush and thus the "chemical grandchild" of Dr. Black, dated 1807, is on the title page of Volume 1. In such a manner does a book collector enjoy direct links to history.
7. A.J. Ihde, *The Development of Modern Chemistry*, Harper & Row, New York, 1964, p. 268.
8. D.S. Tarbell and A.T. Tarbell, *Essays on the History of Organic Chemistry in the United States*, Folio Publishers, Nashville, 1986, pp. 17–23.
9. W. Myles, *Chymia*, Vol. 3, University of Pennsylvania Press, Philadelphia, 1950, pp. 95–113.
10. A. Greenberg, *A Chemical History Tour*, John Wiley and Sons, New York, 2000, pp. 189–191.
11. Smith (1914), op. cit., pp. 5–7.
12. Partington, op. cit., pp. 128–130.
13. J. Black, *Experiments on Magnesia Alba, Quick-Lime, and Other Alcaline Substances; by Joseph Black, M.D., to Which is Annexed, an Essay on the Cold produced by Evaporative Fluids, and of some other means of producing Cold, by William Cullen, M.D.*, William Creech, Edinburgh, 1782, pp. 117–118.
14. Smith (1914), op. cit., pp. 109–118.

♫♪ "FOR IT'S HOT AS HELL . . . IN PHILA-DEL'-PHI-A'" ♪♫ [1]

As already mentioned, the weather was almost unprecedentedly hot; and his laboratory was in sundry places perpetually glowing with blazing charcoal, and red-hot furnaces, crucibles and gun-barrels, and often bathed in every portion of it with the steam of boiling water. Rarely, during the day, was the temperature of its atmosphere lower than from 110° to 115° of Fahrenheit—at times, perhaps, even higher.

Almost daily did I visit the professor in that salamander's home, and uniformly found him in the same condition—stripped to his shirt and summer pantaloons, his collar unbuttoned, his sleeves rolled up above his elbows, the sweat streaming copiously down his face and person, and his whole vesture dripping wet with the same fluid. He, himself, moreover, being always engaged in either actually performing or closely watching and superintending his processes, was stationed for the most part in or near to one of the hottest spots in his laboratory.

My salutation to him on entering his semi-Phlegethon[2] of heat not infrequently was: "Good God, doctor, how can you bear to remain so constantly in so hot a room! It is a perfect purgatory!" To this half interrogatory, half exclamation, the reply received was usually to the same purport. "Hot, sir—hot! Do you call this a hot room? Why, sir, it is one of the coolest rooms in Philadelphia. Exhalation, sir, is the most cooling process. And do you not see how the sweat exhales from my body, and carries off all the caloric? Do you not know, sir, that, by exhalation, ice can be produced under the sun of the hottest climates?"

So writes Benjamin Silliman, the elder during a period (1802–1803) spent attending chemistry lectures and visiting Dr. James Woodhouse, (M.D.), Professor of Chemistry of the University of Pennsylvania.[3] Silliman would become Professor of Chemistry at Yale University. The "salamander" reference evokes the mythical fire-resistant salamander of alchemical lore (see Figure 21). As noted previously, a decade earlier, Woodhouse had founded the Chemical Society of Philadelphia.[4] His lectures were enthusiastic if not riveting, and Silliman notes that "He appeared, when lecturing, as if not quite at his ease, as if a little fearful that he was not highly appreciated,—as indeed he was not very highly."[3] Still, his demonstrations were effective even on "humble" apparatus that had the virtue of affordability. Indeed, Figure 134 depicts "Dr. Woodhouse's Economical Apparatus"—a "Portable Laboratory; containing a Philosophical Apparatus, and a great number of Chemical Agents; by which any person may perform an endless variety of amusing and instructive experiments." The apparatus in Figure 134 appears in the Appendix to the first American edition (Philadelphia, 1802) of *The Chemical Pocket Book*[5] written by Dr. James Parkinson, (M.D.), who characterized the illness of the central nervous system that bears his name. After proclaiming the wide range of experiments available through his "economical apparatus," Dr. Woodhouse points out the inadequacies of the rival portable laboratory of Guyton de Morveau, for example:[6]

It is less expensive. The lamp of Guyton, is one of the worst of the kind, for a Chemical Laboratory. There is no occasion for a number of screws, to elevate or depress the retort or lamp, for a great or low heat may be made, merely by raising or lowering the wick.

Advertising notwithstanding, Dr. Woodhouse was one of the most respected chemists of his day and was the first president of the Chemical Society of

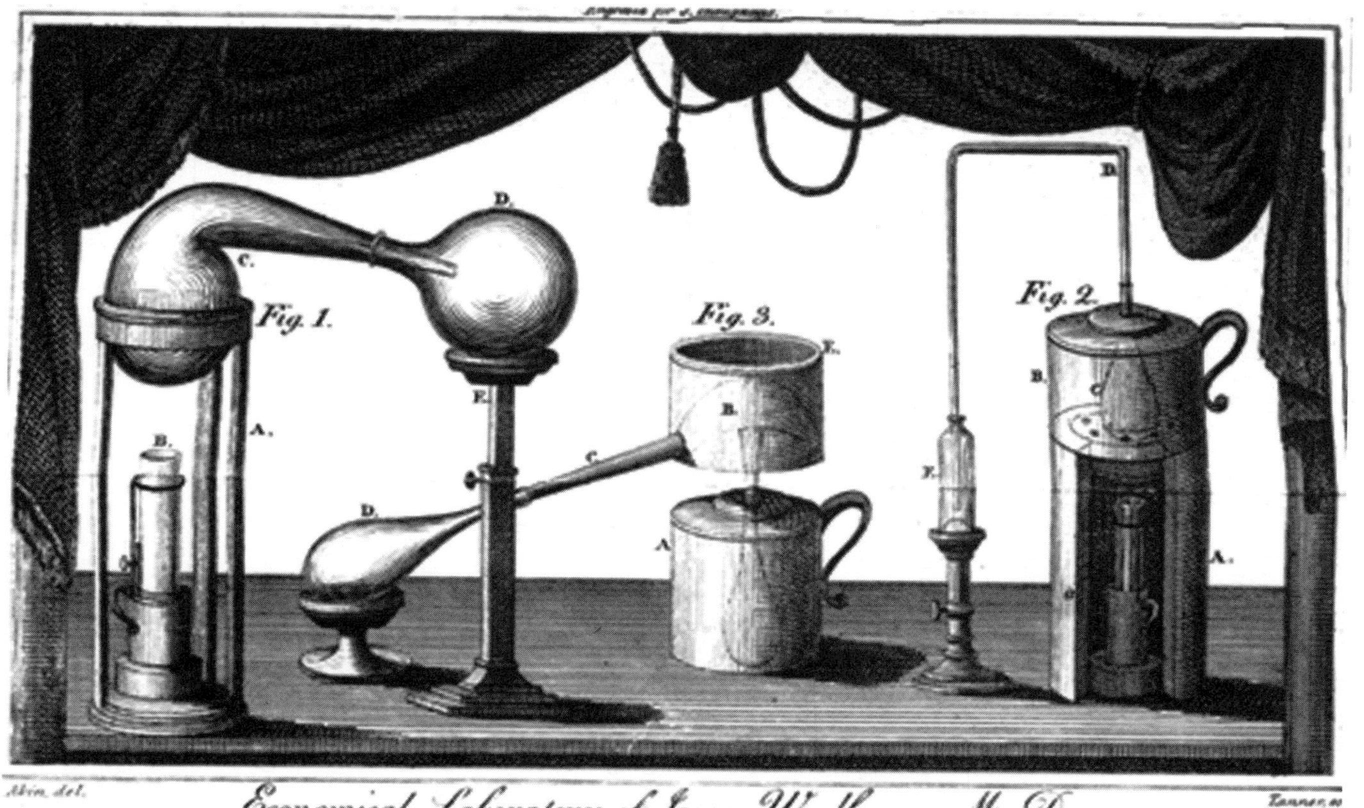

FIGURE 134. ▪ Dr. James Woodhouse, Professor of Chemistry at the University of Pennsylvania, founded the Chemical Society of Philadelphia. He also had a profitable lecture series for which he sold his own chemical apparatus. He proclaimed it to be far superior and more economical than that of Guyton de Morveau in France. (From Parkinson, *The Chemical Pocket-Book*, 1802.)

Philadelphia (1792). As an antiphlogistonist, he was critical of Dr. Joseph Priestley's phlogistic arguments, which were summarized in the latter's 1796 pamphlet, published in Philadelphia and reprinted in the London that he had fled a few short years earlier.[7] For example, Woodhouse noted that when he calcined metals in "pure air" (oxygen), the volume decreased but "I could not find that the air which remained behind was injured."[8] If one takes the phlogistic view, calcinations cause the metal to release its phlogiston to the air until the air is fully phlogisticated and no longer capable of supporting flame or life (the air is "mephitic" or deadly). However, that is clearly not the case since, in an excess of oxygen, the metal takes up a specific quantity of this gas, leaving behind a smaller volume of equally pure oxygen.

However, Woodhouse was not doctrinaire, continued to harbor some uncertainties about the "new" chemistry" emanating from France, and was solicitous of the aging Priestley. In a letter answering an attack by Dr. John Maclean, professor at Princeton University, one paragraph reads[9]

> A judgement may be formed how well you have accomplished your purpose, and what right you have to condemn the experiments of Dr. Priestley in the authoritative manner you have done, having made none yourself, from the following particulars. You are not yet, Doctor, the conqueror of this veteran in philosophy.

In the 1802 *Chemical Pocket-Book*, Dr. Woodhouse[6] writes an Appendix titled: "An Account of the Principal Objections to the Antiphlogistic System of Chemistry." For example, he verifies experimentally a Priestley experiment in which red-hot "scales of iron" (iron oxide devoid of water and, thus, hydrogen) is mixed with red-hot charcoal (likewise devoid of water and, thus, hydrogen) and the result is an inflammable gas. (Hydrogen, i.e., phlogiston? Carbonated hydrogenous gas, i.e., methane?) So the gases produced must have removed some phlogiston (i.e., hydrogen) from the charcoal. (The confusion is that most of the gas produced is carbon monoxide, which *is* flammable.)

The rigors of his work undoubtedly contributed to the untimely death of Dr. Woodhouse, who Silliman noted never made "use of any of the facts revealed by chemistry, to illustrate the character of the Creator as revealed in his works" and Dr. Benjamin Rush, his former teacher, simply called "an open and rude infidel,"[3] in 1809 at the age of 39.

---

1. Homage to the Broadway production of *1776*.
2. In mythology, one of the five rivers of Hades.
3. E.F. Smith, *Chemistry in America*, D. Appleton and Co., New York and London, 1914, pp. 103–106.
4. Smith, op. cit., p. 12.
5. J. Parkinson, *Chemical Pocket-Book*, James Humphreys, Philadelphia, 1802.
6. Parkinson, op. cit., pp. 201–215.
7. J. Priestley, *Experiments and Observations Relating to the Analysis of Atmospherical Air*, Philadelphia, 1796.
8. Smith, op. cit., p. 83.
9. Smith, op. cit., p. 92.

TWELVE CENTS FOR A CHEMISTRY LECTURE

Amos Eaton, A.M. was a busy man. In his 1826 book *Chemical Instructor*[1] he describes himself as "Attorney and Counsellor at Law; Professor of Chemistry and Natural Philosophy in Rensselaer School and in Vermont Academy of Medicine, &c. &c." As an entrepreneurial chemistry lecturer he followed a trail blazed by the likes of Henry Moyes, M.D., whose 1784 twenty-one-part lecture series "Heads of a Course of Lectures on the Philosophy of Chemistry . . ." was advertised for one guinea (or one shilling per lecture)[2] in the *Massachusetts Sentinel.* Here is Mr. Eaton's syllabus (I think we know whose text he used):[3]

COURSE OF LECTURES
TO BE DIVIDED INTO THIRTY-THREE

Lecture	To page	Lecture	To page
1st Lecture to page	22	18th Lecture to page	125
2nd	26	19th	130
3rd	32	20th	137
4th	38	21st	144
5th	42	22nd	152
6th	48	23rd	159
7th	54	24th	16S
8th	60	25th	17S
9th	66	26th	180
10th	70	27th	190
11th	76	28th	199
12th	82	29th	211
13th	88	30th	224
14th	94	31st	239
15th	102	32nd	246
16th	112	33rd	249
17th	119		

For this course charge $4. If it is condensed to 22
lectures, charge $3.

Our Mr. Eaton would never burden his students with esoteric theories about useless compounds:

> It is much to be regretted, that most of the celebrated treatises on chemistry, have so large a proportion of their pages devoted to useless compounds, which can never profit the scholar nor the practical man. Particularly those endless compounds with chlorine and iodine, which may be equally multiplied and extended with any other substance. This is surely trifling with the richest stores of human knowledge. Put such works into the hands of a student, and tell him to place full confidence in the authors, he would form strange views of the science. He would imagine, that the chloridic and iodic theory of Davy constituted the whole science of chemistry; and that all fur-

> ther knowledge of the subject should be pursued as a convenient , though not very important, appendage to chlorine and iodine. And even admitting all Davy's speculations to be well supported, are not those idle speculations as unimportant as any of the smallest mites of human knowledge? I would as soon set a student to commit to memory all the amulets of the dark ages, or the number of ways in which the letters of the alphabet can be arranged.

Well, who said that pandering to tuition-paying students is only a modern phenomenon? Furthermore, we have always been a pragmatic nation—listen to President Thomas Jefferson in 1805:[4,5]

> The chemists have filled volumes on the composition of a thousand substances of no sort of importance to the purposes of life; while the arts of making bread, butter, cheese, vinegar, soap, beer, cider &c remain unexplained.

The "chloridic and iodic theory of Davy" refers to his careful studies some 10 years earlier than the appearance of Eaton's book. These established that chlorine and iodine were pure elements, that hydrochloric (HCl) and hydriodic (HI) acids did not contain oxygen, and that Lavoisier's theory that all acids contain oxygen was incorrect—conclusions perhaps mightier than "the smallest mites of human knowledge." But Eaton was situated on the other side of "the big pond" from Davy and presumably could have defended himself ably if the ailing Davy had pursued a defamation suit. In fact, Davy finished his last days fishing and published *Salmonia: Or Days of Fly Fishing* in 1828, a year before he died, appropriately unaware of the criticisms emanating from the wilds of upstate New York and Vermont.

Actually, Amos Eaton's career took some fascinating twists and turns.[6] Born in Chatham, New York in 1776, Eaton studied law at Columbia College, was first attracted to chemistry by Samuel Latham Mitchill, and passed the bar exam in 1802. He practiced law, went into business, was convicted of forgery on evidence framed by some enemies, was imprisoned in 1811, and was pardoned by Governor Tompkins in 1815 with the pledge of never returning to New York State. He moved to New Haven, learned chemistry from Benjamin Silliman at Yale, became an itinerant chemistry lecturer throughout towns and villages in New England, and even occasionally dared to cross into New York State. His fame spread; he was invited by New York's new Governor, DeWitt Clinton, to give lectures to legislators; and he met the wealthy Stephen van Rensselaer, who founded the Rensselaer School largely to provide a home for the talented lecturer in chemistry and geology. Rensselaer Polytechnic Institute (RPI) is now a widely respected research university with an African-American female physicist, Dr. Shirley Jackson, at its helm as this essay is being composed. RPI has just received a $300 million unrestricted donation. Not bad for a tiny upstate school with a convicted forger as its first chemistry professor.

The chemist-entrepeneur thrived during the early nineteenth century. In *Practical Facts in Chemistry,* Robert Best Ede has written a small instructional book bound as a virtual 193-page preface to his 48-page catalog *Robert Best Ede's Series of Chemical Laboratories and Chests . . .* , published in London in 1837 (see Figure 135).[7] The catalog is complete with four pages of testimonials including magazines, newspapers, and famous chemists such as Thomas Graham, Professor

Robert Best Ede,

In submitting this Catalogue to the Public in general, very respectfully begs to observe, that it is his intention from time to time, to make such an enlargement in his selection both of scientific and domestic articles, as the progressive improvement of the age, and other circumstances may point out, to be best calculated to meet the still advancing taste of the public, for the higher branches of knowledge and science, and secure that popular approbation which it has been, and ever will be his first object to attain.

Feb. 1837.

FOR CHEMICAL STUDENTS, AMATEURS & PROFESSORS.

R. B. EDE'S
SERIES OF
PORTABLE CHEMICAL
Laboratories, Cabinets
AND
BLOWPIPE APPARATUS,
CONTAINING
A CHOICE SELECTION OF THE MOST USEFUL
Tests, Re-Agents & Preparations,
AND AN ORGANIZED COLLECTION OF THE BEST CONTRIVED
MODERN APPARATUS,
Adapted for performing with facility, safety and success,
A Course of Instructive and Entertaining Experiments
AND
For the exhibition of those interesting phenomena in Chemistry, Mineralogy, &c.
WHICH RENDER THE STUDY OF THESE SCIENCES SO FASCINATING;
ALSO,
PRACTICAL FACTS IN CHEMISTRY,
an appropriate
COMPANION TO THE PORTABLE LABORATORIES AND CABINETS;
AND
WARD'S FOOTSTEPS TO CHEMISTRY,
EQUALLY SUITABLE AS A GUIDE TO THE YOUTH'S LABORATORY.

FIGURE 135. ■ Robert Best Ede prefaced his 48-page catalog with a 193-page textbook on chemistry. But selling is the main object—here is the start of his catalog. (From Ede, *Practical Facts in Chemistry*, London, 1839, 1837.)

30

31

APPARATUS FOR CHEMICAL CABINET CONTINUED.

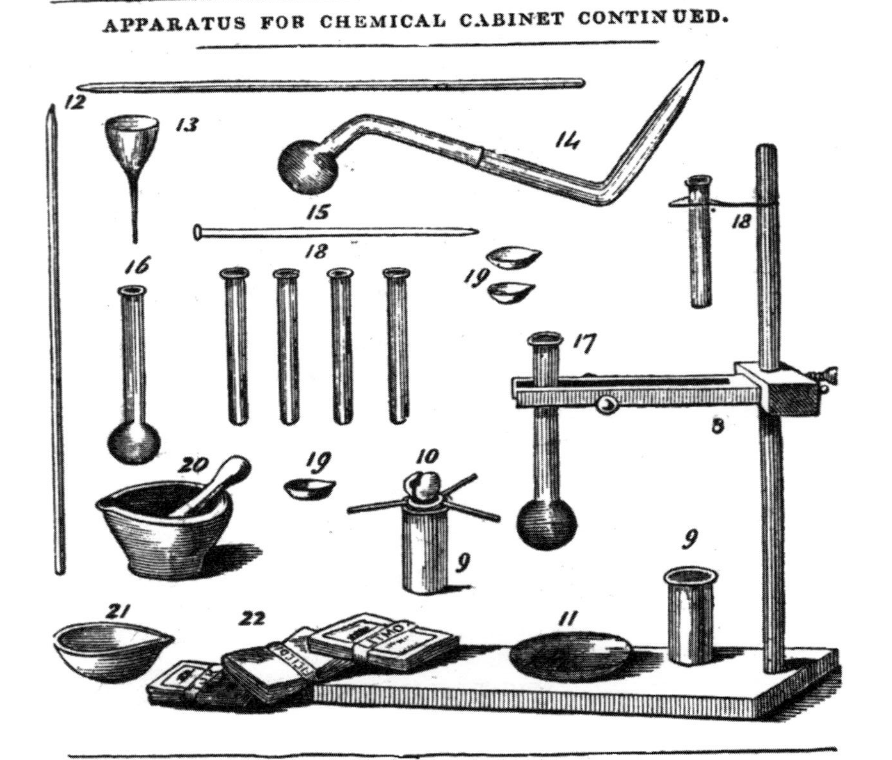

5 Pair of Scales and Set of Weights.
6 Mahogany Test Tube Stand & Brass Triangle.
7 Six Test Tubes.
8 Gay Lussac's Mahogany Test Tube and Retort Holder.
9 Three Phillip's Precipitating Tubes.
10 Glass Triangle for Filters {2 in No.6
11 Green Glass Capsule.
12 Two Tube Rods, and 4 Glass Rods.
13 Blown Glass Decanting Funnel
14 Globe Retort and Kerr's Tube Receiver
15 Suction Tube.
16 Two Bulb Tubes.
17 Two large Green Glass ditto.
18 Six Quill Test Tubes.
19 Three Watch Glasses.
20 Wedgwood's Mortar & Pestle.
21 Two ditto Capsules.
22 Filtering, Litmus, and Turmeric Papers.

of Chemistry at Glasgow ("I have had occasion to look over the contents of Mr. Ede's Portable Laboratory and have formed a high opinion of it") and Thomas Clark, M.D. at Aberdeen ("Mr. Ede's Portable Laboratory, I consider a *cheap* and very *useful selection* for Students in experimental Chemistry")—high praise, indeed, from a Scotsman.

Note the simple functional apparatus in Figure 136. The simple globe retort and tube receiver (item *14*) is an apparatus made from two pieces of glass tubing. The retort is made by slightly bending a tube, sealing it at one end, and blowing a bubble in the sealed end while the tube is heated.[8] The tube receiver is slightly bent and nearly sealed, leaving a pinprick opening for the escape of gases and vapors. The two vessels are connected airtight by paste of linseed meal or a tube of sheet caoutchouc (natural rubber). The use of this apparatus in depicted in Figure 137.[8] The bend in the receiver is immersed in ice water. Lead nitrate [$Pb(NO_3)_2$] is placed in the retort, the apparatus joined, and the retort heated with an alcohol lamp to form mixed oxides, and a solution of nitrous acid is collected in the bend of the tube receiver. When a very small quantity of this liquid is added to a dry test tube and a single drop of distilled water added, a deep blue

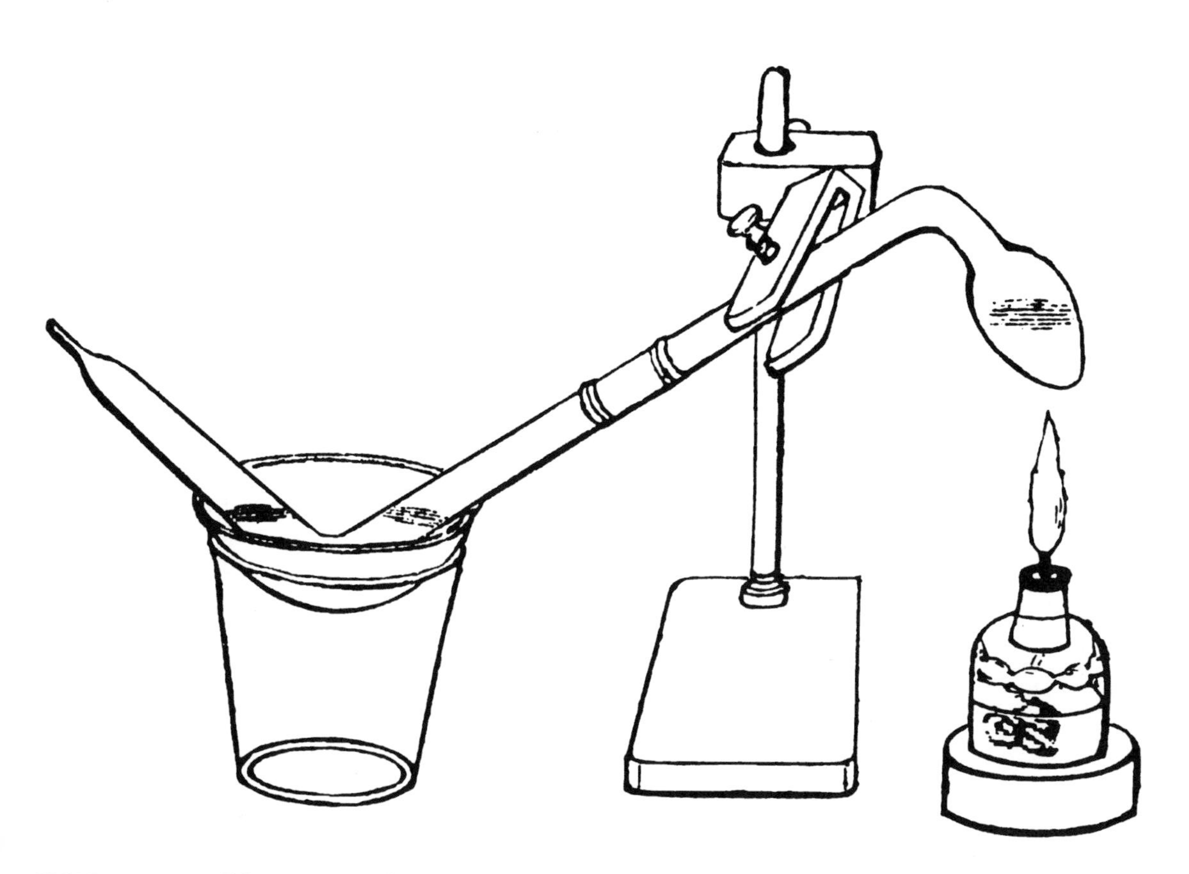

FIGURE 137. ■ Ede's very simple and elegant student experiment for obtaining and trapping at 0°C, the deep blue, highly unstable dinitrogen trioxide (N_2O_3), which decomposes at only 3°C. This would still be a nice experiment today, almost 170 years later. (See Figure 136.)

color is observed, probably due to the presence of dinitrogen trioxide (N_2O_3)—a rather uncommon and unstable substance that decomposes at only 3.5°C.[9] So here we see some pretty elegant chemistry performed in a cheap apparatus assembled using two glass tubes and linseed meal paste.

---

1. A. Eaton, *Chemical Instructor: Presenting a Familiar Method of Teaching the Chemical Principles and Operations of the Most Practical Utility to Farmers, Mechanics, Housekeepers and Physicians; and Most Interesting to Clergyman and Lawyers*, Websters and Skinners, Albany, 1826.
2. H. Moyes, *Heads of a Course of Lectures on the Philosophy of Chemistry*, Boston, 1784.
3. Eaton, op. cit., pp. 3–11.
4. E.F. Smith, *Old Chemistries*, McGraw-Hill, New York, 1927, pp. 50–52, 60–64.
5. A. Greenberg, *A Chemical History Tour*, John Wiley & Sons, New York, 2000, pp. 189–191.
6. H.S. Van Klooster, *Chymia*, Vol. 2, University of Pennsylvania Press, Philadelphia, 1949, pp. 1–15.
7. R.B. Ede, *Practical Facts In Chemistry, Exemplifying the Rudiments and Showing with What Facility the Principles of the Science May Be Experimentally Demonstrated at a Trifling Expense by Means of Simple Apparatus & Portable Laboratories, More Particularly in Reference to Those by Robert Best Ede*, Thomas Tegg, and Simkin, Marshall, and Co., London, 1839. Issued and bound with *Robert Best Ede's Series of Chemical Laboratories and Chests, with Appropriate Companions, Also, Mineralogical Boxes, Labels and Other Select and Approved Articles*, dated February 1837.
8. Ede, op. cit., pp. 144–159.
9. F.A. Cotton and G. Wilkinson, *Advanced Inorganic Chemistry*, fifth edition, John Wiley & Sons, New York, 1988, pp. 320–328.

SECTION VII
SPECIALIZATION AND SYSTEMIZATION

GEODES[1]

The rhythms, rhymes, and imagery of poetry have preserved oral traditions since antiquity and they powerfully fix ideas in our minds. So why not apply this learning tool to contemporary teaching? And so, we have two small volumes of "Werneria" published in 1805 and 1806 by *Terrae Filius* (aka *Terrae Filius Philagricola*—"son of the earth lover Agricola").[2,3] "Werneria" refers to Abraham Werner, a German geologist who believed that all rocks originated in the oceans ("Neptunist school"), but was "a brilliant lecturer in geology."[4] Our mystery author is Reverend Stephen Weston (1747–1830),[5] a poet, a man of letters, having an incredible breadth of interests. Following the death of his young wife around 1790, he devoted the remainder of his life to art and literature.[5] His works included translations from Latin, French, Arabic, Persian, and Chinese languages and discussions of travels and religious thought. It is said that he "lived for some years among the dilettanti in London" and "had a numerous circle of lady admirers who fed his vanity."[5] In his preface to Werneria,[2] Weston quotes Aristotle: "Is it because men, before they discovered the art of writing, sang their laws, that they might not forget them?" and so he applies his own artistic talents to teaching mineralogy.

How about this poetic description of diamond that treats some physical properties and teaches that diamond is pure carbon—its combustion forms carbon dioxide with no other residue remaining?[2]

> In hardness, brilliance, and transparency,
> The diamond every mineral excels,
> Black, yellow, green, blue, brown, or grey, 'tis known,
> And colourless in quartzose sand is found
> In flat, or rounded grains, sometimes cube-shap'd,;
> But oft its form eight-sided is, or twelve:
> In texture laminous, but fibrous too,
> Irregularly so; to solar rays
> Expos'd the diamond is phosphoric;
> Rubb'd, it emits electric sparks: What gem
> But this from rich Golconda's shore, can e'er
> To carbone's acid be converted, and
> Leaves no wreck behind?

And here is the description of lime (CaO, calcium oxide or quicklime), obtained by heating calcium carbonate ($CaCO_3$, limestone)—it was the primary binder in concrete until the early 1800s. It can recombine with carbon dioxide and is strongly basic, but moisture converts it, with evolution of heat, to a milder calcium hydroxide powder:

This earth from carbonate of lime obtained
For various use, by application of
Incessant heat, in form is concrete, or
In powder; in colour white; in taste hot,
Pungent, and caustic; and when in water solved,
Will change the vegetable blue to green:-
When in the concrete state exposed to air,
Cohesion's force is lost; but when the gas
Of carbo from the atmosphere's absorbed,
Then all its pristine hardness is regain'd.

FIGURE 138. ▪ Humphry Davy's apparatus for measuring the amount of limestone in soil. Sulfuric acid releases one equivalent of CO_2 from an equivalent of limestone ($CaCO_3$). The gas fills a balloon, displacing a volume of water that is measured to give the volume of gas generated. (From Davy, *Agricultural Chemistry*, London, 1813.)

Add moisture, and to powder it returns,
Shines in the dark, its caloric evolves,
And doubly mild and temperate becomes.
Per se infusible, all others it can fuse;
With borax, and microsmic salt it melts,
And effervesces not.

Limestone is widely distributed in soils, and in Figure 138 we display the apparatus designed by Humphry Davy[6] to measure its abundance. Sulfuric acid in *B* is added dropwise to the soil in vessel *A*, the carbon dioxide released is carried through C and collected into the balloon in vessel *E*, which is filled with water, and the volume of expansion is measured in cylinder *D*.

---

1. The blame for this title rests on my brother Kenny Greenberg.
2. *Terrae Filius* [i.e., Stephen Weston], *Werneria; or, Short Characters of Earths: with Notes according to the Improvements of Klaproth, Vauquelin and Hauy, C. and R. Baldwin,* London, 1805. I am grateful to chemist and book collector Dr. Roy G. Neville for making me aware of *Werneria* and Stephen Weston.
3. *Terrae Filius Philagricola* [i.e., Stephen Weston], *Werneria, (Part the Second) or, Short Characters of Earths and Minerals According to Klaproth, Kirwan, Vauquelin, and Hauy,* C. and R. Baldwin, London, 1806.
4. *The New Encyclopedia Brittanica,* Vol. 12, Encyclopedia Brittanica, Inc., Chicago, 1986, pp. 582–583.
5. L. Stephen and S. Lee (eds.), *The Dictionary of National Biography,* Oxford University Press, Oxford, 1921 (reprint 1964/65), pp. 1283–1285.
6. H. Davy, *Elements of Agricultural Chemistry, in a Course of Lectures for The Board of Agriculture,* Longman, Hurst, Rees, Orme, and Brown, London, 1813, Figure 15 (facing p. 145).

COLORFUL "NOTIONS OF CHEMISTRY"

Théophile Jules Pelouze (1807–1867)[1] and Edmond Fremy (1814–1894)[1] coauthored one of the most beautifully illustrated chemistry books of the nineteenth century: *Notions Générale de Chimie.*[2] Pelouze was a student of Gay-Lussac at the école Polytechnique in Paris, and Partington describes his living conditions thus: "His lodging was so small that he humorously said he found it necessary to open the window to find space to put on a coat; he dined on bread and water, which he said tended to clear the mind."[1] Pelouze eventually succeeded Gay-Lussac at École Polytechnique, subsequently succeeded Thenard and Dumas at the Collège de France, and, in 1848, became president of the Commission of the Mint. In 1838, Pelouze was the first to react nitric acid with cotton and he produced spontaneously combustible nitrated cellulose. However, it was Christian Friedrich Schönbein who, eight years later, produced highly nitrated cellulose, an explosive commonly called "guncotton," using a mixture of nitric and sulfuric acids.[3]

Fremy began his chemistry career as first assistant to Pelouze at the École Polytechnique. He later was appointed Professor at this institution as well as in the Museum d'Histoire Naturelle. On the very day in 1850 that Pelouze's Chair at Collège de France was to be filled by election, Fremy read a paper attacking the widely favored successor, Auguste Laurent, and the Académie des Sciences did not elect him. Laurent died of poor health less than three years later at the age of 44.[4] Fremy may well have been the first to generate and catch a whiff of fluorine by electrolyzing calcium fluoride in 1854.[5] However, his student Henri Moissan benefitting from Fremy's experiences, isolated fluorine in 1886, and received the 1906 Nobel Prize in Chemistry.[5] Fremy was fascinated by the colors of cobalt salts for which he proposed an original (and long-dead) nomenclature.[6] Perhaps this fascination with colors led to the production of this lovely book.

Figure 1 (Figure 139) depicts Lavoisier's experiment in which a matrass (a type of flask also known as a "bolt head") containing mercury is connected by its long curved neck into a graduated bell jar open over mercury in a basin. Mercury in the matrass is heated just to boiling for five days, until no further reduction of air volume is observed in the bell jar. Further heating occurs for a few days. Lavoisier's finding was that 27% of the gas volume in the bell jar diminished because of the loss of oxygen (the later accepted value was about 21%). The red crystalline substance found floating on the surface of the mercury in the matrass was mercuric oxide (HgO). In *Fig. 2* (Figure 139) an iron wire with a piece of lighted tinder at its end is placed into a jar of pure oxygen. The iron immediately inflames and throws sparks of iron oxide hot enough to melt and penetrate deeply into the glass. In *Fig. 3* (Figure 139) one pound of manganese dioxide (MnO_2) is heated very strongly in an earthen retort to produce oxygen. This is how Gahn first isolated manganese in 1774; less heat is required to produce oxygen from MnO_2 in the presence of sulfuric acid.

In *Fig. 4* (Figure 140) a small plaster cupel sits on a cork that floats, boat-like, on water in a trough. A small piece of phosphorus in the cupel is ignited and a bell jar inverted over the cupel. The gas remaining in the bell jar is nitrogen. *Figure 5* (Figure 140) depicts a flask containing zinc into which water is first added through a funnel followed by sulfuric acid. Hydrogen gas is collected over water. *Figure* 6 (Figure 140) shows Lavoisier's classic decomposition of water using pieces of iron wire in a porcelain tube heated red hot in a furnace. Oxide of iron forms in the tube and hydrogen gas is collected by a receiver inverted in water.

In *Fig. 7 (Figure 141) hydrogen is generated (see* Fig. 5, Figure 140), moves through a drying tube and is combusted with oxygen in common air, and the water condensate drips into a collection bowl. *Figure* 8 (Figure 141) depicts a bell jar containing air and a cylinder of zinc suspended by a wire into acidified water. When the hydrogen formed pushes the acidified water from the bell jar, further reaction ceases. *Figure* 9 (Figure 141) shows a laboratory-scale distillation apparatus. In *Fig. 10* (Figure 142) we see a large-scale distillation apparatus using a copper boiler covered with a hood. The curved condenser tube is called a "worm." *Fig. 11* (Figure 142) shows an amazingly modern-looking laboratory-scale distillation apparatus developed by Gay-Lussac. Cooling water is added through the funnel at the lower right and leaves the condenser at the upper left.

Phosphorus in a cupel, suspended by a wire attached to a cork, burns blindingly in an atmosphere of pure oxygen (*Fig. 12,* Figure 143). Diamonds, known to be pure carbon, are depicted in *Fig. 13* (Figure 143). Heating saltpetre

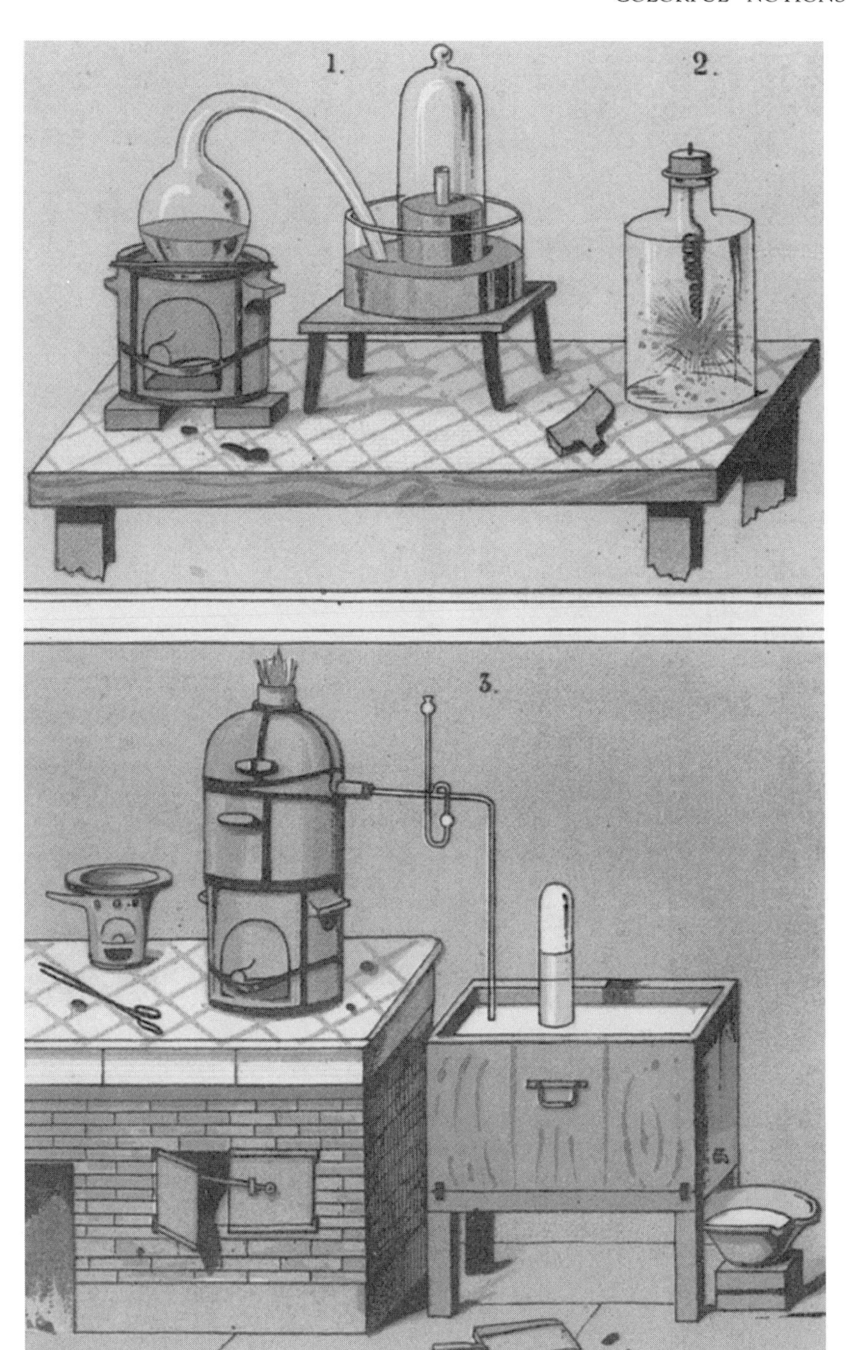

FIGURE 139. ■ Black and white image of a color plate from the American edition (1854) of *Notions Générale de Chimie* by Théophile Jules Pelouze and Edmond Fremy. See color plates. Depicted in *1* is Lavoisier's classic mercury oxidation experiment involving reflux of mercury in air; *2* illustrates the oxidation of an iron wire using a flame in pure oxygen; *3* depicts strong heating of manganese dioxide to release oxygen—an experiment first performed by Scheele.

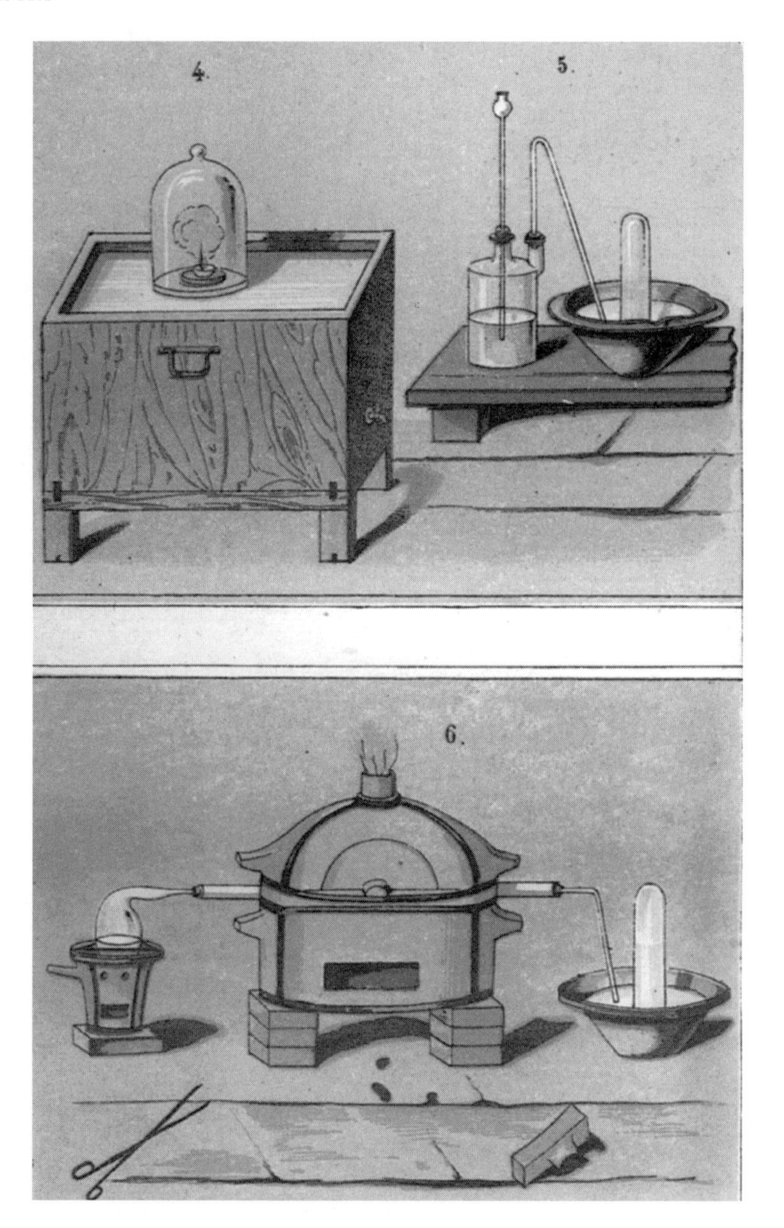

FIGURE 140. ■ Black and white image of a color plate from the 1854 American edition of *Notions Générale de Chimie* by Pelouze and Fremy. See color plates. Depicted in *4* is the combustion of phosphorus in air leaving unreacted nitrogen; *5* shows the generation of hydrogen through reaction of zinc and sulfuric acid—work first published by Henry Cavendish in 1766. In *6*, we see Lavoisier's decomposition of water using an iron wire in a porcelain tube heated red hot in a furnace (see Figure 114).

(KNO_3) and sulfuric acid in a glass retort and distillation produces "azotic" (nitric) acid (*Fig. 14*, Figure 143). Large-scale production of nitric acid (*Fig. 15*, Figure 143) employs niter ($NaNO_3$), which is cheaper to produce than saltpetre. Quantities of 100–150 kg (kilograms) of niter are heated in earthen cylinders to which sulfuric acid is added periodically. The distillate is received by a series of 12–15 three-necked flasks.

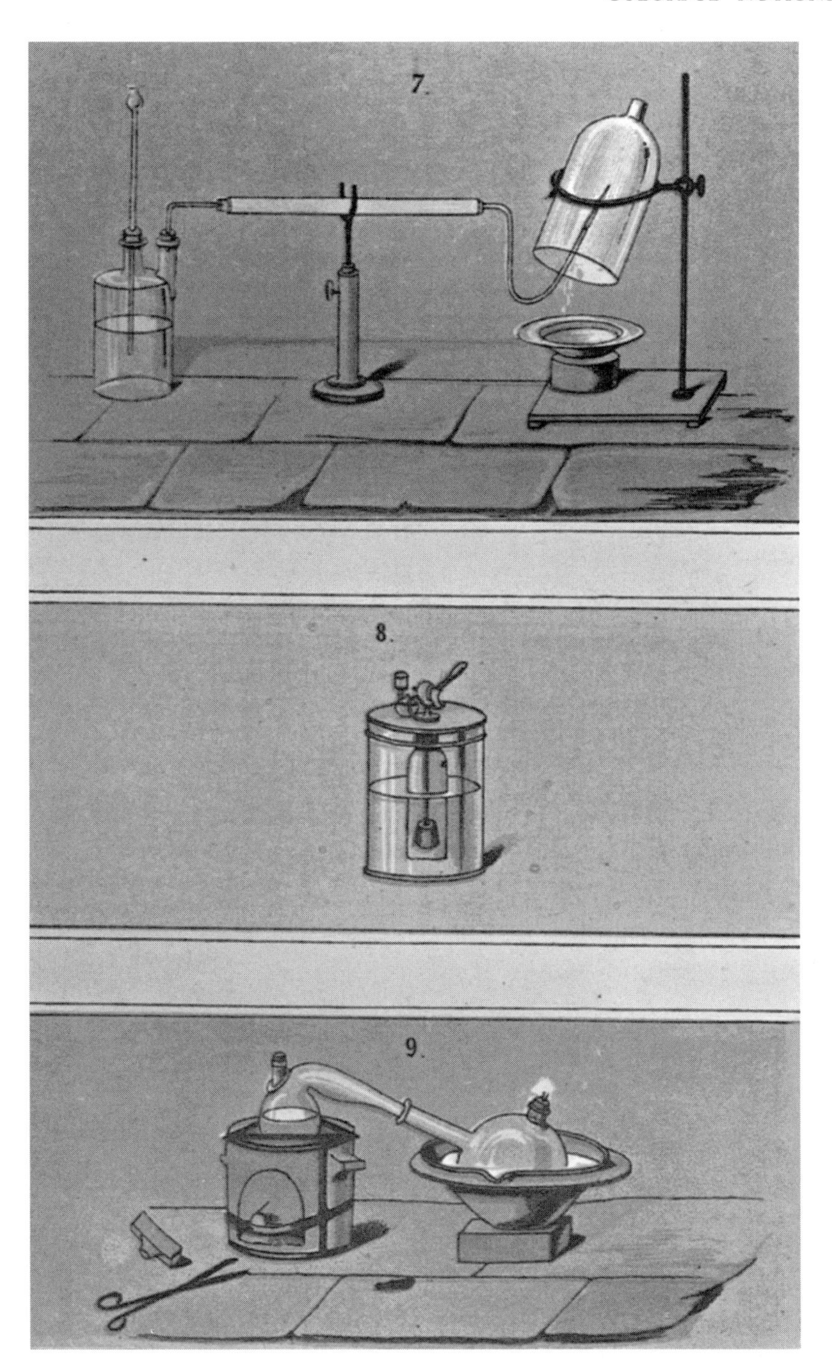

FIGURE 141. ▪ Black and white image of a color plate from the 1854 American edition of *Notions Générale de Chimie* by Pelouze and Fremy. See color plates. Depicted in 7 is a synthesis of water involving generation of hydrogen and its combustion in air; 8 shows a clever self-controlling hydrogen gas generator; 9 illustrates a laboratory-scale distillation apparatus appropriate for students.

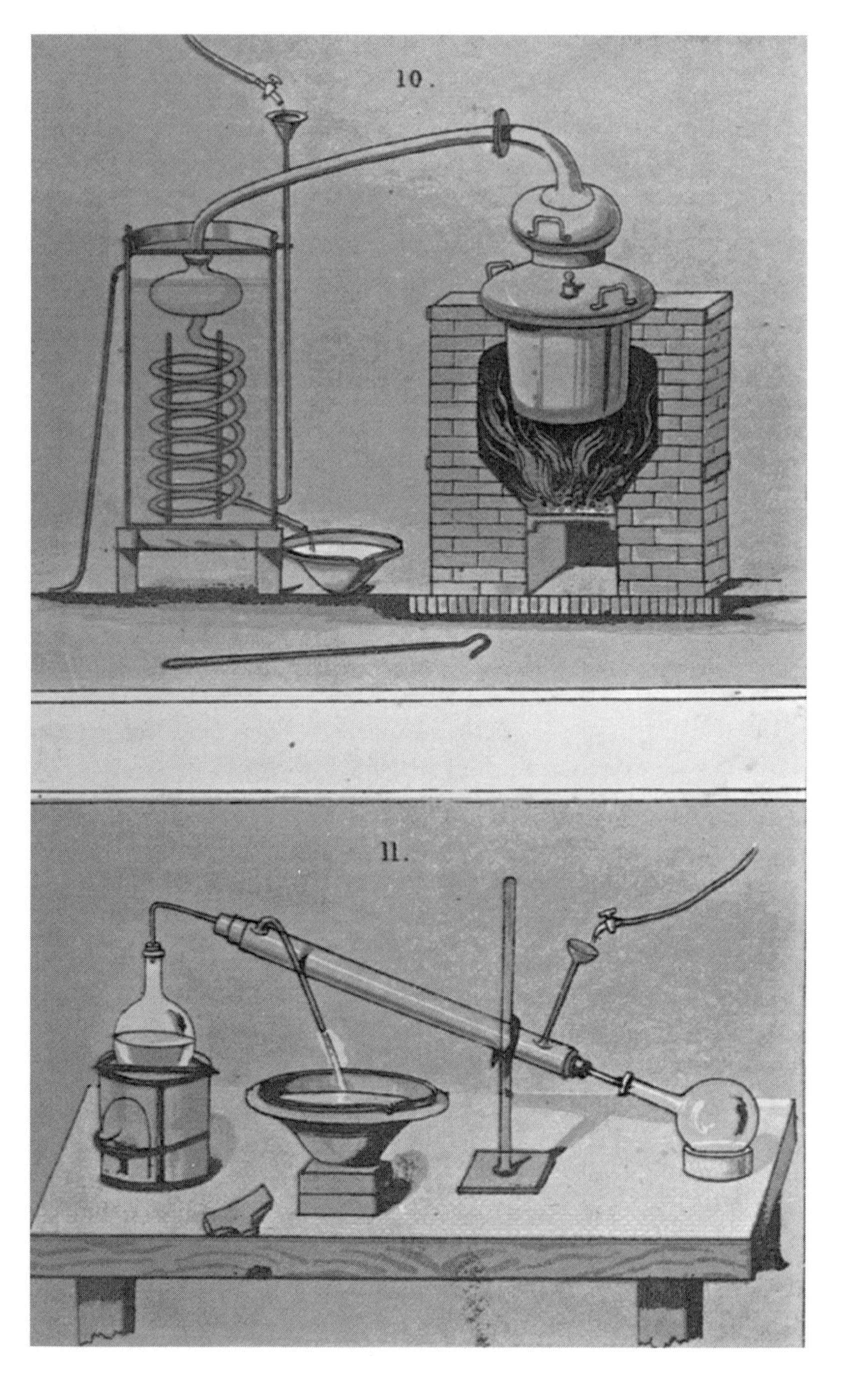

FIGURE 142. ▪ Black and white image of a color plate from the 1854 American edition of *Notions Générale de Chimie* by Pelouze and Fremy. See color plates. Depicted in *10* is an industrial-scale still consisting of a copper boiler covered with a hood; *11* displays the very modern-looking water-cooled distillation apparatus designed by Gay-Lussac.

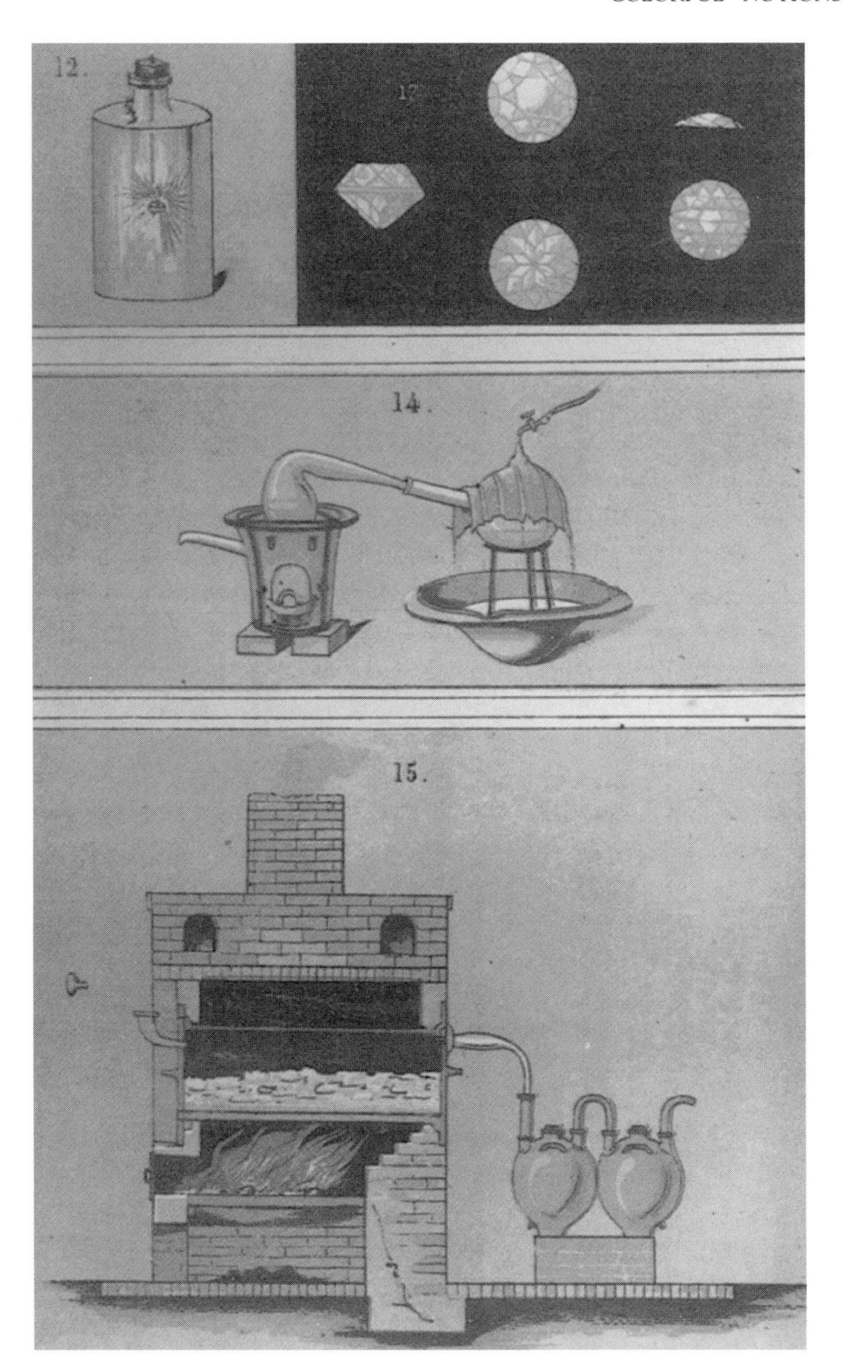

FIGURE 143. ▪ Black and white image of a color plate from the 1854 American edition of *Notions Générale de Chimie* by Pelouze and Fremy. See color plates. In *12* phosphorus, suspended by a wire, burns with blinding brightness in pure oxygen; diamonds (*13*) are composed of pure carbon just like the humble mineral plumbago (graphite), another carbon allotrope, and totally different from ruby and other gemstones; heating saltpetre (KNO_3) and sulfuric acid produces nitric acid (*14*) but on an industrial scale (*15*) it is cheaper to use niter ($NaNO_3$).

FIGURE 144. ▪ A wonderfully surrealistic rendition of stalactites (suspended from the ceiling, in case you forgot) and stalagmites composed of limestone ($CaCO_3$) formed by contact between the CO_2 dissolved in groundwater and lime (CaO) in the soil. See color plates. This image seems to anticipate the artistic style of René Magritte some 45 years before his birth. (From the 1854 American edition of *Notions Générale de Chemie* by Pelouze and Fremy.)

Figure 27 (Figure 144) is a rather surrealistic rendition of stalactites (suspended from the ceiling) and stalagmites that rise to meet them in a cave that looks like the open maw of some hideous beast. Stalactites and stalagmites are composed of calcium carbonate ($CaCO_3$) arising from contact of carbonic acid dissolved in water with lime (CaO) in the earth's surface. The wide sky, strange clouds, and bizarre cave seem to anticipate the style of modernist René Magritte (1898–1967).

---

1. J.R. Partington, *A History of Chemistry*, Vol. 4, Macmillan & Co., Ltd., London, 1964, pp. 395–396.
2. T.J. Pelouze and E. Fremy, *Notions Generales de Chimie. Avec un Atlas de 24 Planches en Couleur, en 2 Volumes*, Victor Masson, Paris, 1853. The plates shown here are from the American edition: *General Notions of Chemistry*, Lippincott, Grambo & Co., Philadelphia, 1854.
3. A.J. Ihde, *The Development of Modern Chemistry*, Harper & Row, New York, 1964, p. 451.
4. Partington, op. cit., p. 376.
5. Ihde, op. cit., p. 367.
6. W.H. Brock, *The Norton History of Chemistry*, W.W. Norton & Co., New York, 1993, pp. 577–578.

WHAT ARE ORGANIC CHEMISTS GOOD FOR?

Until the middle of the nineteenth century the dyes used in textiles and other commercial applications had their origins in plant and animal matter.[1] Indeed, indigo dyes from three species of snails were the basis of the ancient dye *tekhelet*, specified by Moses for coloring blue the fringes of the Hebrew prayer shawl or *tallit*.[2] This interesting history is related by Roald Hoffmann and Shira Leibowitz Schmidt, who note that *tekhelet* was likely a mixture of two closely related indigo dyes.[2] The Hebrews, according to their lore, lost the art of making *tekhelet* by the year 760. Since that time, the fringes have been white since no substitutes were allowable under religious law. Despite the subsequent discovery of plant sources for these dyes and modern chemical techniques that definitively validate their identities, the modern tradition of white fringes remains firm. And since, as the authors note, "there is no authentic Hebrew textile dyed in *tekhelet* that has survived," no attempt to re-create *tekhelet* is likely to be acceptable.[2]

In the middle of the nineteenth century, the precocious William Henry Perkin (1838–1907) entered the Royal College of Chemistry at the age of 15 and soon became an assistant to its Director, Professor August Wilhelm Hofmann.[3,4] By that time, coal tar had became an unwanted waste product and while commercial benzene and toluene had been obtained from coal tar by distillation, it was still considered a massive nuisance.[5] Working in his home laboratory in London in 1856, young Perkin tried unsuccessfully to synthesize the drug quinine but obtained instead dark tars. A modification, using the coal-tar component aniline, provided another dark substance that was found, again quite by accident, to be an excellent purple dye, that Perkin named *mauve*. Perkin left the university, much to Hoffmann's dismay, and built a factory to manufacture mauve financed by his father. Suddenly, a synthetic dye industry emerged and coal tar became a commodity rather than a waste product.[1,5]

Figure 145 is a family tree showing the development of synthetic dyes for about the first 75 years following the discovery of mauve by Perkin.[6] The limb branching to the left of mauve includes a series of chemically related dyes, some of which are named in Figure 146a. There is a fairly smooth transition in color from the purplish-red fuchsine through a series of three violet dyes to the two blue dyes whose structures are shown in this figure. In Figure 146a, fuchsine is employed as the "core dye" and the five others differ slightly through substitutions of the highlighted groups for hydrogen atoms. The custom synthesis and fine tuning of the colors of these dyes illustrate one of the fundamental strengths of organic chemistry. Organic chemists are experts at "tweaking" the properties of complex molecules by substituting atoms or groups of atoms for each other. The difference between fuchsine and methyl violet B is a rather simple replacement of five hydrogen atoms by five methyl groups.

It is interesting to note that each of the six compounds in Figure 146a has one nitrogen atom forming five bonds with other atoms. Indeed, the beautiful book, published in 1935, containing the structures in Figure 146a also includes some rather "precious" cartoons of atoms with hands signifying valences.[6] Figure 146b depicts a water molecule in which one oxygen and two hydrogen atoms hold hands (rather like the "water fairies" depicted elsewhere[7]). In Figure 146c we note illustrations of the valence of each atom including a valence of 5 for nitrogen. The confusion may be illustrated for ammonium chloride (NH_4Cl). Since the valences of hydrogen and chlorine are commonly known to be 1, the only reasonable arrangement appears to have nitrogen forming single bonds with the four hydrogen atoms as well as with the fifth atom, chlorine. It had previously been known that NH_4Cl separated into ammonia (NH_3) and hydrogen chloride (HCl) upon heating. The theory of ionic solutions developed by Svante Arrhenius[8] in the 1880s laid the basis for understanding the true nature of ammonium chloride, an ionic salt of formula $NH_4^+Cl^-$ that behaves as two distinct particles when the salt is dissolved in water, rather than one NH_4Cl molecule. The four bonds attached to nitrogen in the NH_4^+ ion were completely consistent with the octet rule and Lewis structures (1916).[9] However, consolidation of ionic theory and the Lewis octets really occurred only in the 1920s,[10] barely a decade before Figure 146 was published.

The molecular fine tuning and "tweaking" by organic chemists found early application in the pharmaceutical industry. For example, the differences between morphine, codeine, and heroine are not very significant structurally but enormously significant pharmaceutically. The accidental discovery of penicillin in 1928 by Alexander Fleming stimulated a two-decade search for its chemical structure, ultimately obtained in the mid–1940s by the crystallographer Dorothy Crowfoot (later Hodgkin) (1910–1994),[11] who received the 1964 Nobel Prize in Medicine and Physiology for her determination of the structure of Vitamin B_{12}.[12] Once the penicillin structure was known, pharmaceutical chemists synthesized thousands of derivatives looking to increase efficacy, lower cost, and diminish undesired side effects, such as allergic responses.

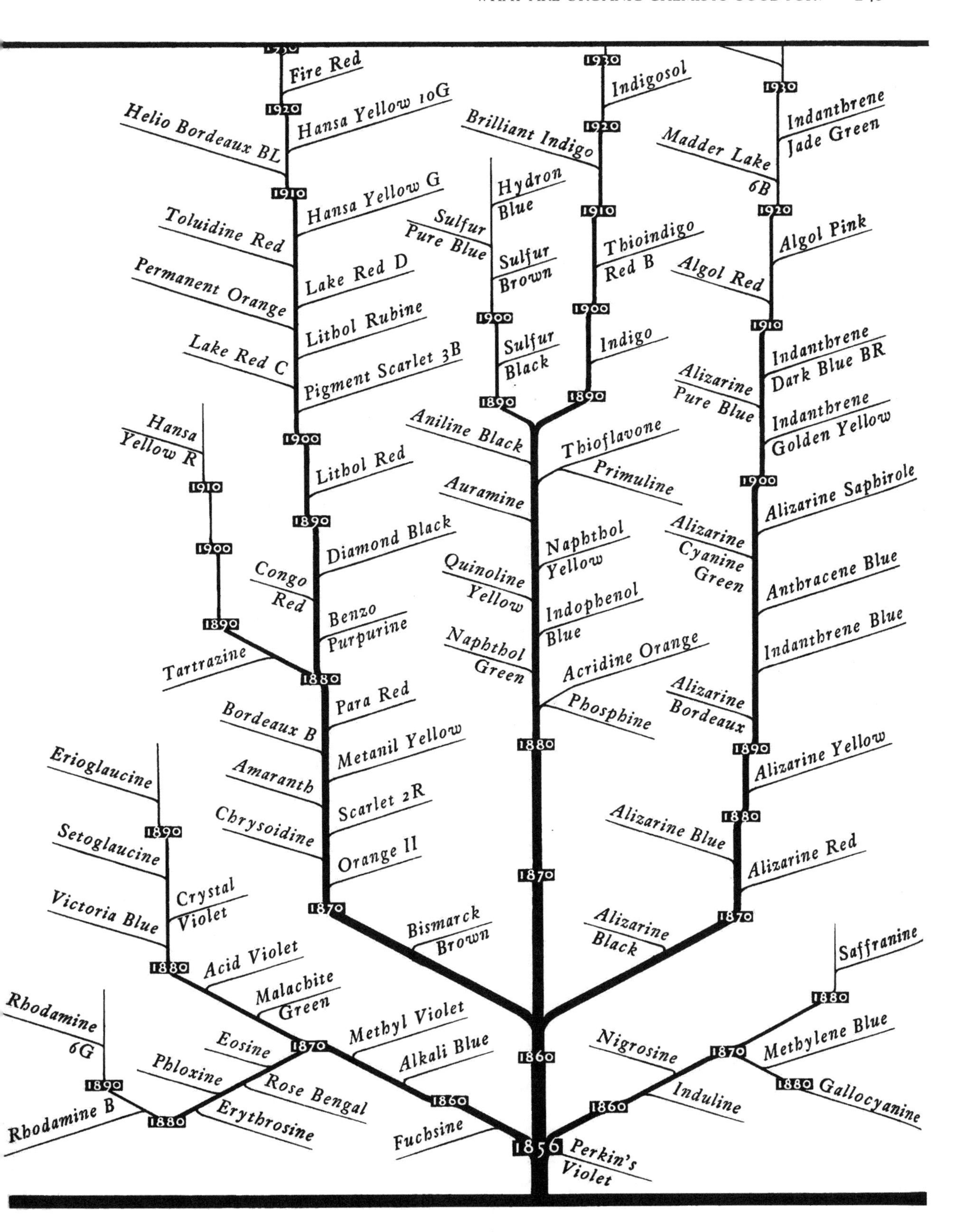

FIGURE 145. ■ The Tree of Dyes (?), showing the evolution of synthetic organic dyes through the early twentieth century following William Henry Perkin's discovery in 1856, at the age of 18, of "mauve," termed here "Perkin's Violet" (from *Color Chemistry*, No. 1, courtesy Ms. Lynne Crocker).

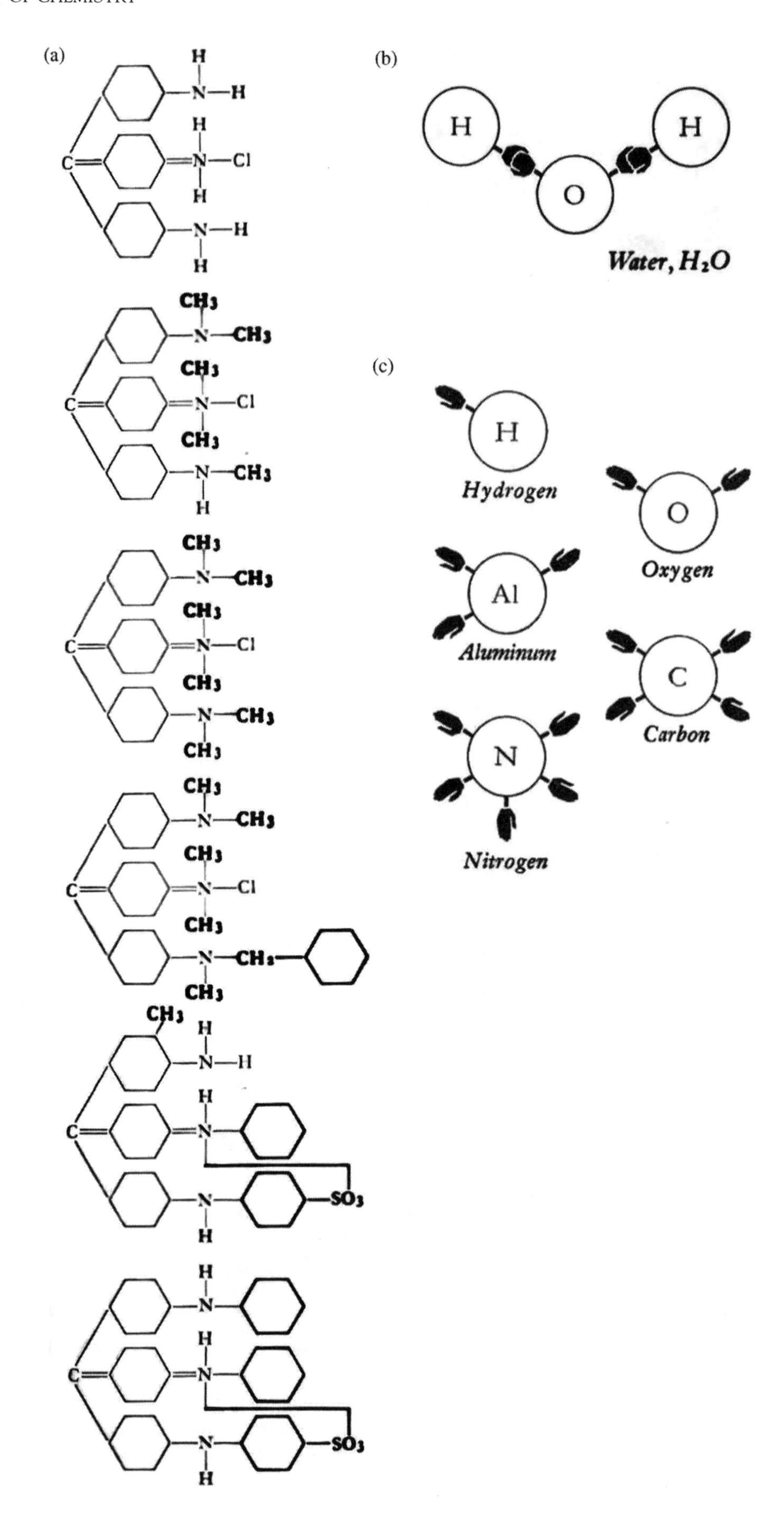
(a)
(b)
H
O
H
Water, H₂O
(c)
H
Hydrogen
O
Oxygen
Al
Aluminum
C
Carbon
N
Nitrogen

FIGURE 146. ▪ In (a) we see the synthetic organic chemist's skill in "tweaking" a molecular framework in order to fine-tune desired properties; substitution of the hydrogen atoms in the top structure (the purplish-red dye fuchsine) by methyl (CH_3) and other groups smoothly varies color through shades of violet, violet-blue, and then blue. This approach has long been used to modify penicillins and other drugs and now to synthesize laser dyes. Diagram (b) depicts two hydrogen atoms joining their friend oxygen atom to make water. Diagram (c) the hands (valences or combining "power") of five common atoms; the confusion about the valence of nitrogen [sometimes 3, sometimes 5—see the nitrogen atoms in (a)] was finally settled in the 1920s. (From *Color Chemistry*, No. 1, courtesy Ms. Lynne Crocker.)

---

1. W.H. Brock, *The Norton History of Chemistry*, W.W. Norton & Co., New York, 1993, pp. 297–301.
2. R. Hoffmann and S. Leibowitz Schmidt, *Old Wine New Flasks—Reflections on Science and Jewish Tradition*, W.H. Freeman and Co., New York, 1997, pp. 159–174.
3. J.R. Partington, *A History of Chemistry*, Macmillan and Co. Ltd., London, Vol. 4, 1964, pp. 772–774; 791–793.
4. S. Garfield, *Mauve: How One Man Invented a Color that Changed the World*, Faber and Faber Ltd., London, 2001.
5. A.J. Ihde, *The Development of Modern Chemistry*, Harper & Row, New York, 1964, pp. 452–458.
6. *Color Chemistry. Number One of a Series of Monographs on Color*, The Research Laboratories of the International Printing Ink Corporation and Subsidiary Companies, New York, 1935. I thank Ms. Lynne Crocker, Portsmouth, New Hampshire, for supplying this book.
7. A. Greenberg, *A Chemical History Tour*, John Wiley and Sons, New York, 2000, pp. 236–238.
8. Partington, op. cit., pp. 672–681.
9. Greenberg, op. cit., pp. 271–274.
10. G.B. Kauffman and I. Bernal, *Journal of Chemical Education*, Vol. 66, pp. 293–300, 1989.
11. D. Crowfoot, C.W. Bunn, B.W. Rogers-Low, and A. Turner Jones, in *The Chemistry of Penicillin*, H.T. Clarke, J.R. Johnson, and R. Robinson (eds.), Princeton University Press, Princeton, 1949, pp. 310–381.
12. G. Ferry, *Dorothy Hodgkin: A Life*, Granta Books, London, 1998.

NEVER SMILE AT A CACODYL

In 1835 Friedrich Wöhler called organic chemistry a "primeval forest of the tropics"[1] and the metaphor, of an unimaginably complex living system, was seemingly an apt one. Organic compounds seemed to be isolable only from living creatures—plants and animals. Often, they had to be extracted from enormously complex matrices and were challenging to isolate pure. Even urine, a clear liquid, is extraordinarily complex. The simple organic compound urea (later understood to be CH_4N_2O) was reported in 1773 by Hilaire Martin Rouelle (it had earlier been described by Boerhaave).[2] The substance was impure, but its "soapy" texture and ease of decomposition upon "distillation" marked it as notably different from typical inorganic salts, which were crystalline and usually heat-stable. Although Wöhler quite accidentally synthesized urea from inorganic compounds in 1828, he retained, at the time, the prevalent opinion that organic compounds were imbued with a "vital force" and could thus never be made artificially.[3] Some two decades later Hermann Kolbe "killed vitalism" by quite effectively synthesizing acetic acid ("vinegar") from its chemical elements.[3] There were three prevalent systems of relative atomic weights in common use by the midnineteenth century. These have been summarized by Aaron J. Ihde as follows:[4]

	H	C	O
Berzelius	1	12	16
Liebig	1	6	8
Dumas	1	6	16

Further confusion was caused by the difficulties in precise chemical analysis. Early nineteenth-century analyses involved measurements of the volume of CO_2 generated during combustion. This limited analytical samples to quite small quantities and had the effect of magnifying small errors. Even Liebig's *kaliapparat,* which captured carbon dioxide in condensed form for weighing, greatly increasing sample size and accuracy, did not completely solve these problems. This is well illustrated[5] by the disparity between the formula of cholic acid ($C_{48}H_{39}O_9$, Liebig atomic weights) reported by the famous chemist Adolph Strecker working in Liebig's Giessen laboratory in 1848, and the present-day formula ($C_{24}H_{40}O_5$). The disparities, albeit small, are highly significant since use of the Berzelius atomic weights for C, H, and O (very close to the modern values) would have given the formula $C_{24}H_{19.5}O_{4.5}$. This was, of course, incompatible with whole atoms as well as the rules of valence that were still about a decade into the future.

Organic compounds also posed numerous problems for early chemical theory. Davy's electrolytic studies, which produced electropositive potassium at one electrode and electronegative chlorine at the other, led Berzelius to postulate a theory of dualism. Hydrogen was clearly electropositive—it formed water and hydrogen chloride with the electronegative elements oxygen and chlorine. Electropositive carbon also formed compounds with oxygen and chlorine. However, organic compounds did not seem to fit this theory. How could one explain methane—a compound of carbon and hydrogen—both electropositive elements? How could an electronegative element such as chlorine fully replace the four electropositive methane hydrogen atoms to form CCl_4?

Among the early significant discoveries that helped clarify and systematize organic chemistry was the notion of a radical ("from the root") that had its earliest origins with Lavoisier: acid = radical + oxygen (where the radical could be the element sulfur whose combination with oxygen formed "sulfuric acid"—really SO_3).[6] This crude concept was followed by much more refined studies that disclosed the existence of the cyano radical (CN). It was Scheele who first treated the pigment Prussian Blue, which consists of iron compounds of ferrocyanide [today $Fe(CN)_6$], possibly in the presence of alkali metals or ammonia {e.g., $NH_4Fe[Fe(CN)_6]$—modern formula, of course}. Treating potassium ferrocyanide with sulfuric acid, he obtained "prussic acid" ("hydrocyanic acid" or hydrogen cyanide, HCN), and it was remarkable that he did not kill himself testing its odor. Scheele obtained potassium cyanide (KCN), mercury cyanide [$Hg(CN)_2$], and silver cyanide (AgCN).[7] In 1787, Berthollet reacted "prussic acid" with chlorine and discovered cyanogen chloride (ClCN).[8] In 1815, Gay-Lussac discovered cyanogen [$(CN)_2$] from his work on "prussic acid."[9] Thus, a body of evidence suggested that the CN radical acts virtually like an atom (i.e., "Cy") in passing unchanged from compound to compound.[10] Even more exciting was the 1832 publication by Liebig and Wöhler of the benzoyl radical ("$C_{14}H_{20}O_2$"—actually $C_7H_{10}O$), a stable unit containing three different types of atoms.[10]

Figure 147 is from the 1857 edition of Youmans' *Chemical Atlas*.[11] Radicals were initially considered to be stable "superatoms," which were joined, separated, and recombined to form molecules. The first radical in Figure 146 is ethyl, depicted as C_4H_5 (using the Liebig convention and thus C_2H_5 in modern terms). If we look at the third structure on the top, we see diethyl ether, which we recognize today as $C_4H_{10}O$ rather than C_4H_5O (or C_2H_5O using Liebig atomic weights). From a dualistic viewpoint, the C_4H_5O "molecule" is composed of the "ethyl" radical ("C_4H_5") as the electropositive part and O as the electronegative part. Seemingly, addition of water ("OH") to ethyl ether forms its hydrate, also known as "ethyl alcohol" (here as $C_4H_5 \cdot OH \cdot O$; really C_2H_6O using modern, rather than Liebig, atomic weights). In any case, the early theory of compound radicals (top third of Figure 147) depicted simple exchange of stable radicals to form different organic molecules.

This brings us to cacodyl, an early name for the awful smelling, spontaneously flammable, colorless liquid tetramethyldiarsine, obtained by heating arsenious oxide and potassium acetate:[12]

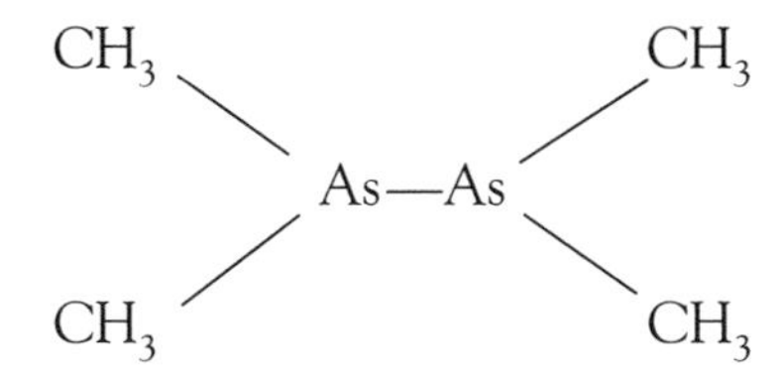

The cacodyl radical [$(CH_3)_2As$][13] also appeared to Bunsen to be a stable "superatom" that could be exchanged amongst other radicals. Many cacodyl compounds are explosive as well as spontaneously flammable. One of these, cacodyl cyanide [$(CH_3)_2AsCN$], exploded during Bunsen's exploratory studies, and he lost his right eye.[13]

In one of his studies, Bunsen synthesized cacodyl oxide from cacodyl and converted it to the chloride. Upon reaction with zinc, chlorine was lost and the pure arsenic, carbon, hydrogen compound remaining was thought by Bunsen to be the free cacodyl radical $(CH_3)_2As$. In fact it was liquid cacodyl [$(CH_3)_2As—As(CH_3)_2$].[14] Similarly, reaction of ethyl iodide with zinc freed the organic molecule of iodine and was thought to produce the free radical "ethyl" but in fact yielded butane ($C_2H_5—C_2H_5$)—the dimer of ethyl. This work by Edward Frankland (1825–1899) actually produced some diethylzinc, $(C_2H_5)_2Zn$, a spontaneously flammable volatile substance, which heralded the beginning of organometallic chemistry.[14] Searches for these free radicals were fruitless, and they were thus assumed to be incapable of isolation until the unexpected observation of the triphenylmethyl radical by Moses Gomberg.[15] Gomberg reacted triphenylmethyl chloride with zinc dust expecting to obtain hexaphenylethane:[16]

PLATE VI.

ILLUSTRATION OF THE THEORY OF COMPOUND RADICALS.

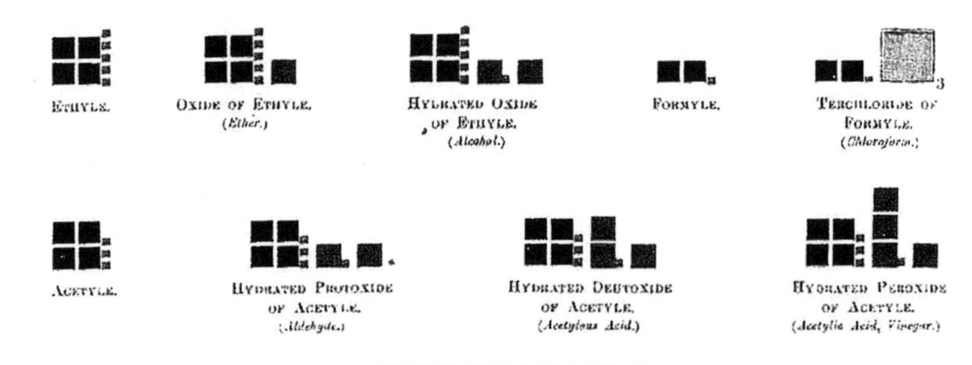

THEORY OF CHEMICAL TYPES—DOCTRINE OF SUBSTITUTION.

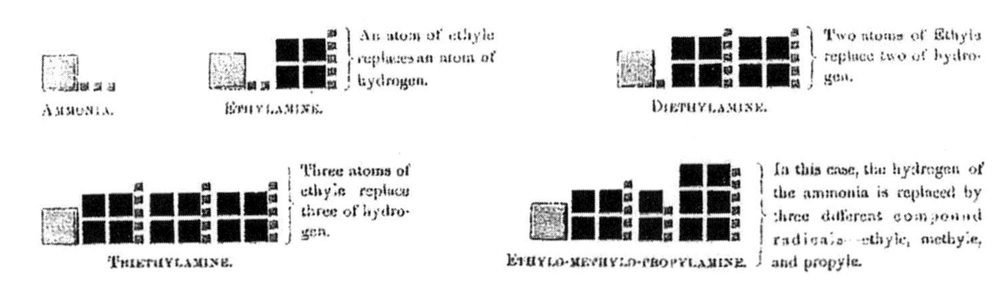

THEORY OF PAIRING—EXAMPLE OF COUPLED ACIDS.

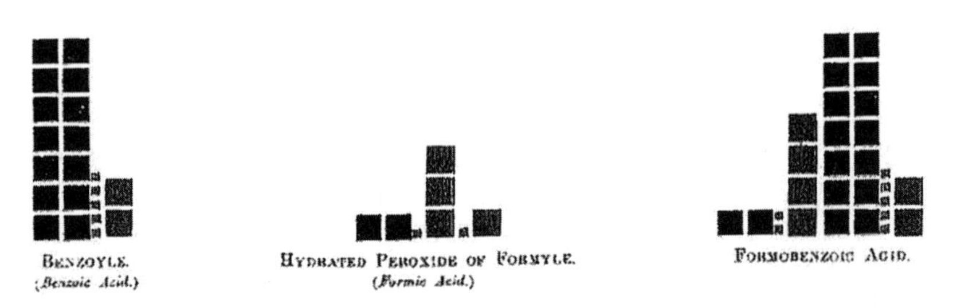

FIGURE 116 ■ For full description, see page xv.

FIGURE 147. ■ Illustration of mid-nineteenth-century theories of organic chemistry (see text); the original color figure is found in E. Youmans, *Chemical Atlas*, New York, 1857. The cacodyl radical $[(CH_3)_2As]$ was postulated early by Robert Bunsen, who lost an eye when cacodyl cyanide $[(CH_3)_2As–CN]$ exploded in his laboratory.

$$2\ \Phi_3C\text{—}Cl + Zn \xrightarrow{??} \Phi_3C\text{—}C\Phi_3 + ZnCl_2 \qquad (\text{where } \Phi = \text{phenyl or } C_6H_5)$$

What occurred was a surprising reaction forming a colored solution. Addition of iodine, for example, produced triphenylmethyl iodide and a colorless solution. Gomberg had generated a stable, yet reactive, free radical—triphenylmethyl radical:

$$2\ \Phi_3C\text{—}Cl + Zn \rightarrow ZnCl_2 + 2\ \Phi_3C. \xrightarrow{I_2} 2\ \Phi_3C\text{—}I$$

Interestingly, although the dimer of triphenylmethyl radical was thought to be the expected hexaphenylethane for some 60 years, we now know the correct structure of the dimer that exists in equilibrium with triphenylmethyl radical—it is not an ethane derivative.[16] Indeed, although pentaphenylethane ($C\Phi_3$—$CH\Phi_2$) exists and has abnormally long C—C bonds,[17] almost a century after Gomberg's discovery, hexaphenylethane continues to elude clever modern chemists.[18]

It is a wonderful irony to note that there is indeed a truly stable (actually, "persistent") cacodyl free radical.[19,20] Some 20 years ago it was discovered that mild heating of the compound $[(CH_3)_3Si)_2CH]_2As$—$As[CH(Si\{CH_3\}_3)_2]_2$ produces two $As[CH(Si\{CH_3\}_3)_2]_2$ radicals that are stable and observable for indefinite periods in solution at 25°C.[17,18] The trick here is the group of four huge $[(\{CH_3\}_3Si)_2CH]$ groups that hinder recombination of the radicals and formation of the weak[21] As—As bond.

The middle part of Figure 147 depicts the type theory in which, for example, an amine type could be replaced in turn by alkyl radicals in a series clearly related to ammonia. This was a positive contribution since it recognized families of related compounds (functional groups). The bottom section in Figure 147 illustrates the "pairing" ("copula") theory advanced by Berzelius in one final attempt to rescue dualism. As Ihde illustrates, acetic acid ($C_2H_4O_2$, modern formula) could be rationalized (using "double formulas") as a combination of an electropositive part (C_4H_6), an electronegative part (O_3), and water (H_2O). Trichloroacetic acid ($C_2HCl_3O_2$) was troubling to Berzelius. The substitution theory of the period visualized simple replacement of the hydrogens in the C_4H_6 radical by chlorines. The resulting formula, $C_4Cl_6 + O_3 + H_2O$, now has a serious "charge imbalance" since the carbon-containing radical is much less electropositive if not "downright electronegative" while the O_3 part is now fully electronegative. Berzelius felt that a dramatic rearrangement was needed such that the chlorinated carbon part (now C_2Cl_6) was "coupled" ("copulated") with "oxalic acid" (C_2O_3) with water a recognizable unit: $C_2Cl_6 + C_2O_3 + H_2O$. In this way, increased electronegativity in the C_2Cl_6 part was balanced by decreased electronegativity in the C_2O_3 part. The bottom section of Figure 147 depicts this kind of rearrangement of atoms in the "combination" of benzoyl radical with formic acid to produce formylbenzoic acid. However, a variety of chemical investigations clearly showed that acetic acid (CH_3COOH) and trichloroacetic acid (CCl_3COOH) were very closely related chemically and dualism was forced to disappear as a viable theory in organic chemistry. Resolution would begin to occur only with the development of the valence concept.

1. A. Greenberg, *A Chemical History Tour,* John Wiley and Sons, New York, 2000, pp. 194–199.
2. J.R. Partington, *A History of Chemistry,* Macmillan and Co. Ltd., London, 1962, Vol. 3, p. 78.
3. Greenberg, op. cit., pp. 201–204.
4. A.J. Ihde, *The Development of Modern Chemistry,* Harper & Row, New York, 1964, p. 191.
5. W.H. Brock, *The Norton History of Chemistry,* Norton, New York, 1993, pp. 194–207.
6. J.R. Partington, *A History of Chemistry,* Macmillan and Co. Ltd., London, 1964, Vol. 4, pp. 142–177.
7. Partington (1962), op. cit., pp. 233–234.
8. Partington (1962), op. cit., p. 511.
9. Partington (1964), op. cit., pp. 253–254.
10. Ihde, op. cit., pp. 184–189.
11. E. Youmans, *Chemical Atlas, Appleton, New York, 1857.*
12. *D.H. Hey (ed.),* Kingzett's Chemical Encyclopedia, ninth edition, Baillière, Tindall and Cassell, London, 1966, p. 149.
13. Partington (1964), op. cit., pp. 283–286.
14. J. Hudson, *The History of Chemistry,* Chapman & Hall, New York, 1992, pp. 114–116.
15. Ihde, op. cit., pp. 619–621.
16. F.A. Carroll, *Perspectives on Structure and Mechanism in Organic Chemistry,* Brooks/Cole Publishing Co., Pacific Grove, CA, 1998, pp. 257–258.
17. R. Destro, T. Pilati, and M. Simonetta, *Journal of the American Chemical Society,* Vol. 100, pp. 6507–6509, 1978.
18. C.R. Arkin, B. Cowans and B. Kahr, *Chemistry of Materials,* Vol. 8, pp. 1500–1503, 1996.
19. M.J.S. Gynane, A. Hudson, M.F. Lappert, P.P. Power and H. Goldwhite, *Journal of the Chemical Society Dalton Transactions,* pp. 2428–2433, 1980.
20. P.R. Hitchcock, M.F. Lappert, and S.J. Smith, *Journal of Organometallic Chemistry,* Vol. 320, pp. C27–C30, 1987.
21. Pauling provides a value of only 32.1 kcal/mol (kilocalories per mole) for the As—As bond energy, compared to 83.1 kcal/mol for a typical C—C bond (L. Pauling, *The Nature of the Chemical Bond,* third edition, Cornell University Press, Ithaca, 1960, p. 85.) Actually, there does not appear to be data allowing a good determination or estimate of the As—As bond in "cacodyl" according to J.F. Liebman, J.A. Martinho-Simões, and S.W. Slayden, in *The Chemistry of Organic Arsenic, Antimony and Bismuth Compounds,* S. Patai (ed.), John Wiley and Sons, Chichester, 1994, pp. 153–168. Interestingly, the carbon–carbon bond in cyanogen is quite strong (134.7 kcal/mol using National Institute of Standards data—see *http://nist.gov*). In this case, as strong as the C—C bond is, it is much weaker than the carbon–nitrogen triple bonds in cyanogen that maintain the integrity of the cyano "radical."
22. Ihde, op. cit., p. 198.

MENDELEEV'S "COSMIC STAIRCASE" TO A "PYTHAGOREAN HEAVEN"

In 1945, 12-year-old Oliver Sacks was first entranced by the giant periodic table at the Science Museum in South Kensington as he examined individual samples of each element.[1] More than half a century later, a distinguished career as physician and author only seems to have enhanced his love of chemistry, metals in particular, and his ecstatic passion for the periodic table. The autobiographical *Uncle Tungsten* recounts young Sacks mentally spiraling the "sober, rectangular table" into "a sort of cosmic staircase or a Jacob's ladder, going up to, coming down from, a Pythagorean heaven."[2] The ancient Pythagoreans imagined a universe governed by pure mathematics, and Dr. Sacks perceived "Mendeleev's garden," the periodic table, as part of a cosmology vibrant with natural harmony.[1]

Dmitry Ivanovich Mendeleev (1834–1907)[3] was the youngest of 14 children. The heroism of his mother in obtaining a suitable education for him under tragic circumstances is well known.[3,4] Mendeleev obtained his master's degree at St. Petersburg in 1856, and Figures 148–150 show the title page and four additional pages from his eight-sheet (15-page) dissertation.[5] Chemist/book collector Roy G. Neville has noted hints of the periodic law in the masters dissertation[6] and Mendeleev's interest in atomic masses and the relationships between elements is certainly apparent. Almost 40 years earlier, Eil-

ПОЛОЖЕНІЯ,

ИЗБРАННЫЯ ДЛЯ ЗАЩИЩЕНІЯ НА СТЕПЕНЬ МАГИСТРА ХИМІИ

Д. Менделѣевымъ.

9 Сентября 1856 года.

САНКТПЕТЕРБУРГЪ.

ВЪ ТИПОГРАФІИ ДЕПАРТАМЕНТА ВНѢШНЕЙ ТОРГОВЛИ.

1856.

FIGURE 148. ▪ The title page of young Dmitri Mendeleev's 15-page master's dissertation (courtesy The Roy G. Neville Historical Chemical Library).

— 6 —

ковая разность въ составѣ опредѣляетъ весьма различныя величины въ разности объемовъ. Каждая изъ названныхъ теорій, изъясняя нѣсколько фактовъ, противорѣчитъ большей части ихъ.

12) Попытки Шрёдера, Гмелина, Гросгана и Дана согласить удѣльные объемы твердыхъ и жидкихъ тѣлъ съ числомъ паевъ совершенно безуспѣшны до сихъ поръ.

13) Магнитные элементы имѣютъ меньшій удѣльный объемъ, чѣмъ діамагнитные, что подтверждаетъ теорію Фелпча.

14) Предположеніе Авогардо о томъ, что электроположительныя тѣла имѣютъ бóльшій удѣльный объемъ (или его кратное), чѣмъ электро–отрицательныя — согласуется съ бóльшею частію точно извѣстныхъ фактовъ.

15) Близость кристаллическихъ формъ (изоморфизмъ и гомёоморфизмъ) независитъ отъ близости или кратности удѣльныхъ объемовъ, какъ думали Коппъ, Дана и Гунтъ; потому что:

16) a) Тѣла съ близкими формами, но неаналогическимъ составомъ (гомёоморфныя) не имѣютъ близкихъ или кратныхъ удѣльныхъ объемовъ.

17) b) Иногда и изоморфныя тѣла (т. е. близкихъ формъ и аналогическаго состава) не имѣютъ близкихъ удѣльныхъ объемовъ, что особенно ясно надъ простыми тѣлами ромбоедрической системы (по опред. Густ. Розе): осмій R (уголъ ромбоедра) $= 84°52'$,

— 7 —

$V = 63,4$; иридій $R = 84°52'$, $V = 56,7$; мышьякъ $R = 85°4'$, $V = 328$; теллуръ $R = 86°57'$, $V = 64,7$; сурьма $R = 87°35'$, $V = 115,7$ и висмутъ $R = 87°40'$, $V = 270,1$.

18) c) Многія изоморфныя тѣла имѣютъ близкіе удѣльные объемы только потому, что они сходственны между собою (по составу и свойствамъ); ибо, и безъ изоморфизма.

19) Сходственныя тѣла очень часто имѣютъ близкіе удѣльные объемы. Напримѣръ a) $Ni^2 — 42,4$; $Co^2 — 42,7$; $Cu^2 — 44,7$; $Fe^2 — 44,6$; $al^2 — 44,1$. b) $Mn^2 — 46,4$; $Cr^2 — 46,8$. c) $Ag — 128,3$; $Au^2 — 127,1$. d) $NaCl — 170$; $SrCl — 172$; $BaCl — 175$. e) $AgJ — 264$; $hgJ — 265$. f) $Pb^2SO^4 — 301,6$; $Sr^2SO^4 — 300,9$. g) $Cr^2O^3 — 234$; $V^2O^3 — 234$. h) Оловянный камень sn^2O — 63, брукитъ $ti^2O — 61$, анатазъ $ti^2O — 65$; рутилъ $ti^2O — 59$. i) $Cd^2S — 190$, $Pb^2S — 198$.

20) Многія сходственныя тѣла имѣютъ удѣльные объемы (V) постепенно увеличивающіеся съ увеличеніемъ пая (П). Напримѣръ: a) $Li^2 — V = 136$, $П — 81$; $Na^2 — V = 404$, $П = 289$; $K^2 — V = 561$, $П = 488$. b) $Be^2 — V = 41$, $П = 87$; $al^2 — V = 44$, $П = 114$; $Mg^2 — V = 88$, $П = 158$; $Ca^2 — V = 158$, $П = 250$; $Sr^2 — V = 215$, $П = 546$; $Ba — V = 231$, $П = 854$. c) $Hg^2 — V = 92$, $П = 1250$; $Pb^2 — V = 114$, $П = 1294$; $Ag^2 — V = 128$; $П = 1350$. d) $Sb^2 — V = 115,7$, $П = 778$; $Bi^2 — V = 270,1$, $П = 2660$. e) $S^6 — V = 581$, $П = 1200$; $Se^6 — V = 671$, $П = 2946$. f) $MgCl — V = 137$,

FIGURE 149. ■ Statement 17 on pages 6 and 7 of Mendeleev's master's dissertation discusses the angles measured in crystals of related substances. The Periodic Law, published by Mendeleev in 1869, would explain the similarities between isomorphic crystals, first noted by Eilhard Mitscherlich some 40 years before Mendeleev's master's dissertation was published. (Courtesy The Roy G. Neville Historical Chemical Library.)

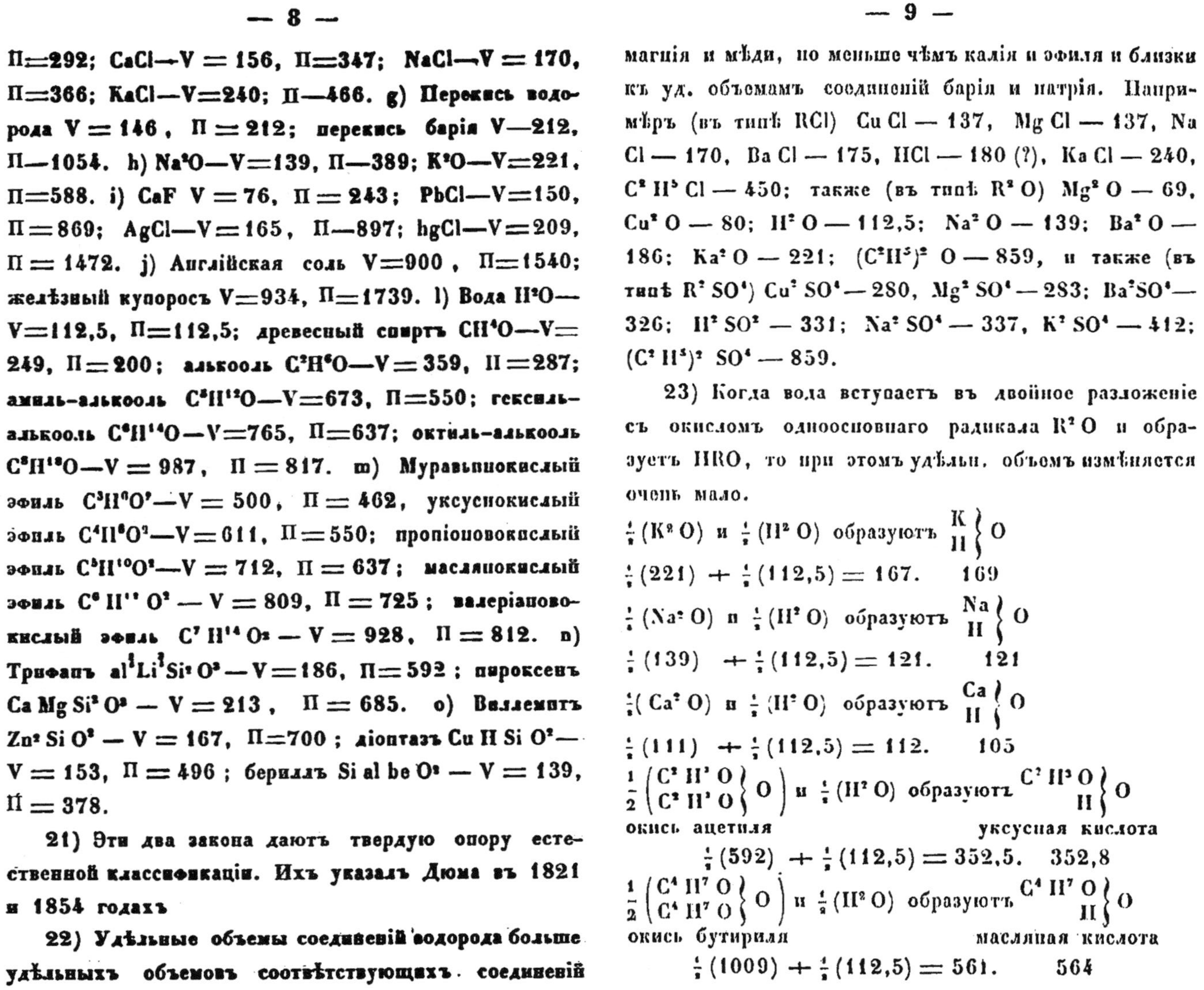

П=292; CaCl—V = 156, П=347; NaCl—V = 170, П=366; KaCl—V=240; П—466. g) Перекись водорода V = 146, П = 212; перекись барія V—212, П—1054. h) Na^2O—V=139, П—389; K^2O—V=221, П=588. i) CaF V = 76, П = 243; PbCl—V=150, П = 869; AgCl—V = 165, П—897; hgCl—V=209, П = 1472. j) Англійская соль V=900, П=1540; желѣзный купоросъ V=934, П=1739. l) Вода H^2O—V=112,5, П=112,5; древесный спиртъ CH^4O—V= 249, П=200; алкооль C^2H^6O—V=359, П=287; амиль-алкооль $C^5H^{12}O$—V=673, П=550; гексиль-алкооль $C^6H^{14}O$—V=765, П=637; октиль-алкооль $C^8H^{18}O$—V = 987, П = 817. m) Муравьинокислый эфиль $C^3H^6O^2$—V = 500, П = 462, уксуснокислый эфиль $C^4H^8O^2$—V=611, П=550; пропіоновокислый эфиль $C^5H^{10}O^2$—V = 712, П = 637; масляцокислый эфиль $C^6H^{12}O^2$ — V = 809, П = 725; валеріановокислый эфиль $C^7H^{14}O^2$ — V = 928, П = 812. n) Трифанъ $al^{\frac{4}{3}}Li^{\frac{2}{3}}Si^2O^3$—V=186, П=592; пироксенъ Ca Mg Si^2O^3 — V = 213, П = 685. o) Виллемитъ Zn^2SiO^3 — V = 167, П=700; діоптазъ Cu H Si O^3— V = 153, П = 496; бериллъ Si al be O^3 — V = 139, П = 378.

21) Эти два закона даютъ твердую опору естественной классификаціи. Ихъ указалъ Дюма въ 1821 и 1854 годахъ

22) Удѣльные объемы соединеній водорода больше удѣльныхъ объемовъ соотвѣтствующихъ соединеній

магнія и мѣди, но меньше чѣмъ калія и эфиля и близки къ уд. объемамъ соединеній барія и натрія. Напримѣръ (въ типѣ RCl) Cu Cl — 137, Mg Cl — 137, Na Cl — 170, Ba Cl — 175, HCl — 180 (?), Ka Cl — 240, C^2H^5Cl — 450; также (въ типѣ R^2O) Mg^2O — 69, Cu^2O — 80; H^2O — 112,5; Na^2O — 139; Ba^2O — 186; Ka^2O — 221; $(C^2H^5)^2O$ — 859, и также (въ типѣ R^2SO^4) Cu^2SO^4 — 280, Mg^2SO^4 — 283; Ba^2SO^4— 326; H^2SO^2 — 331; Na^2SO^4 — 337, K^2SO^4 — 412; $(C^2H^5)^2SO^4$ — 859.

23) Когда вода вступаетъ въ двойное разложеніе съ окисломъ одноосновнаго радикала R^2O и образуетъ HRO, то при этомъ удѣльн. объемъ измѣняется очень мало.

$\frac{1}{2}(K^2O)$ и $\frac{1}{2}(H^2O)$ образуютъ $\left.\begin{matrix}K\\H\end{matrix}\right\}O$

$\frac{1}{2}(221) + \frac{1}{2}(112,5) = 167.$ 169

$\frac{1}{2}(Na^2O)$ и $\frac{1}{2}(H^2O)$ образуютъ $\left.\begin{matrix}Na\\H\end{matrix}\right\}O$

$\frac{1}{2}(139) + \frac{1}{2}(112,5) = 121.$ 121

$\frac{1}{2}(Ca^2O)$ и $\frac{1}{2}(H^2O)$ образуютъ $\left.\begin{matrix}Ca\\H\end{matrix}\right\}O$

$\frac{1}{2}(111) + \frac{1}{2}(112,5) = 112.$ 105

$\frac{1}{2}\left(\left.\begin{matrix}C^2H^3O\\C^2H^3O\end{matrix}\right\}O\right)$ и $\frac{1}{2}(H^2O)$ образуютъ $\left.\begin{matrix}C^2H^3O\\H\end{matrix}\right\}O$

окись ацетиля — уксусная кислота

$\frac{1}{2}(592) + \frac{1}{2}(112,5) = 352,5.$ 352,8

$\frac{1}{2}\left(\left.\begin{matrix}C^4H^7O\\C^4H^7O\end{matrix}\right\}O\right)$ и $\frac{1}{2}(H^2O)$ образуютъ $\left.\begin{matrix}C^4H^7O\\H\end{matrix}\right\}O$

окись бутириля — масляная кислота

$\frac{1}{2}(1009) + \frac{1}{2}(112,5) = 561.$ 564

FIGURE 150. ■ Statement 20 on pages 7 (see Figure 149) and 8 deals with the relative atomic weights so vital to the development of the Periodic Law. The relative mass of water ("H^2O" on page 8), is 112.5, and this is based on the widely employed assumed relative mass = 100 for the oxygen atom. (Courtesy The Roy G. Neville Historical Chemical Library.)

hard Mitscherlich (1794–1863)[7] had originated the concept of isomorphism, noting the great similarities in form and measured angles between certain crystalline substances. For example, sodium hydrogen phosphate hydrate ($Na_2HPO_4 \cdot 12H_2O$) and the analogous arsenate ($Na_2HAsO_4 \cdot 12H_2O$) form isomorphic crystals. Crystalline ammonium sulfate [$(NH_4)_2SO_4$] and potassium sulfate (K_2SO_4) are also isomorphic (Figure 151). In the master's dissertation, the 22-year-old Mendeleev applied his interest in isomorphism to exploring re-

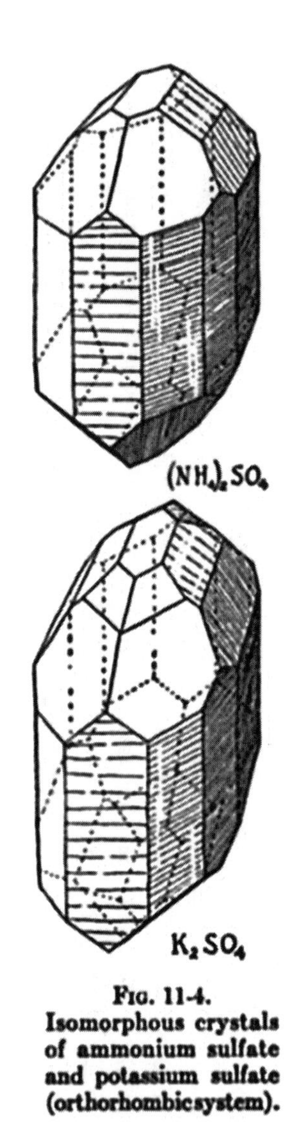

FIGURE 151. ▪ Isomorphous crystals of ammonium sulfate and potassium sulfate (illustration by artist Roger Hayward in the first edition of his friend Linus Pauling's text, *General Chemistry*, 1947, p. 202. © 1947 by Linus Pauling. Used with permission of W.H. Freeman and Company). Mitscherlich had discovered that some phosphate and arsenate crystals were isomorphic, and these observations were explained by Mendeleev as arising from arsenic and phosphorus being in the same chemical family. We now know that ammonium (NH_4^+) and potassium (K^+) ions are similarly congruent, thus accounting for the isomorphism of their sulfate crystals.

lationships between the elements. This is demonstrated in statement 17 on page 6 (Figure 149), where he compared angles in crystals of related substances. Thirteen years later (in 1869), Mendeleev would place phosphorus and arsenic in the same chemical family and thus provide an understanding of the isomorphism of the aforementioned phosphate and arsenate. We now recognize that ammonium (NH_4^+) and potassium (K^+) are monovalent, positive ions of similar size and this is the basis for the isomorphism of their sulfates (Figure 151). Statement 19 on page 7 (Figure 149) describes similarities in specific volume between chemically related substances.

In statement 20 on pages 7 and 8 (Figures 149 and 150), we observe atomic weight relationships, noted by Mendeleev, on a scale where the relative weight of oxygen is taken as 100. The numbers that follow "V" in statement 20 refer to relative volume, while the data following "II" refer to relative mass. It was well known that in water, 100 grams of oxygen were combined with 12.5 grams of hydrogen and, thus, the relative mass for "H^2O" (see page 8 in Figure 150) on this scale is 112.5. If we employ the relative mass of "Na^2O," we see that a total mass of 289 combines with 100 of oxygen. Therefore, each sodium atom has a relative mass of 289/2 or 144.5. If we multiply 144.5/100 by 16.0 (the modern atomic mass of oxygen), we would obtain a value of 23.1 for sodium, in decent agreement with the modern value.

Despite the agreements of these particular relative atomic masses with modern data, there were considerable uncertainties during the 1850s in atomic masses and equivalent weights. At the international chemical congress in Karlshuhe, Germany, during 1860 some of the greatest chemists of the day gathered to debate these issues.[8] Critical insights were provided by Stanislao Cannizzaro who enlightened the gathering by making the 50-year-old hypothesis of his fellow countryman Amedeo Avogadro accessible within the new chemical knowledge of the midnineteenth century.[8] The youthful Mendeleev took a leisurely trip to Karlsruhe accompanied by another young chemist, Aleksandr Porfirevich Borodin (1833–1887).[9] Borodin became, of course, one of the world's great composers, and his opera *Prince Igor* is still immensely popular. But this precocious musician (self-taught on the cello; he composed musical pieces at the age of 14) was also a precocious chemist (a homebuilt laboratory and youthful efforts at making explosives).[9] It is fun to imagine a film depicting the 26-year-old Mendeleev and the 27-year old Borodin making their leisurely way to Karlsruhe, delighting in the music from the giant pipe organ at Freiburg.[9] Both had powerful Russian mothers[10] who lovingly guided their gifted sons' educations and even established lodging near their colleges. Although music would be his most powerful legacy, Borodin made some fundamental discoveries in organic chemistry. Today's overburdened college sophomores who must remember that the Aldol Condensation may be followed by dehydration have Borodin to blame for this additional fact.[11]

The Karlsruhe congress settled many controversies and placed atomic masses and equivalents on firm footing in the chemical community. Attempts to organize these data were not long in coming, and early periodic tables were proposed by, among others, John Newlands (1865), William Odling (1865), and Julius Lothar Meyer (1868).[12] "Eureka moments" are actually very rare in science, and an evolution of ideas is more typical. As noted earlier, attempts to organize the elements are apparent in Mendeleev's 1856 master's thesis. Figure 152

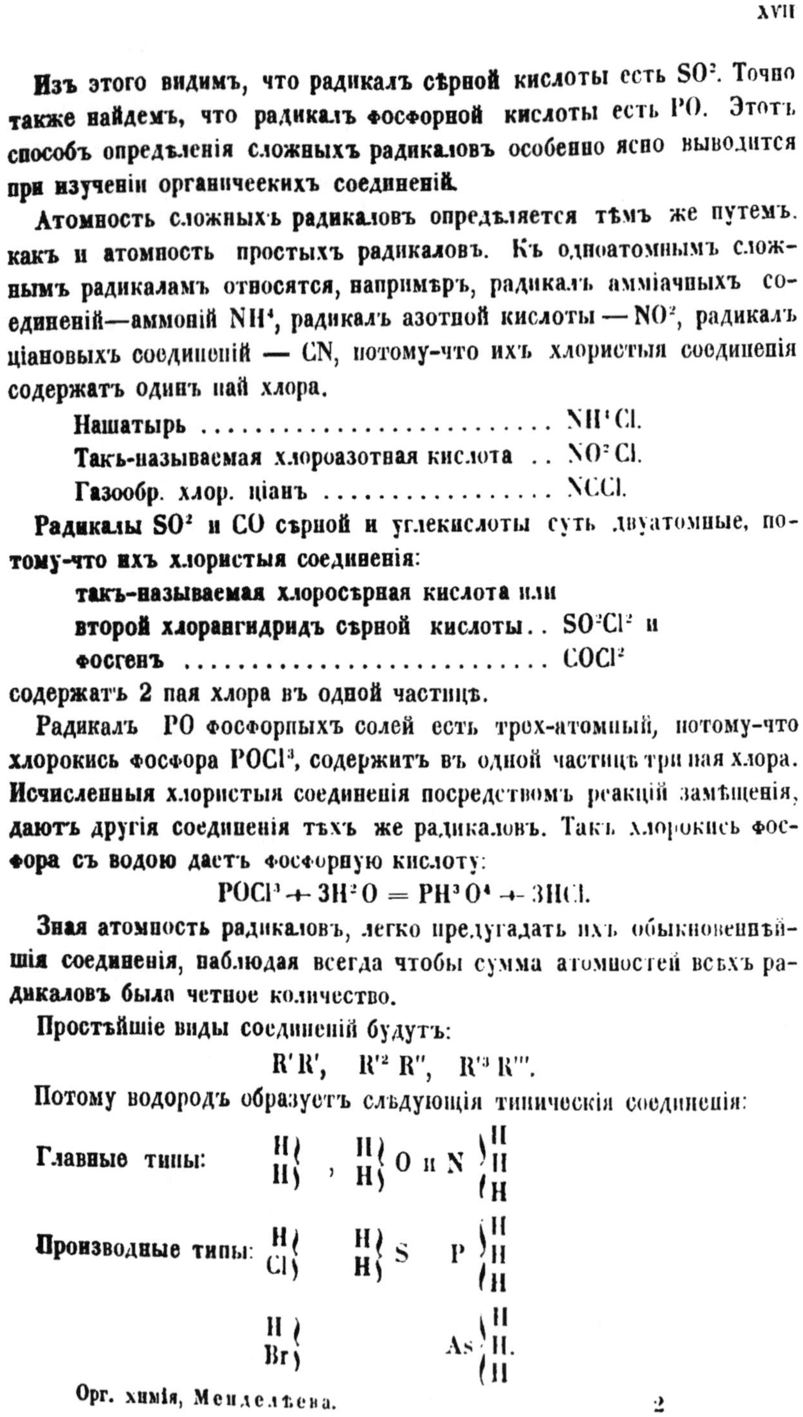

XVII

Изъ этого видимъ, что радикалъ сѣрной кислоты есть SO^2. Точно **также найдемъ, что радикалъ фосфорной** кислоты есть PO. Этотъ **способъ опредѣленія сложныхъ радикаловъ** особенно ясно выводится **при изученіи органическихъ соединеній.**

Атомность сложныхъ **радикаловъ опредѣляется тѣмъ** же путемъ, **какъ и атомность** простыхъ **радикаловъ.** Къ одноатомнымъ слож**нымъ радикаламъ** относятся, **напримѣръ,** радикалъ амміачныхъ со**единеній—аммоній** NH^4, **радикалъ азотной кислоты** — NO^2, **радикалъ** ціановыхъ соединеній — CN, потому-что ихъ хлористыя соединенія **содержатъ одинъ** пай хлора.

Нашатырь .... NH^4Cl.
Такъ-называемая хлороазотная кислота .. NO^2Cl.
Газообр. хлор. ціанъ .... $NCCl$.

Радикалы SO^2 **и** CO **сѣрной и** углекислоты суть двуатомные, по**тому-что ихъ** хлористыя **соединенія:**

такъ-называемая хлоросѣрная кислота или
второй хлорангидридъ сѣрной кислоты.. SO^2Cl^2 и
фосгенъ .... $COCl^2$

содержатъ 2 пая хлора въ одной частицѣ.

Радикалъ PO фосфорныхъ солей есть трех-атомный, потому-что **хлорокись фосфора** $POCl^3$, **содержитъ** въ одной частицѣ три пая хлора. **Исчисленныя** хлористыя соединенія посредствомъ реакцій замѣщенія, **даютъ другія соединенія тѣхъ же** радикаловъ. Такъ хлорокись фос**фора съ водою** даетъ фосфорную кислоту:

$$POCl^3 + 3H^2O = PH^3O^4 + 3HCl.$$

Зная атомность радикаловъ, легко предугадать ихъ обыкновеннѣй**шія соединенія, наблюдая** всегда чтобы сумма атомностей всѣхъ ра**дикаловъ была** четное количество.

Простѣйшіе виды соединеній будутъ:

$$R'R', \quad R'^2R'', \quad R'^3R'''.$$

Потому водородъ образуетъ слѣдующія типическія соединенія:

Главные типы: $\left.\begin{matrix}H\\H\end{matrix}\right\}$, $\left.\begin{matrix}H\\H\end{matrix}\right\}O$ и $N\left\{\begin{matrix}H\\H\\H\end{matrix}\right.$

Производные типы: $\left.\begin{matrix}H\\Cl\end{matrix}\right\}$ $\left.\begin{matrix}H\\H\end{matrix}\right\}S$ $P\left\{\begin{matrix}H\\H\\H\end{matrix}\right.$

$\left.\begin{matrix}H\\Br\end{matrix}\right\}$ $As\left\{\begin{matrix}H\\H\\H\end{matrix}\right.$.

Орг. химія, Менделѣева. 2

FIGURE 152. ▪ See the bottom of page XVII in Mendeleev's text on organic chemistry published in St. Petersburg in 1863. You will note this early organization anticipating the periodic law that Mendeleev will publish six years hence. (From Mendeleev, *Organischeskaja Khimia*, 1863.)

is from his 1863 text[13] on organic chemistry. It is thrilling to see an embryonic periodic table here.[4] These ideas truly crystallized in Mendeleev's mind in 1868 and his first periodic table (Figure 153) was published in 1869.[14] We are accustomed to seeing a "horizontal" periodic table and this one was "vertical." Mendeleev ordered the elements according to relative atomic mass and noted a periodicity in properties. Thus, the alkali metals cesium (Cs), rubidium (Rb), potassium (K), and sodium (Na) mimicked lithium (Li) as metals that would react violently, even explosively, with water and form salts with chlorine, all sharing the formula MCl. Similarly, we see listed in one family the nonmetals fluorine (F), chlorine (Cl), bromine (Br), and iodine (I), which all form MX salts,

Ueber die Beziehungen der Eigenschaften zu den Atomgewichten der Elemente. Von D. Mendelejeff. — Ordnet man Elemente nach zunehmenden Atomgewichten in verticale Reihen so, dass die Horizontalreihen analoge Elemente enthalten, wieder nach zunehmendem Atomgewicht geordnet, so erhält man folgende Zusammenstellung, aus der sich einige allgemeinere Folgerungen ableiten lassen.

			Ti = 50	Zr = 90	? = 180
			V = 51	Nb = 94	Ta = 182
			Cr = 52	Mo = 96	W = 186
			Mn = 55	Rh = 104,4	Pt = 197,4
			Fe = 56	Ru = 104,4	Ir = 198
			Ni = Co = 59	Pd = 106,6	Os = 199
H = 1			Cu = 63,4	Ag = 108	Hg = 200
	Be = 9,4	Mg = 24	Zn = 65,2	Cd = 112	
	B = 11	Al = 27,4	? = 68	Ur = 116	Au = 197?
	C = 12	Si = 28	? = 70	Sn = 118	
	N = 14	P = 31	As = 75	Sb = 122	Bi = 210?
	O = 16	S = 32	Se = 79,4	Te = 128?	
	F = 19	Cl = 35,5	Br = 80	J = 127	
Li = 7	Na = 23	K = 39	Rb = 85,4	Cs = 133	Tl = 204
		Ca = 40	Sr = 87,6	Ba = 137	Pb = 207
		? = 45	Ce = 92		
		?Er = 56	La = 94		
		?Yt = 60	Di = 95		
		?In = 75,6	Th = 118?		

1. Die nach der Grösse des Atomgewichts geordneten Elemente zeigen eine stufenweise Abänderung in den Eigenschaften.
2. Chemisch-analoge Elemente haben entweder übereinstimmende Atomgewichte (Pt, Ir, Os), oder letztere nehmen gleichviel zu (K, Rb, Cs).
3. Das Anordnen nach den Atomgewichten entspricht der *Werthigkeit* der Elemente und bis zu einem gewissen Grade der Verschiedenheit im chemischen Verhalten, z. B. Li, Be, B, C, N, O, F.
4. Die in der Natur verbreitetsten Elemente haben *kleine* Atomgewichte

FIGURE 153. ▪ The first version of Mendeleev's 1869 periodic table; note the question marks in some places, notably following aluminum and silicon. The daring aspect of Mendeleev's table was the presence of these gaps. *Eka*-aluminum (i.e., gallium) and *eka*-silicon (i.e., germanium) were predicted by Mendeleev to exist and were discovered shortly thereafter. This is one of the most excellent illustrations of the power of the scientific method in human history. (From *Zeitschrift für Chemie;* with permission from The Edgar Fahs Smith Collection.)

(a)

Tabelle I.

			K = 39	Rb = 85	Cs = 133	—	—
			Ca = 40	Sr = 87	Ba = 137	—	—
			—	?Yt = 88?	?Di = 138?	Er = 178?	—
			Ti = 48?	Zr = 90	Ce = 140?	?La = 180?	Th = 231
			V = 51	Nb = 94	—	Ta = 182	—
			Cr = 52	Mo = 96	—	W = 184	U = 240
			Mn = 55	—	—	—	—
			Fe = 56	Ru = 104	—	Os = 195?	—
	Typische Elemente		Co = 59	Rh = 104	—	Ir = 197	—
			Ni = 59	Pd = 106	—	Pt = 198?	—
= 1	Li = 7	Na = 23	Cu = 63	Ag = 108	—	Au = 199?	—
	Be = 9,4	Mg = 24	Zn = 65	Cd = 112	—	Hg = 200	—
	B = 11	Al = 27,3	—	In = 113	—	Tl = 204	—
	C = 12	Si = 28	—	Sn = 118	—	Pb = 207	—
	N = 14	P = 31	As = 75	Sb = 122	—	Bi = 208	—
	O = 16	S = 32	Se = 78	Te = 125?	—	—	—
	F = 19	Cl = 35,5	Br = 80	J = 127	—	—	—

der chemischen Elemente. 149

(b)

Tabelle II.

Reihen	Gruppe I. — R^2O	Gruppe II. — RO	Gruppe III. — R^2O^3	Gruppe IV. RH^4 RO^2	Gruppe V. RH^3 R^2O^5	Gruppe VI. RH^2 RO^3	Gruppe VII. RH R^2O^7	Gruppe VIII. — RO^4
1	H=1							
2	Li=7	Be=9,4	B=11	C=12	N=14	O=16	F=19	
3	Na=23	Mg=24	Al=27,3	Si=28	P=31	S=32	Cl=35,5	
4	K=39	Ca=40	—=44	Ti=48	V=51	Cr=52	Mn=55	Fe=56, Co=59, Ni=59, Cu=63.
5	(Cu=63)	Zn=65	—=68	—=72	As=75	Se=78	Br=80	
6	Rb=85	Sr=87	?Yt=88	Zr=90	Nb=94	Mo=96	—=100	Ru=104, Rh=104, Pd=106, Ag=108.
7	(Ag=108)	Cd=112	In=113	Sn=118	Sb=122	Te=125	J=127	
8	Cs=133	Ba=137	?Di=138	?Ce=140	—	—	—	— — — —
9	(—)	—	—	—	—	—	—	
10	—	—	?Er=178	?La=180	Ta=182	W=184	—	Os=195, Ir=197, Pt=198, Au=199.
11	(Au=199)	Hg=200	Tl=204	Pb=207	Bi=208	—	—	
12	—	—	—	Th=231	—	U=240	—	— — — —

11*

der chemischen Elemente. 151

FIGURE 154. ▪ Mendeleev's 1871 versions (actually published in 1872) of his periodic tables in (a) vertical form, and (b) horizontal form (from *Annalen der Chemie und Pharmazie*, 1872).

such as NaCl, with the alkali metals. It is noteworthy that fluorine, known to be a unique element as it could not be separated from its compounds, would first be isolated as an element by Henri Moissan in 1886, almost 20 years after Mendeleev's first table. In Figure 154a we see Mendeleev's vertical periodic table published in German in 1872.[15] It is obvious that he was still "playing" with his arrangement and this one has eight columns ("periods") and 17 rows compared to six columns and 19 rows in the 1869 version (Figure 153). Figure 154b shows the 1872 version[15] of Mendeleev's horizontal periodic table, first published in 1871.[3]

The truly singular and daring aspect of Mendeleev's periodic table was the gaps deliberately left for elements yet undiscovered. Prominently featured in Figure 153 is the heavier element in the aluminum family ("eka-aluminum") and that in the silicon family ("*eka*-silicon"). Mendeleev predicted the existence of these elements as well as their properties. Within only six years, the first of these, gallium, was discovered by Paul Émile (dit François) Lecoq de Boisbaudran, and 11 years later the other, germanium, by Clemens Alexander Winkler.[3] Scientific theories are developed to explain natural phenomena and can really be tested only by the predictions they make—the more daring, the better. Mendeleev's prediction of the existence and properties (chemical *and* physical) of hitherto unknown (and unsuspected) elements ranks among the most powerful of all scientific achievements.

---

1. O. Sacks, *Uncle Tungsten—Memories of a Chemical Boyhood*, Alfred A. Knopf, New York, 2001.
2. Sacks, op. cit., pp. 187–211.
3. J.R. Partington, *A History of Chemistry*, Macmillan and Co. Ltd., London, Vol. 4, 1964, pp. 891–899. Partington notes here that the Russian chemist spelled his name "Mendeleeff" when he signed the register of the Royal Society and Ramsay recommended "Mendeléeff." We will employ the more commonly used transliteration "Mendeleev".
4. A. Greenberg, *A Chemical History Tour*, John Wiley & Sons, New York, 2000, pp. 214–216.
5. D.M. Mendeleev, *Polozhenija, izbrannya dlja zachschishchenija na stepen' magistra khimii*, St. Petersburg, 1856. I am grateful to The Roy G. Neville Historical Chemical Library (California) for supplying these images.
6. The Roy G. Neville Historical Chemical Library (California), catalog in preparation. I am grateful to Dr. Neville for making me aware of his views on Mendeleev's early thoughts on the relationships between elements manifested in his master's thesis.
7. Partington, op. cit., pp. 205–211.
8. Greenberg, op. cit., pp. 212–214.
9. S.A. Dianin, *Borodin*, Oxford University Press, London, 1963, pp. 12–13; 22–29.
10. The roots of my mother, Bella Greenberg, are similarly Russian. She first took me as a young child to see the Halls of Dinosaurs at the American Museum of Natural History, encouraged my early interest in reading, and gently helped remove hundreds of newborn praying mantids from the wall of my bedroom and placed them in the backyard (see the essay "A Natural Scientist" in the Epilogue of this book).
11. C.C. Gillispie (ed.), *Dictionary of Scientific Biography*, Charles Scribner's Sons, New York, Vol. II, 1970, pp. 316–317.
12. Partington, op. cit., pp. 883–891.
13. D.M. Mendeleev, *Organischeskaja Khimia*, St. Petersburg, 1863, p. xvii.
14. Partington, op. cit., p. 895.
15. D. Mendelejeff, *Annalen der Chemie Und Pharmacie*, VIII. Supplementband, Leipzig und Heidelberg, 1872, pp. 133–229. (This paper was received from St. Petersburg in August 1871).

THE ELECTRIC OXYGEN

Renewal and rebirth—the bracing "electric" aroma of the seaside air following a thunderstorm; the evocative smell of an electric train set bringing back warm memories of childhood; even the New York City subway; traces of gaseous ozone, produced by electric arcs and sparks, remain in our memories. This "electric oxygen" has been used for over 100 years to purify drinking water and remove unpleasant odors. The century-old bottle in Figure 155 proclaims "Ozone is life" and probably contained treated water from which the ozone was long gone. And why not use this magical-sounding name to advertise a product having nothing

FIGURE 155. ▪ Renewal and rebirth, the bracing aroma of the sea, and purifier of water—the lore of ozone was quickly marketed. Here we see a century-old "Ozone Is Life" bottle for ozone-treated water produced in Canada. We have filled the bottle with life-giving milk in harmony with this theme (and also to improve photographic contrast). (Photograph courtesy of Ms. Susan J. Greenberg.)

whatsoever to do with ozone—say, soap (Figure 156)? Ozone Soap—rebirth and renewal—flowers; a happy, healthy baby.

It is truly remarkable that most of the significant facts about ozone, including its formula (O_3), were known by about 1872. Ozone is an allotrope of oxygen. Allotropes are different forms of an element in the same state. Diamond, graphite, and fullerenes (such as the "soccer ball" C_{60}) are carbon allotropes just as red phosphorus and white phosphorus are allotropes. However, ozone is typically found at only ppm (parts per million) levels in air [1 cubic meter of air contains about 1 mg (milligram) of ozone]. Moreover, it is much higher in energy than oxygen, is a much more reactive substance (stronger oxidizer), and readily decomposes thermally or after absorbing ultraviolet (UV) light.[1] When concentrated in liquid form, it is explosive. So how was so much learned so early?

In 1785, a decade following the discovery of dephlogisticated air (oxygen) by Scheele and then Priestley, M. van Marum sparked a sample of this gas trapped in a tube above mercury.[2,3] He observed that the mercury in contact with this electrified oxygen tarnished.[4] In contrast, "plain oxygen" typically reacts with mercury at temperatures exceeding 300°C.[5] He further noted that the gas in the tube possessed the characteristic sulfurous odor that was already associated with electricity. However, it took over 50 years for Christian Friedrich Schönbein to postulate a distinct new substance and provide a name (derived from *ozon* = "I smell") and another 20 years to understand that it was composed solely of the element oxygen.[2,3] Indeed, in the latter part of the nineteenth century ozone–oxygen mixtures were liquefied[5] and concentrated by fractional distillation from oxygen. The compression of samples containing concentrated ozone had to be done with great care since heating upon compression could cause an explosion.[2] Pure ozone is a deep blue, explosive liquid boiling at –112°C.[1]

Figure 157 depicts a late-nineteenth-century ozone generator.[2] *BB* is an iron tube through which cold water is passed (tube CC). Glass cylinder *AA* has a slightly larger diameter than *BB*, and the small space between these two cylinders is filled with oxygen introduced through tube *DD*. Part of the outer (glass) cylinder is covered with tinfoil (GG). The outer tin jacket and the inner iron tube are connected at points *E* and *F* to an induction coil. This apparatus was designed to produce a "silent discharge" since sparks are also known to decompose ozone to oxygen. The early characteristic test for ozone was its ability to turn potassium iodide/starch-impregnated filter paper blue:[2,3]

$$O_3 + 2KI + H_2O = O_2 + I_2 + 2\ KOH \qquad \text{(iodine–starch complex = blue)} \quad (1)$$

It is noteworthy that this reaction and other many other reactions of ozone produced oxygen as a by-product (1 molecule of gas yielding 1 molecule of gas—i.e., no volume change). However, it was known that components in turpentine "absorbed" ozone completely:[3,6]

$$O_3 + \text{turpentine} = \text{(turpentine–ozone "compound")} \quad (2)$$

The ozone generator in Figure 157 does not produce complete conversion to oxygen, since decomposition to starting material is very significant. Ten percent ozone/oxygen mixtures are usually attained.[1] So the problem becomes one of es-

FIGURE 156. ■ Early advertising cards for Ozone Soap. See color plates. While we strongly doubt that even the minutest traces of ozone were ever to be found in Ozone Soap, it was still a wonderfully evocative trade name.

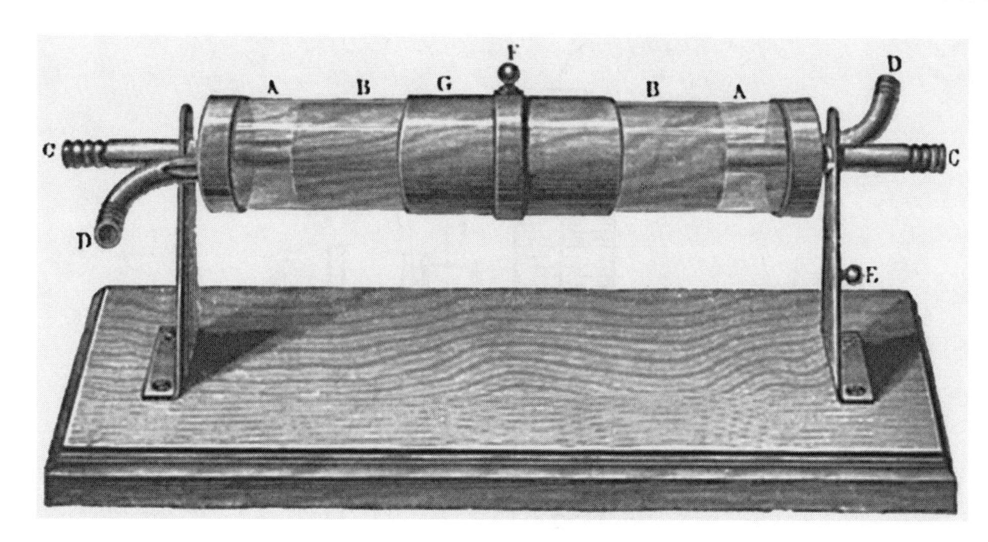

FIGURE 157. ▪ Illustration of a late-nineteenth-century ozone generator. The apparatus was designed to expose flowing oxygen gas to a "silent discharge" of electricity since sparks decompose ozone (1 mol O_3) back to oxygen (1.5 mol O_2). (From Roscoe and Schorlemmer, *A Treatise on Chemistry*, 1894.)

tablishing the formula of this highly reactive substance, present in only 10% quantity:

$$2O_3 = 3O_2 \quad (3)$$

The problem was solved by J.L. Soret in 1872 using the apparatus schematically depicted in Figure 158.[2] The solution in the concentric vessel on the left-hand side of Soret's apparatus contains dilute sulfuric acid or copper sulfate with a wire dipping into it. (Not shown is the bath vessel containing another wire dipped into ice water; the wires are connected to a source of electricity.) Oxygen is introduced into the concentric space, which also contains a sealed thin glass tube filled with turpentine. There is airtight communication of this glass tube with the outside so that it may be broken when desired. A manometer containing concentrated sulfuric acid with indigo dye is placed in series with the left-hand vessel.

Thus, if a 10% ozone mixture is produced, complete absorption by turpentine will reduce 100 cm^3 (cubic centimeters) of gas to 90 cm^3. On the other hand, if this 10% mixture is heated in order to decompose ozone to oxygen, the 10 cm^3 of ozone present yields 15 cm^3 of oxygen for a total gas volume of 105 cm^3. In other words, if reactions (2) and (3) hold (i.e., the formula of ozone is O_3), then the diminution of volume upon complete reaction with ozone must be twice the expansion of volume upon thermal decomposition of ozone. The reproducibility of this experiment was verified and allowed the assignment of the formula despite its relatively low abundance in the mixture.

Since the concept of valence was more than a decade old and oxygen was assigned the valence 2, the most reasonable structural formula in 1872 was **1** (remember, gentle reader, there was no octet rule or lone pairs of electrons and no appreciation of ring strain in 1872). This structure was favored at least through the late 1920s.[3] However, experimental structural chemistry of solids (via X-ray

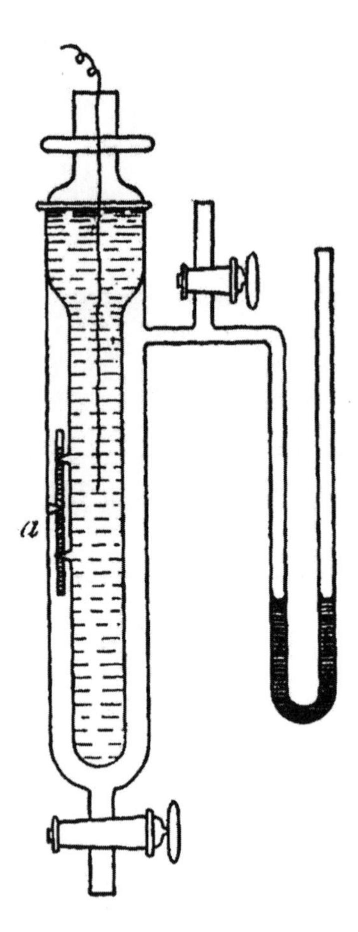

FIGURE 158. ▪ Schematic of the apparatus used by J.L. Soret in 1872 to elucidate the formula (O_3) of ozone—see text for explanation (from Partington, *Everyday Chemistry*, 1929, Macmillan and Co., Ltd, London). It is quite amazing that the formula of this highly energetic, highly reactive trace component of the atmosphere was solved more

diffraction) and gases (via electron diffraction) began in the 1920s and 1930s, and ozone was found to be a bent molecule[1] (not an equilateral triangle) with an O—O—O angle of almost 117° and O—O bond lengths (1.28 Å) intermediate between single (1.49 Å) and double (1.21 Å). This structure was nicely rationalized by the resonance theory of Pauling in the 1930s as a hybrid of the two canonical Lewis structures, **2A** and **2B** (which obey the octet rule):[7]

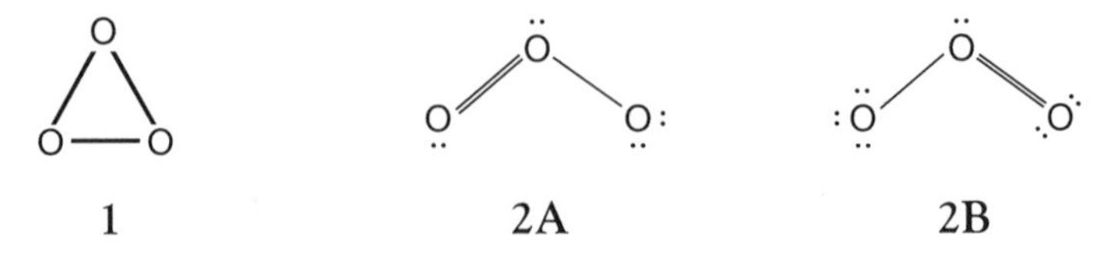

Just as the ozone molecule is a resonance hybrid of two contributing (though identical) Lewis structures, so are ozone's properties a kind of hybrid of our conventional views of "good" and "bad." Ozone remains today an effective drinking water purification agent. However, its concentration in the lower atmosphere in human-made smog poses a significant health risk, particularly to asthmatics and the elderly. On the other hand, stratospheric ozone absorbs harmful ultraviolet light and lowers our risk of skin cancer. We have great cause to be concerned about the current decrease in stratospheric ozone caused by chlorofluorocarbons (CFCs) long used in aerosol cans. So to quote from a rock-and-roll era song—"Ozone," are you "Devil or Angel?"[8]

---

1. *F.A. Cotton and G. Wilkinson, Advanced* Inorganic Chemistry, fifth edition, John Wiley & Sons, New York, 1988, pp. 452–454.
2. H.E. Roscoe and C. Schorlemmer, *A Treatise On Chemistry*, Vol. 1, Macmillan and Co., London, 1894, pp. 235–243.
3. J.R. Partington, *Everyday Chemistry*, Macmillan and Co., Ltd, London, 1929, pp. 285–288.
4. This is mercury(I) oxide (Hg_2O). Heat will disproportionate it to Hg and HgO.
5. The liquefaction of "permanent gases" will be described in the essay on the discovery of neon.
6. Ozone reacts with alkenes (olefins) through addition to the double bond. The initially formed ozonides decompose further. Schõnbein appears to have performed the first such identified ozonolysis (on ethylene). Turpentine is a complex mixture of olefinic terpenes (see P.S. Bailey, *Ozonation in Organic Chemistry*, Academic Press, New York, 1978, pp. 1–4, for a brief historical perspective).
7. L. Pauling, *The Nature of the Chemical Bond*, Cornell University Press, Ithaca, 1939.
8. Or is ozone simply an oxymoron—"a human-made, naturally occurring, environment-protecting hazardous pollutant?" Actually, "oxygen" is derived from "acid maker"—Lavoisier's incorrect conclusion that all acids contain oxygen. Acidic properties are considered by early definition to be "sharp." "Oxymoron" means literally "sharp–dull"—an internal contradiction in a single word. But this would hardly surprise the ancients, who recognized contraries (male–female, good and evil) in all earthly things.

"CHEMISTRY COMPRESSED"

Chemia Coartata (Figure 159) is Latin for "chemistry compressed."[1] This book's "singular" oblong shape, which one reviewer noted "more nearly approaches that of a cheque-book than of any ordinary volume,"[2] allows long horizontal tables starting on the verso of one leaf and ending on the *recto* of the next leaf. It is a surprisingly uncommon book—either its "jazzy" title didn't stimulate many sales, or its devoted adherents were buried with their copies so that they could cope with the chemical rigors of their final destinations. From the Preface,[2] "The main object of the author has been to compress into as small a space as possible everything connected with the study that deserves attention, and to give no more explanatory matter than is actually required to render each subject perfectly intelligible." Indeed, I had a friend in junior high school who could have even further "coartated" this entire 111-page book onto one small sheet of paper just before the chemistry final exam (see "A Natural Scientist," the first essay in the Epilogue of the present book). But I digress. According to the author Kollmyer the target audiences include[3]

1. Students intending to present themselves for examinations ("to whom, as a rule, we should offer the general advice—'don't!'" sayeth one book reviewer in 1876)[2]

CHEMIA COARTATA;

OR,

THE KEY TO MODERN CHEMISTRY

BY

A. H. KOLLMYER, A.M., M.D.

PROFESSOR OF MATERIA MEDICA AND THERAPEUTICS AT THE UNIVERSITY OF BISHOP'S COLLEGE; PROFESSOR OF MATERIA MEDICA AND PHARMACY AT THE MONTREAL COLLEGE OF PHARMACY; AND LATE PROFESSOR OF CHEMISTRY, &c.

PRINTED AND PUBLISHED BY

J. STARKE & CO., 54 ST. FRANCOIS XAVIER STREET

MONTREAL, CANADA.

—18—

SUBSTANCE.	SYNONYMS.	HISTORY.	OBTAINED FROM.	EQUATIONS.
HI	Hydrogen Iodide. Hydriodic Acid. Iodohydric Acid.		(1) Phosphorus, Iodine, Water, Glass. (2) Iodine, Sulphuretted Hydrogen & Water.	(1) $P_4+I_{10}+16H_2O=4H_3PO_4+10HI$. The glass only moderates the action of the P upon the I. (2) I_2 suspended in $H_2O+H_2S=S$ deposited and 2HI remains in solution.
NI.	Nitrogen Iodide.		Iodine and Solution of Ammonia.	$6I+4NH_3=3NH_4I+NI_3$
Bromine	*Bromos*, a bad odor.	Balard. 1826 It exists in sea water as Magnesium Bromide. It is also found in certain saline springs as in that of Kreutznack, in Prussia.	Sea water. Chlorine. Ether. Solution of Potash. Manganese dioxide. Sulphuric'.	Evaporate sea water and remove the less soluble salts, pass Cl through to free the Br. from the Magnesium; add Ether which brings the Br. to the surface, separate the Ethereal solution by a pipette; add the KHO and evaporate to dryness; then add the MnO_2 and H_2SO_4 (a) $MgBr_2+Cl_2=MgCl_2+Br_2$ (b) $6KHO+3Br_2=3H_2O+KBrO_3+5KBr$ (c) $2KBr+MnO_2+2H_2SO_4=K_2SO_4+MnSO_4+2H_2O+Br_2$
Fluorine	*Fluo* to flow.	Exists as Fluor-spar.	It is used as a flux.	Has never been isolated.
HF	Hydrogen Fluoride. Hydrofluoric Acid. Fluohydric Acid.	Scheele.	Fluor Spar and Sulphuric'.	$CaF_2+H_2SO_4=CaSO_4+2HF$
H_2SiF_6	Hydrofluosilicic Acid. Hydrogen Fluosilicate		Fluor Spar, Sulphuric', Glass, and Water.	(a) $CaF_2+H_2SO_4=CaSO_4+2HF$ (b) $4HF+SiO_2=2H_2O+SiF_4$ (c) $3SiF_4+4H_2O=SiH_4O_4+2H_2SiF_6$

—19—

PROPERTIES.			TESTS.
Sp. gr.	4.443	An acid gas, fumes, soluble in water. The solution spoils in a very short time, the Iodine being set free.	Two oxides of Iodine exist, but are unimportant; they are I_2O_5 and I_2O_7. Compounds are also formed with Nitrogen and Chlorine.
		A dark powder, explodes, often spontaneously.	The slightest touch will cause it to explode even under water.
Symb. Comb. weight Sp. gr. " " vapor 1 litre weighs	Br. 80 2.976 5.41 6.99 grms.	A brownish-red liquid, suffocating odor, produces symptoms of Influenza or Catarrh for several days after it has been inhaled. It is poisonous, bleaches, is disinfectant, soluble in 33 parts of water, but freely in Alcohol, Ether, and Chloroform; freezes at—7°C (19°F.), boils at 63°C (143°F.) Ether will separate Bromine from any of its watery solutions and rises to the surface of the liquid with it, acquiring a yellow or red color from it; Chloroform does the same, but sinks to the bottom of the fluid. It stains the skin yellow.	1. Silver Nitrate gives a light *yellow* precipitate, but sparingly soluble in Ammonia. 2. Solution of Starch gives an *orange* Bromide of Amiden. 3. Antimony burns in it. 4. Phosphorus and Potassium unite with it with explosive violence. The compounds with Oxygen are unimportant Hydrobromic' may be prepared like Hydriodic'. Bromic' and Hypobromous' are prepared by the same means adopted for getting the corresponding Chlorine compounds.
Symb. F.	Weight 19	Supposed to be a gas. Theoretical density 1.313	It attacks and destroys all vessels used to obtain it.
Sp. gr.	1.0609	A corrosive fluid, fumes very irritating, used in the arts to etch on glass. If dropped upon the skin it produces deep and extensive ulcers.	Its etching on glass is a delicate test for its presence. No compound of Fluorine and Oxygen is known.
		The solution is separated by filtration and is kept to precipitate the Salts of Potash.	Any salt of Potash is thrown down by this acid as an insoluble *Hydrofluosilicate.*

FIGURE 159. ■ One of the tables from the rare 1875 *Chemia Coartata* ("Chemistry Compressed"). No, Virginia, it isn't possible to cram for a chemistry final or "refresh" (readers') "memories without doing so at the expense of their own engagements."

2. Persons who have learned the *old* notation and wish to become acquainted with the *modern system*
3. Those who desire to keep themselves posted on this subject, and who can thus easily refresh their memories without doing so at the expense of their other engagements

This last promise has the appeal of one of those late-night ads that promise weight loss while you eat anything you want as you watch TV. *Plus ça change, plus c'est la même chose.*

It's interesting to examine the entry on fluorine (Figure 159), only 11 years before its isolation in 1886 by Henri Moissan. There were tantalizing hints of a new element over a century earlier when Carl Wilhelm Scheele reported, in 1771, the results of adding fluorspar (calcium fluoride, CaF_2) to oil of vitriol (sulfuric acid) followed by distillation. The remarkable result was the corrosion of the glass distilling retort (by the hydrofluoric acid, HF, formed) and consequent formation of a gas (SiF_4) that would yield gelatinous silica upon contact with water.[4] It still took another 115 years for Moissan to finally free fluorine, using electrochemistry, from any other chemical partner.[5]

---

1. "Coartate" is a variation of "coarctate" ("compressed") *The Oxford English Dictionary*, second edition, Clarendon Press, Oxford, 1989, Vol. 11, pp. 391–392.
2. *Chemical News*, Vol. 34 (Dec. 22, 1876), pp. 271–272. I am grateful to Professor William D. Williams for this reference and helpful discussions about this book.
3. A.H. Kollmyer, *Chemia Coartata; or the Key to Modern Chemistry*, J. Starke & Co., Montreal, 1875, Preface.
4. J.R. Partington, *A History of Chemistry*, Macmillan & Co., Ltd., London, 1962, p. 214.
5. Partington, op. cit., Vol. 4, 1964, pp. 911–915.

LÆVO-MAN WOULD ENJOY THE "BUZZ" BUT NOT THE TASTE OF HIS BEER

Louis Pasteur's brilliant insight that the chirality ("handedness") of crystals has its origins in the underlying molecular structure stood unexploited for a quarter of a century.[1] However, following the totally independent postulations by Joseph Achille Le Bel and Jacobus Henricus van't Hoff of tetrahedral carbon in 1874,[1] chemical investigator's moved rapidly into the third dimension. Pasteur had first discovered in 1848 that some crystalline substances were "chiral" or "handed" (like a left hand and a right hand in that some crystals were mirror images that could not be superimposed). When he dissolved "right-handed" and "left-handed" crystals of potassium ammonium tartrate, derived from winemaking, in separate vessels of water, the resulting transparent solutions were optically active in equal but opposite senses. Pasteur concluded that this "handedness" or dissymmetry was present in the molecules that made up the crystal but were free in solution. This abstract idea was formulated some ten or so years before the concept of valence, so Pasteur could not have had a clue about the rules governing how atoms were connected.

The next 25 years witnessed one of those rapid and dramatic revolutions so often seen in science—the concept of valence explained formulas and isomers; the realization that chemical behavior is often governed by molecular structure and finally the expansion of chemical theory into the third dimension. Figure 160 is derived from the first German edition (1877)[2] of van't Hoff's work on stereochemistry and shows some of the cutouts for his "fold your own" cardboard molecular models—a kind of "molecular origami" consistent with the artistic aspirations of the book you are *now* reading. The top two forms in Figure 160 (*Fig.* 39 and *Fig.* 40) are cutouts for two tetrahedra in which the corners are painted different colors (red, blue, white, and yellow), each color representing a different type of atom or group of atoms bonded to a central carbon (*Cwrsb*). *Figures 41* and *42* are cutouts for two tetrahedra in which the four triangular faces, rather than the corners, are painted four different colors. Assemble these two pairs of figures, and you will discover that they form two pairs of three-dimensional mirror-image figures that are nonsuperimposable and therefore chiral or "handed." (Color photographs of van't Hoff's original cardboard models may be found in the attractive book co-authored by Edgar Heilbronner and Jack D. Dunitz.[3]) Actually, the two pairs of figures shown in two dimensions at the top of Figure 160 themselves are obviously mirror images that cannot be superimposed by any movement or rotation in *two* dimensions. In "Flatland,"[4] they would be nonsuperimposable mirror images [as would be two-dimensional (2D) tracings on paper of your left and right hands]. If the colored areas were the same on each side of these four 2-D figures, then we three-dimensional (3D) people could cut one out, rotate it by 180° and superimpose it on the related figure. Similarly, if each of your hands were identical top and bottom (10 knuckles per hand—a "benefit" allowing delivery of both forehanded and backhanded punches), the 2D mirror images would be superimposable in three dimensions. The same principle indicates that we spatial people cannot superimpose two (3D) tetrahedra of opposite chirality or, for that matter, our left and right hands. However, a person living in four-dimensional space could clearly have lots of fun with us. For starters, it might be fun to imagine a right-handed barber setting his scissors down for a brief moment, having them projected into the fourth dimension, and returned to him in a flash as left-handed scissors. The results would probably not amuse the barber or the client.

Figure 161 is from a series of four articles published in 1901 for the purpose of enlightening those working in the arts and manufacturing about the breakthroughs in stereochemistry.[5] The author, William Jackson Pope, was Professor of Chemistry at Cambridge, and a remarkable contributor to the field of stereochemistry.[6] The three pictures in Figure 161a depicts the three-dimensional, tetrahedral structure of methane,[5] and the tetrahedral structures of the two enantiomers (nonsuperimposable mirror images) of lactic acid (the central carbon sits in the center of the tetrahedron).[5] The central carbon in lactic acid is bonded to four different substituents (atoms or groups of atoms). This asymmetric carbon center is a sufficient although not necessary condition for chirality. A helix (e.g., a spring or a screw) is also "handed."

Lactic acid played a central role in the development of stereochemistry.[1] It was first isolated by Scheele from fermented milk in 1770. Berzelius isolated lactic acid from muscles in 1807. Following the development of polarimetry in the early nineteenth century, Scheele's lactic acid was found to be optically inactive

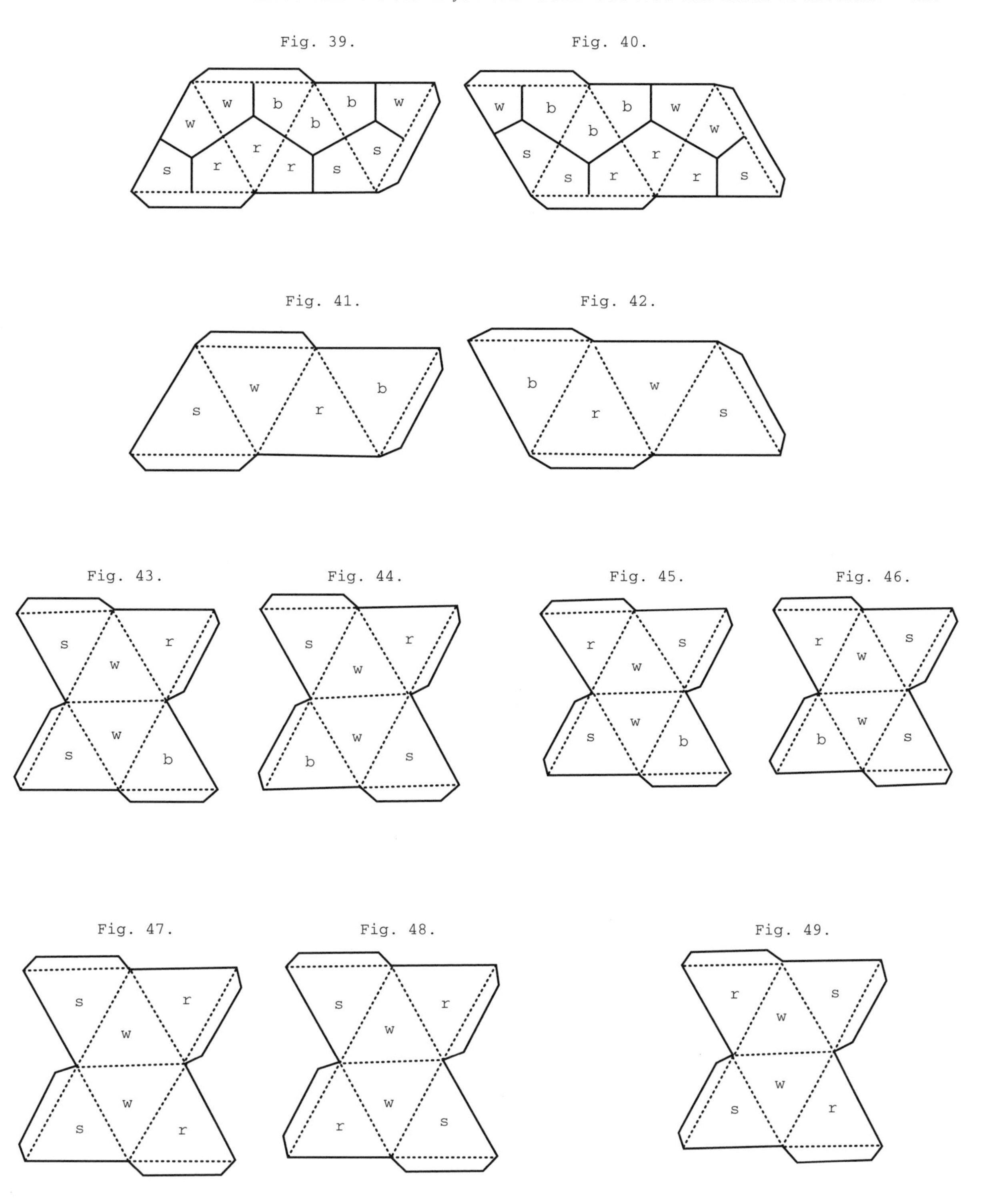

FIGURE 160. ■ Cutouts for paper or cardboard molecular models from the first German edition of van't Hoff's *The Arrangement of Atoms in Space* (*Die Lagerung Der Atome Im Raume*, 1877). Color photographs of the assembled models are found in Heilbonner and Dunitz, *Reflections on Symmetry*, 1993).

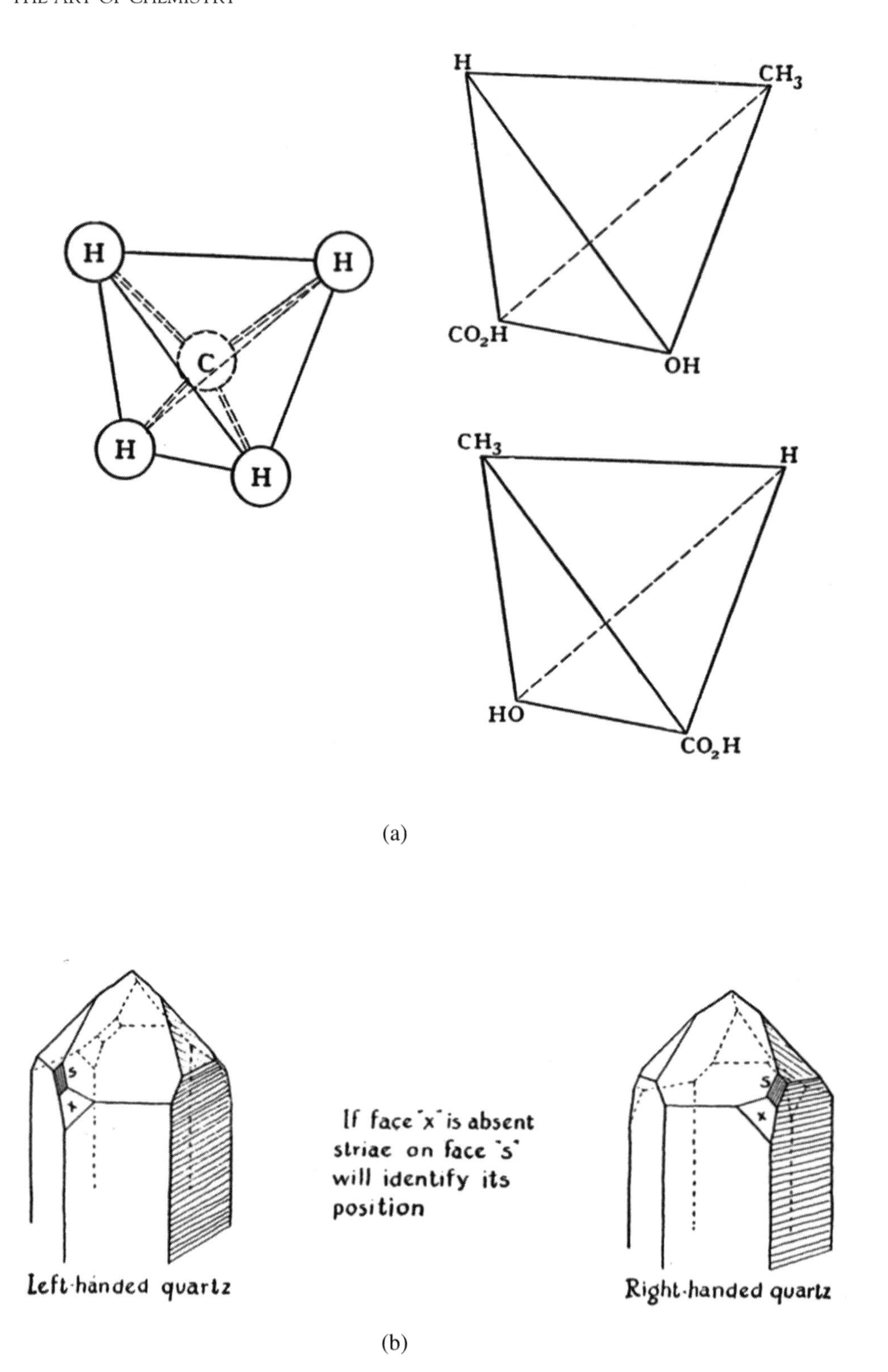

FIGURE 161. ▪ (a) Depictions of tetrahedral methane as well as the nonsuperimposable mirror images (enantiomers) of lactic acid (from Pope, *Journal of the Society of the Arts*, 1901); (b) enantiomeric crystals of quartz, arising from its hidden helical structure. (From *General Chemistry* by Linus Pauling © 1947 by Linus Pauling. Used with the permission of W.H. Freeman and Company.)

while Berzelius' lactic acid, identical with Scheele's in all other respects, was optically active. Van't Hoff and LeBel both explained these phenomena by postulating that Berzelius' lactic acid, derived from muscles, contained only one enantiomer while Scheele's lactic acid was the racemic mixture—both enantiomers in exactly equal quantities.

Pasteur had first noted the chirality of crystalline potassium ammonium tartrate, an organic substance, which he laboriously separated by hand into "left-handed" and "right-handed" crystals. Many naturally occurring minerals also exhibit macroscopic "handedness," and this was duly noted by Pasteur. The two structures in Figure 161 are drawings of "left-handed" and "right-handed" crystals of quartz.[7] Pasteur also discovered that while the physical and chemical properties of enantiomers appeared to be identical in virtually all ways, yeasts, molds, and bacteria could differentiate them. Thus, he incubated racemic tartaric acid with *Penicillum glaucum* and found that the mold "resolved" the mixture by metabolizing one enantiomer while passing the other.[5] Pasteur thus realized that the chemistry of life was chiral even though he could understand it only in the most abstract manner. Tartaric acid contains two attached asymmetric carbons. In principle, such an arrangement could allow 2 × 2 or four stereoisomers. *Figures 43–46* in Figure 160 provide molecular models that can be assembled to provide all four stereoisomers of a molecule of generalized formula srwC-Cwbs. However, only three stereoisomers (*Figs. 47–49*) are possible for srwC-Cwrs. Assemble the figures and try them out. Pasteur's tartrates correspond to two of these three possible structures. Figure out which ones.

As Pope noted in 1901,[5] the observed optical activity of quartz crystals allowed late nineteenth-century scientists to reason that its underlying molecular structure is helical. They did understand at that time that the bonding connections in quartz involved no chiral centers, yet the crystal was chiral. The crystalline nature of quartz demanded a regular, periodic structure. The helix is the only regular, chiral structure that can meet these demands. While we now know that quartz is indeed composed of long helical structures[8] and clearly understand the logic of the chemists who discerned its molecular structure, I remain awed by these remarkably prescient predictions that predated X-ray crystallography by decades. The origin of optical activity on earth remains to this day a mystery. One view is that crystals such as quartz or calcite (calcium carbonate) formed chiral templates that, by pure chance, formed an excess of "handed" molecules of only one type on the planet.[9]

Pope informs his readers that, by the end of the nineteenth century, it was known that the sugars and amino acids that compose our bodies are specifically "right-handed" and "left-handed" respectively. He then has some fun by imagining the sudden appearance of a "lævo-man" ("left-handed" person) based upon "left-handed" sugars and "right-handed" amino acids.[5] Pope concludes his essay series rather grimly:

> if a human being enantiomorphously related to ourselves—the lævo-man of whom we have spoken—made his appearance on our planet, he would in all probability quickly starve to death, owing to his inability to assimilate the foodstuffs we were able to place before him.

To be a bit less grim about it, we might imagine lævo-man's first day on the planet starting nicely enough. Happily, he "gets up on the wrong side of the bed," has a refreshing glass of water, and takes a bracing shower.[10] Brewing coffee, he notes an off-odor and samples his first cup—black as usual. It tastes awful.[10] One teaspoon of sugar is followed by another, another, and then two more.[10] There is no improvement in the taste of the coffee. His toast tastes like chocolate-flavored mashed potatoes.[10] Agitated now, he pours a cold beer and gulps it down. Utterly awful.[10] Next, a stiff vodka, and this begins to taste familiar and starts to sooth his nerves.[9] By late morning a headache begins to develop, but two aspirins and a glass of water quickly soothe it.[10] Back to bed for a short nap, and then he suddenly awakens, abruptly "gets up on the right side of the bed," and immediately realizes that the rest of the day will be a disaster.

---

1. A. Greenberg, *A Chemical History Tour,* John Wiley and Sons, New York, 2000, pp. 247–250.
2. J.H. van't Hoff, *Die Lagerung Der Atome Im Raume,* Friedrich Vieweg Und Sohn, Braunshweig, 1877, p. 48.
3. E. Heilbronner and J.D. Dunitz, *Reflections on Symmetry,* VCH-Wiley, New York, 1993, pp. 72–73.
4. E.A. Abbott, *Flatland: A Romance of Many Dimensions* (with Foreward by Isaac Asimov), Barnes & Noble, New York, 1983.
5. W.J. Pope, *Journal of the Society of Arts (London),* Vol. 49, pp. 677–683, 690–697, 701–708, 713–718.
6. Greenberg, op. cit., pp. 251–252.
7. L. Pauling, *General Chemistry,* W.H. Freeman and Co., San Francisco, 1947, p. 521.
8. A.F. Wells, *Structural Inorganic Chemistry,* fifth edition, Clarendon Press, Oxford, 1984, pp. 1004–1006.
9. R.M. Hazen, *Scientific American,* Vol. 284, No. 4, pp. 76–85, April 2001.
10. Obviously we are joking when we talk about "left-hand and right-hand sides of the bed." Right-handed people and left-handed people all contain the same "right-handed" sugars and "left-handed" amino acids. However, the chirality or nonchirality of substances should affect, sometimes dramatically, how they interact with living beings. Fortunately for lævo-man, the oxygen and nitrogen in air as well as water are achiral (not "handed"). Ethyl alcohol and aspirin (acetylsalicylic acid) are also achiral. However, table sugar is chiral, as are the flavor and aroma ingredients in coffee and beer. Chocolate mashed potatoes is a secret family recipe best kept secret for the sake of humanity. It appears that Oliver Sacks' family had an old secret family recipe for Passover "matzoh balls of an incredible tellurian density, which would sink like little planetismals below the surface of the soup" (see O. Sacks, *Uncle Tungsten—Memories of a Chemical Boyhood,* Alfred A. Knopf, New York, 2001, p. 97).

WHAT ELSE COULD A WOMAN WRITE ABOUT?

Don't be fooled by the quaint title of Ellen Henrietta Swallow Richards' *The Chemistry of Cooking and Cleaning* (Figure 162),[1] published in 1882. Richards was the first female student at Massachusetts Institute of Technology (B.S., 1873), became an instructor at MIT and founded its women's laboratory. She bridged pure and applied chemistry with social science and founded the field of scientific home economics. She was a co-founder in 1882 of what would eventually become the American Association of University Women.[2–6]

Born to teachers in rural Massachusetts in 1842 (d. 1911), the precocious Ellen Swallow received a rural education, taught locally, and saved sufficient money to enter an experimental school for women's higher education in Poughkeepsie, New York. Her interest in analytical chemistry was stimulated by Professor A.C. Farrar. Notes Swallow after her first laboratory exercise: "Prof. Farrar encourages us to be very thorough there, as the profession of an analytical chemist is very profitable and means very nice and delicate work, fitted for ladies' hands."[5] She was a member of the first graduating class of Vassar College in 1870 and is honored by a plaque in Blodgett Hall.[5] The new graduates promised "cheerful submission to authority, compliance, diligence, and lady-like deportment."[5] In 1871 she entered MIT, excelled in her studies, and met Professor Robert Hallowell Richards, whom she married in 1875 after she had become a member of the faculty. A children's book dramatizes her student days at MIT with a quaint scene in which she wins the acceptance of the males in her class by baking cookies for them.[6] Richards' early work in the analytical chemistry of minerals and water earned her wide respect, but the work in bringing sanitary chemistry into the home eventually won her worldwide renown.

The Chemistry of Cooking and Cleaning is slim, economical, and very effective in its straightforward presentation to its female audience. Here is Mrs. Richards on the perfidy of manufacturers whose claims about "secret ingredients" she debunks:

> There is, lingering in the air, a great awe of chemistry and chemical terms, an inheritance from the age of alchemy. Every chemist can recall instances by the score in which manufacturers have asked for recipes for making some substitute for a well-known article, and have expected the most absurd results to follow the simple mixing of two substances. Chemicals are supposed by the multitude to be all-powerful, and great advantage is taken of this credulity by unscrupulous manufacturers.

Well, even Ivory Soap is only "ninety-nine and forty-four-one-hundredths percent pure." And what would she have said about Ozone Soap (see the earlier essay in this book, p. 260)?

Discussion of the chemical composition of foods is accompanied by analyses of their energy contents and the dietary habits of different cultures. It is duly noted that rice, a carbohydrate, is much lower in energy content than fat, explaining why the former is the dietary staple of tropical cultures while the latter is an important staple in arctic climates. Indeed, an astute woman, noting that her children (or husband) might be accumulating too much "residue" from their diet, can chemically "titrate" them with oxygen to burn off the excess as CO_2 and H_2O through outdoor activity:

> Cooking has thus become an art worthy the attention of intelligent and learned women. The laws of chemical action are founded upon the law of definite proportions, and whatever is added more than enough, is in the way. The head of every household should study the condition of her family,

FIGURE 162. ▪ Title page from Ellen Henrietta Swallow Richards' ("Ellen Richards") 1882 book *The Chemistry of Cooking and Cleaning*. Ms. Richards was in the first graduating class at Vassar, inaugurated the women's chemistry laboratory at MIT, and was the founder of the field of sanitary chemistry, a co-founder of the American Association of University Women, and an environmentalist many decades ahead of her time. She would not have been amused by the deceptive advertising for Ozone Is Life bottled water (Figure 155) or Ozone Soap (Figure 156).

> and tempt them with dainty dishes if that is what they need. If the ashes have accumulated in the grate, she will call a servant to shake them out so that the fire may burn. If she sees that the ashes of the food previously taken are clogging the vital energies of her child, she will send him out into the air, with oxygen and exercise to make him happy, but she will not give him more food.

The 1910 MIT convocation address written by Ellen Swallow Richards was cited over 60 years later by Yale University scientist Bill Hutchinson as an early clarion call to conservation and respect for the environment:[5,7]

> The quality of life depends on the ability of society to teach its members how to live in harmony with their environment—defined first as the family, then with the community, then with the world and its resources.

Ellen Swallow Richards was an early pioneer in the education of women in chemistry. Figure 163 shows a photograph, from the Frank Lloyd Wright Foundation, taken between 1900 and 1910.[8] The young women are from the Hillside Home School, Spring Green, Wisconsin. The famous architect's aunts were supporters of the school, and when it was closed, it eventually became part of the Frank Lloyd Wright Estate. The picture was probably taken routine-

FIGURE 163. ■ Photograph of a chemistry class for women at the Hillside Home School in Spring Green, Wisconsin near the beginning of the twentieth century (courtesy The Frank Lloyd Wright Archives, Scottsdale, Arizona).

ly in the school and eventually became part of the estate's holdings. Another early pioneer was Dr. Edgar Fahs Smith, Professor of Chemistry, University of Pennsylvania. Ten women completed their Ph.D.s with Professor Smith between 1894 and 1908, a number of whom became college faculty members.[9] Smith's collection of chemical books and artwork presently forms the core of the University of Pennsylvania History of Chemistry collection. His sublime book *Old Chemistries*[10] is a work of warmth and erudition that helped inspire the present book. It includes a fine discussion of Jane Marcet and her "conversations in chemistry."

---

1. E.H. Richards, *The Chemistry of Cooking and Cleaning—a Manual for Housekeepers*, Estes & Lauriat, Boston, 1882.
2. C.L. Hunt, *The Life of Ellen H. Richards*, Whitcomb and Barrows, Boston, 1912.
3. R. Clarke, *Ellen Swallow: The Woman Who Founded Ecology*, Follett Publishing Co., Chicago, 1973.
4. *The New Encyclopedia Britannica*, Vol. 10, Encyclopedia Britannica, Inc., Chicago, 1986, p. 45.
5. *http://departments.vassar.edu/~anthro/bianco/hidden/ellen.html* (6/14/01).
6. E.M. Douty, *America's First Woman Chemist—Ellen Richards*, Julian Messner, Inc., New York, 1961.
7. B. Hutchinson, "Swallow Warned Us All Years Ago," *Miami Herald*, January 17, 1974.
8. This photograph was provided courtesy The Frank Lloyd Wright Archives, Scottsdale, AZ. I am also grateful for conversations with Ms. Margo Stipe, The Frank Lloyd Wright Archives.
9. J.J. Bohning, *Chemical Heritage*, Vol. 19, No. 1, pp. 10–11, 38–44, Spring 2001.
10. E.F. Smith, *Old Chemistries*, McGraw-Hill Book Co., Inc., New York, 1927.

SEARCHING FOR SIGNS OF NEON

The discovery of the noble gas argon was made in 1894, and the scientific details are provided by Rayleigh and Ramsay in their Hodgkin's Prize–winning 1896 essay[1] and briefly summarized elsewhere.[2] As often happens in science, a tiny but very real discrepancy led to a momentous series of discoveries. Briefly, physicist Lord Rayleigh (John William Strutt) and chemist William Ramsay had communicated about the curious observation that the density of "atmospheric nitrogen" (1.2572 grams per liter) was about 0.6% higher than that of "chemical nitrogen" (1.2505 grams per liter). "Chemical nitrogen" could be synthesized, for example, by heating pure crystalline ammonium chloride to release ammonia and then reacting this gas with oxygen over red-hot copper. In their study of "atmospheric nitrogen," Ramsay chemically removed water, carbon dioxide and other trace gases, and subsequently oxygen (via exposure to red-hot copper). The full complement of nitrogen was removed by reaction with red-hot magnesium to form the powdery nitride. What remained was less than 1% of an unreactive residue of gas that they named "argon" ("lazy"). They noted with great respect that Henry Cavendish had obtained this same 1% residue more than a century earlier by exhaustively sparking atmospheric nitrogen with oxygen. The sheer totality of their discovery included their finding that argon was monoatomic despite the conceptual difficulties entailed by an atomic weight of 40; that it was totally unreactive, like the recently discovered helium, and, with helium, probably defined an entirely new family in the periodic table.[3]

It is important to emphasize that argon's density would be nicely explained if it were a diatomic (Ar_2) just like all the other elemental gases then known. An atomic weight of 20 would fit nicely between that of fluorine (19) and sodium (23). But 40 would appear to place it after potassium (39) and equal to calcium. Although a similar "problem" was known between the atomic weights of iodine and tellurium, it was felt that this single anomaly would someday be corrected. In 1896, there was no concept of nuclei or isotopes and, most importantly, the reality of atomic number—the parameter that truly sets the sequence of elements (rather than atomic weight)—would not be discovered for another two decades. So the fact that argon was out of order truly threatened to break the periodic law.

The discoveries of argon and the rest of the rare gases are conveyed on a very personal level by Morris W. Travers, who, three decades earlier, was a young graduate student of Ramsay's at Bristol.[4] I see an ironical aspect noted early in Travers' book. He quotes Van't Hoff, the first chemistry Nobel laureate, from a contemporary Dutch review as follows:[5]

> How then was this discovery made? Year after year Lord Rayleigh! Poor Rayleigh! had been weighing nitrogen: nitrogen from urea, nitrogen from ammonium nitrate, nitrogen from the air, ever did he find the latter heavier: 1.2572 against 1.2505 gram per litre of nitrogen. Nitrogen from the air was thus somewhat different, it contained something different from chemical nitrogen; and starting from the latter supposition Ramsay removed all possible substances from the air, and there remained the celebrated little bubble of gas of Cavendish, colourless, without taste or smell.

Travers finds this article to be "in a bitterly sarcastic vein throughout, and makes light of the work of both Lord Rayleigh and Ramsay."[5] Indeed, Travers notes that Rayleigh did not "make a career" of weighing nitrogen, and his study of the relative densities of gases was part of a broad investigation of the relative atomic weights of the elements with, I might now add over a century later, unseen yet profound future implications for the understanding of the nuclear structures of atoms.[6] But wasn't this the same van't Hoff who, as a 26-year-old instructor at the Veterinary College at Utrecht some 22 years earlier, postulated tetrahedral carbon and chemistry in three-dimensional space? The same Van't Hoff who was famously "trashed" by the nearly 60-year-old doyen of German organic chemistry, Professor Doctor Adolph Wilhelm Hermann Kolbe?[7,8] Ah, the foibles of gifted humans. Happily, Travers describes his erstwhile mentor Ramsay as a kind, fatherly mentor, solicitous of his students and moderate and fair in argumentation.

A new series of spectroscopic lines were observed by Norman Lockyer during the solar eclipse of 1868. He recognized a new element and named it "helium" (from the Greek *helios*, the sun). In 1888, an unreactive gas obtained from the mineral clevite, which contains uranium, was isolated by Dr. W.F. Hildebrande of the U.S. Geological Survey, who erroneously characterized it as nitrogen. In 1895, a year after the discovery of argon, Ramsay, following a suggestion from Mr. Henry Miers of the British Museum, examined clevite as well as some other minerals containing uranium, collected the gaseous components as described above, and found an inert, monoatomic gas of mass 4 as the major component.[9] William Crookes, who had obtained the first spectrum of argon in collaboration with Rayleigh and Ramsay, discovered that the spectrum of the new light gas was identical with that of Lockyer's helium.[9] As noted above, Ramsay postulated that helium and argon formed a new periodic family. At this point, it is vital to remark that Ramsay's work on helium occurred about one year before Henri Becquerel's discovery of the phenomenon of radioactivity using a uranium salt; three years before the Curies coined the term *radioactivité* and isolated polonium and radium; and some eight years before alpha particles started to be generally recognized as helium lacking two electrons.[10] So helium was apparently to be found in certain exotic minerals containing uranium and thorium. There was no association with radioactivity since it was unknown.

And now arose a powerful challenge for Ramsay—the new family of noble gases had a member in the first row of the periodic table (helium) and in the third row (argon). Missing was the second-row noble gas calculated to have an atomic weight of 20. Missing, too, were possible heavier inert gases, but the primary goal was to fill the Mendeleevian gap. Attempts to find new noble gases in the atmosphere were initially unsuccessful. Mysterious and exotic minerals such as Norwegian clevite, meteorites, and gases from the bowels of the earth bubbling out of hot springs in Iceland and elsewhere were investigated to no avail.[9,11]

And now, very briefly, we need to learn how to liquefy (even solidify) gases at extremely low (cryogenic) temperatures. Hints of very low levels of at least another noble gas in argon motivated Ramsay and his co-workers to investigate the possibility of liquefying air (or selected fractions) followed by fractional distillation. Again, it is remarkable to note that temperatures very close to absolute zero [zero degrees Kelvin (0 K) or –273.16°C or about –460°F] were achieved before the end of the nineteenth century. The important principle here is called the Joule–Thomson effect,[12] discovered in the mid-nineteenth century. If a gas expands from a vessel into a vacuum, the remaining gas (and other matter in the flask) will be cooled. If a gas is first compressed, the gas will heat up, but a cooling jacket can remove this heat, leaving cold compressed gas, which can cool down further when exposed to reduced pressure. Indeed, Michael Faraday accidentally condensed chlorine gas into a green, nasty liquid by injecting it via syringe into a closed tube.[13] Carbon dioxide can be compressed into liquid form at pressures above 5.11 atmospheres (atm), but no degree of coldness will condense it to a liquid under atmospheric pressure. A. Thilourier discovered in 1835 that when pressurized, liquefied carbon dioxide is quickly exposed to atmospheric pressure, the cooling expansion (which also "steals" the heat of vaporization from the remaining material and surroundings) forms solid CO_2 or dry ice.[13] Now gas samples could be condensed in a "thermodynamic sink" of dry ice in diethyl ether maintained at –78°C. It wasn't long before oxygen [boiling point (bp) –183°C] and nitrogen (bp –196°C) and atmospheric air could be liquefied and even solidified. And boiling liquid oxygen could condense helium and even hydrogen under pressure. The masters of this technology were two Polish scientists, Olszewski and Wroblewski.[14,15] James Dewar, working in England, developed the vacuum vessels that today bear his name, and while he kept this apparatus secret for years, by 1892 it was widely disclosed.[15] The path was now clear for W. Hampson, who developed a process for liquefying air in collaboration with Mr. K.S. Murray, the managing director of the British Oxygen Company,[14] to supply Ramsay with liquid air sufficient to provide many liters (!) of gaseous argon.

At the end of May, 1898, Hampson brought a 750 cm^3 sample of liquid air, and at Ramsay's suggestion, over the course of about a week Travers allowed it to boil away until only about 25 cm^3 of gas (i.e., perhaps 0.025 cm^3 of the original liquid) remained. Travers describes his interplay with a young friend and colleague who gently teased him:[16] "'It will be the new gas this time, Travers.' 'Of course it will be,' I replied, and passed on upstairs to Ramsay's room. I was beginning to think that the discovery of the new gas would correspond with the Greek *Kalends*;[17] but Ramsay still had faith in the periodic law, and perhaps I had even stronger faith in Ramsay. However, we had to put up with a good deal of kindly chaff both within and without the department."

But later that day, following the usual residual cleanup of the remaining gas, a small quantity was introduced into a Plucker tube, the electricity turned on, and the light viewed using direct-vision spectroscopes. Lo! A distinctly new yellow band appeared—a new element—in a *less volatile* fraction derived from argon. The gas was not, of course, the missing atomic weight 20 species, but was a monoatomic gas of weight close to 80. The new element was named krypton (from the Greek *kryptos* = "hidden"). The dry spell was broken. Another large sample of liquid air was fractionally distilled and chemically treated to give high-boiling and low-boiling distillation cuts containing argon. The low-boiling cut was carefully fractionated and provided a gas that did not require the subtlety of spectroscopic glasses to disclose its secret, but let's allow Travers to tell it:[18]

> we each picked up one of the little direct-vision spectroscopes which lay on the bench. But this time we had no need to use the prism to decide whether or not we were dealing with a new gas. The blaze of crimson light from the tube told its own story, and it was a sight to dwell upon and never to forget.

Holy Krypton, Batman! A Neon Sign for Neon![19] Neon is derived from the Greek *neos* ("new"). And in a very short time xenon (Greek xenos = "strange") was similarly discovered [and five years later in a collaboration between Ramsay and Frederick Soddy—radon (originally called "niton"), a by-product of the decay of radium was identified].[20] In summary, the percent by volume of the noble gases in the atmosphere are now known: argon, 0.93%; neon, 0.0018%; krypton, 0.0011%; helium, 0.00052%; xenon, 0.0000087%.[21] Small wonder that neon, krypton, and xenon required very large quantities of liquid air in order to detect their presence. But it is fascinating to note, too, that every cubic meter of air contains nearly 10 grams of argon; each adult inhales almost 200 grams of argon per day. Yet we were unaware of argon's existence until 1894.

Ramsay was awarded the 1904 Nobel Prize in Chemistry, and Rayleigh was awarded the 1904 Nobel Prize in Physics. In Figure 164, we see an early twentieth-century caricature of the fatherly Ramsay blissfully posing with his own periodic family.

FIGURE 164. ■ An early-twentieth-century caricature in *Vanity Fair* of William Ramsay pointing with fatherly pride to his chemical family—the rare gases. See color plates.

1. Lord Rayleigh and Professor William Ramsay, *Argon, a New Constituent of the Atmosphere*, Smithsonian Institution, Washington, DC, 1896.
2. A. Greenberg, *A Chemical History Tour*, John Wiley & Sons, New York, 2000, pp. 255–256.
3. A.J. Ihde, *The Development of Modern Chemistry*, Harper & Row, New York, 1964, p. 374.
4. M.W. Travers, *The Discovery of the Rare Gases*, Edward Arnold & Co., London, 1928.
5. Travers, op. cit., pp. 1–4.
6. In 1815 William Prout postulated, on the basis of the small number of atomic weights known, that all were simple whole-number multiples of the lightest element—hydrogen. At one level, this seemed to be an incredibly prescient hypothesis. We have known for about 100 years that hydrogen has one proton [mass based on carbon–12 is 1.0073 atomic mass unit (amu)] and one electron of relatively negligible mass (0.0005486 amu). However, careful determinations by Berzelius, Dumas, and Stas throughout the midnineteenth century indicated significant discrepancies. And our modern understandings include the occurrence of isotopes, the fact that the neutron is slightly heavier than the proton (1.0087 amu), departures from ideal-gas behavior at higher pressures, nuclear binding energies, and numerous other flaws in the hypothesis. But it was a useful construct and remains conceptually helpful today in a very simplistic way.
7. J.E. Marsh, *Chemistry in Space, from Professor J.H. van't Hoff's "Dix Années Dans l'Histoire d'Une Théorie,"* Clarendon, Oxford, 1891, p. 16.
8. Greenberg, op. cit., pp. 247–250.
9. Travers, op. cit., pp. 56–57.
10. J.R. Partington, *A History of Chemistry*, Vol. 4, Macmillan & Co. Ltd., London, 1964, pp. 936–947.
11. Travers, op. cit., pp. 82–84.
12. W. Kauzmann, *Thermodynamics and Statistics: With Applications to Gases*, W.A. Benjamin, Inc., 1967, pp. 53–58.
13. Partington, op. cit., pp. 105–108.
14. Travers, op. cit., p. 87.
15. Partington, op. cit., pp. 904–906.
16. Travers, op. cit., p. 90.
17. *Kalends* = the first day of the month in the ancient Roman calendar.
18. Travers, op. cit., pp. 95–96.
19. This essay is dedicated to my brother Kenny, who immersed himself in *Superman* and *Batman* comics folklore as a boy, became a neon artist, and is the proprietor of *Krypton Neon* in New York City. I have tried, unsuccessfully, to convince him to post a sign "Krypton Neon Argon" when he goes to lunch.
20. Travers, op. cit., pp. 105, 110, 126.
21. F.A. Cotton and G. Wilkinson, *Advanced Inorganic Chemistry*, fifth edition, John Wiley & Sons, New York, 1988, p. 588.

A "GROUCH" OR A "CRANK"?

Did Mendeleev and Priestley Become Scientific Grouches in Old Age?

The discovery of the noble gases between 1894 and 1898 presented Dmitri Mendeleev the opportunity to apply his periodic law to develop a fully chemical explanation of the universal ether, the all-penetrating imponderable medium surrounding and imbuing all matter.[1] Although the Michaelson–Morley experiment in 1887 disproved the existence of the ether, many important scientists, including a number of prominent physicists, still accepted its existence at the start of the twentieth century. Part of the problem was that the experimental result—

that the velocity of light is equal in all directions, although accepted, could not be explained by current theory. In that sense, it could be considered to be an "anomaly."[2,3] Einstein's relativity theory would eventually furnish the explanation. Mendeleev stretched his periodic law to postulate the existence of an undiscovered inert gas of atomic mass (relative to $H = 1.0$) on the order of 0.00000096 to 0.000000000055, that would form the substance of the ether.[1] However, in order to postulate this ethereal element, he also needed to postulate one additional inert gas of atomic mass 0.4. There was no serious evidence for this element, which would represent an *extrapolation* rather than the interpolations that had worked so brilliantly in predicting new elements during the 1870s and 1880s. Perhaps it is unfair to say this, "hindsight always being 20 : 20," but we may view the aging Mendeleev as becoming a bit "grouchy," scientifically, with age.

There are other precedents for great scientists who "grouchily" persisted in retaining theories beyond their useful lives. A prominent example is Dr. Joseph Priestley,[4] whose discoveries of new gases including oxygen were so critical to the development of chemistry. Priestley was an early adherent of phlogiston theory, and his final chemical publication, *The Doctrine of Phlogiston Established and that of the Composition of Water Refuted,* was published in 1800 (Figure 165),[4] and the second edition was published in 1803, a year before he died and two decades after discovery of the true composition of water sank phlogiston theory. (Partington terms the prominent Swedish chemist Anders Retzius, who died in 1821—"probably the last phlogistonist."[5]) Such conservatism is not necessarily an unhealthy thing for science. It protects scientific theories from rapidly shifting with the prevailing winds and demands stronger proof and even generational change before what philosopher of science Thomas Kuhn calls a "paradigm shift"[6] is widely accepted.

Rejecting Atomic Theory and Dismissing Continental Drift

In this light, it is important to note that Dalton's atomic theory, which we blithely inform our students was introduced and accepted at the start of the nineteenth century, was resisted by some prominent chemists (and many physicists) until the first decade of the twentieth century when Einstein and later Jean Perrin explained the molecular basis of Brownian motion. Thomas Sterry Hunt,[7] Professor of Geology at the Massachusetts Institute of Technology, member of the National Academy of Sciences (1873), President of the American Association for the Advancement of Science (1871), and two-time President of the American Chemical Society (1879, 1888), wrote a book in 1887 that totally rejected atomic theory. He retained this belief until he died five years later. The great German chemist Friedrich Wilhelm Ostwald (1853–1932) firmly resisted atomic theory throughout the first four decades of his scientific career. However, in 1909, the year he was awarded the Nobel Prize in Chemistry, Ostwald finally admitted that the work of Perrin as well as J.J. Thomson "justify the most cautious scientist in now speaking of the experimental proof of the atomic nature of matter."[8]

The boundary line between a "scientific grouch" and a "scientific crank" is

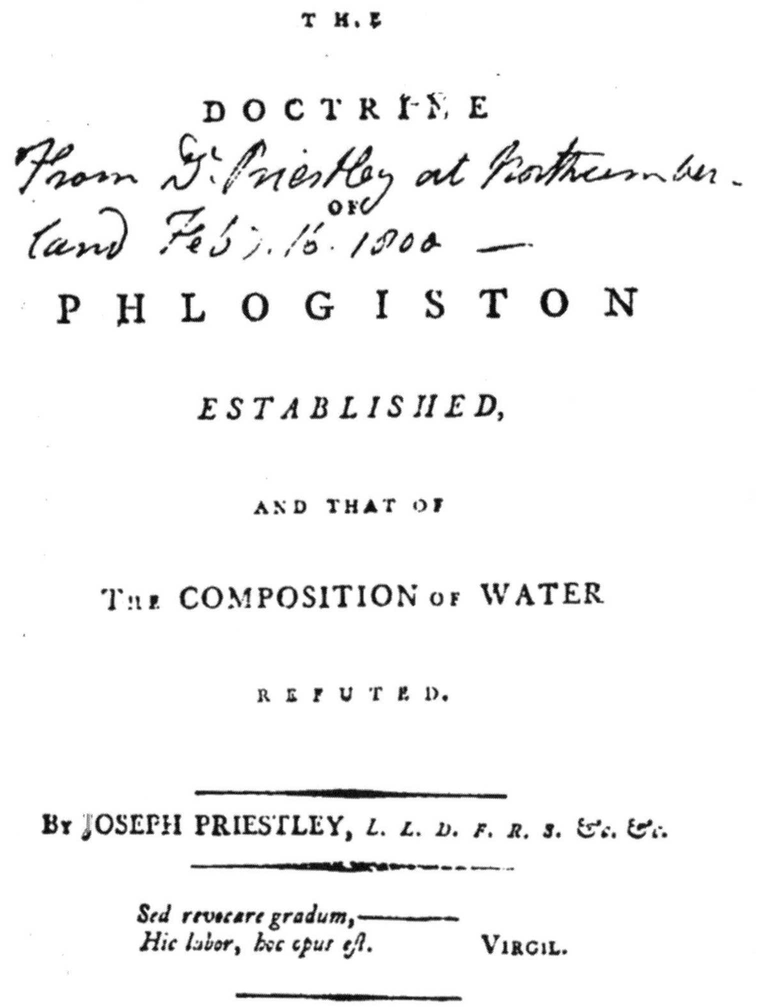

THE

DOCTRINE

From Dr. Priestley at Northumberland Feb. 16. 1800 —

OF

PHLOGISTON

ESTABLISHED,

AND THAT OF

THE COMPOSITION OF WATER

REFUTED.

By JOSEPH PRIESTLEY, *L. L. D. F. R. S.* &c. &c.

Sed revocare gradum,——
Hic labor, hoc opus est. VIRGIL.

FIGURE 165. ▪ Joseph Priestley "grouchily" retained his belief in phlogiston theory through the end of his life. Here is a copy of his spirited defense that he signed and presented to an acquaintance. (Courtesy The Roy G. Neville Historical Chemical Library.)

sometimes not a very clear one. Simply imagining a person stubbornly resisting the prevailing views of the scientific establishment and "howling alone in the desert" is not sufficient reason to label the "infidel" a "crank." For example, countless people, including children, who have viewed maps of the world have undoubtedly noted the apparent jigsaw puzzle fit between the South American and African continents. It remained a curious, but not very informative, observation since no mechanism for explaining this "anomaly" existed until fairly recently. However, in 1912 the German geophysicist Alfred Wegener (1880–1930) noted similarities in fossils collected in the two continents, combined this with geophysical data, and postulated the theory of "continental drift."[2] He was isolated, and his views were considered "highly controversial," a description sometimes applied diplomatically to the work of "cranks," until he was vindicated by the theory of plate tectonics developed in the late 1960s.[2]

A Crank Who "Trashed" Phlogistonists and Antiphlogistonists with Great Gusto

I believe that a "crank" adheres to an ideology and it is this ideology rather than an inherent scientific conservatism or radicalism that defines a "crank" (i.e., unless conservatism or radicalism for its own sake is the ideology). Partington[9] clearly labels Robert Harrington,[10] an English surgeon, a "crank." Although a believer in phlogiston theory, Harrington's "ideology" seemed based on a scientific enterprise in which he was correct and all others wrong. He believed in "equal opportunity" and, in a pamphlet published in 1786, joyfully "trashed" phlogistonists of the English school (Priestley, Cavendish, and Richard Kirwan) and antiphlogistonists of the French school (Lavoisier):[10]

> Letter . . . to Dr. Priestley, Messrs. Cavendish, Lavoisier, and Kerwan . . . to prove that their . . . opinions of Inflammable and Dephlogisticated Airs forming Water, and the Acids being compounded of different Airs are fallacious, London, 1786.

And one must appreciate the titles of two of his subsequent works:[10]

> The Death-warrant of the French Theory of Chemistry . . . with a Theory fully . . . accounting for all the Phenomena. Also a full . . . Investigation of . . . Galvanism, and Strictures upon the Chemical Opinions of Messrs. Weiglet, Cruickshanks, Davy, Leslie, Count Rumford, and Dr. Thompson; likewise Remarks upon Mr. Dalton's late Theory and other Observations, 1804.

Or perhaps, "How To Make Friends and Influence Chemists" and finally, and most modestly, Harrington on Harrington:[10]

> An Elucidation and Extension of the Harringtonian System of Chemistry, explaining all the Phenomena without one single Anomaly, London, 1819.

Let us briefly examine Harrington's "hash" with his fellow phlogistonists. In his 1785 book on different kinds of air, he notes that it appears to be appropriate to "conclude that phlogiston is fire and light, or a certain subtile elastic fluid, upon the modifications of which the phænomena of heat and light immediately depend."[11] But he then observes that fresh air is required to receive the phlogiston from the combustible body until the air is "injured" by the fire and incapable of further supporting combustion. Harrington sees a "timing" issue here. If fresh air must first attract phlogiston and then fire and light follow, how could phlogiston actually be fire and light? Indeed, he ridicules other phlogistonists by indicating the logical extension of loss of phlogiston followed by heat and light: "For if a body be saturated with water it will not burn. But as the air attracts the water from the burning body it will burn, and agreeable to its quick or slow attraction of this moisture."[11] Thus, rather than being firelike, phlogiston would seem to be, idiotically enough, waterlike.

Here is another problem for phlogistonists less savvy than Dr. Harrington. Inflammable air (actually H_2) and nitrous air (actually NO) are known to contain equal quantities of phlogiston. Let us trace this correct phlogistonist logic for a moment using modern chemical equations. Aqueous sulfuric acid will "dissolve" metals because the acids hydrogen ions (H^+) are readily reduced by all metals except the most inert:

$$2H^+ + SO_4^{2-} + Cu\ (\text{metal}) \rightarrow H_2\ (\text{gas}) + Cu^{2+} + SO_4^{2-}$$

However, in dilute nitric acid, less active metals such as copper and iron will reduce the nitrate group rather than the hydrogen ion to produce nitric oxide ("nitrous air"):

$$3Cu\ (\text{metal}) + 2NO_3^- + 8H^+ \rightarrow 3Cu^{2+} + 2NO\ (\text{gas}) + 4H_2O$$

In both cases, calxes of copper remain, so it is "clear" that the phlogiston has been carried off in "flammable air" in the first case and in "nitrous air" in the second case. Now, in comparing "flammable air" and "nitrous air," it is "clear" using eudiometry that each contains the same quantity of phlogiston since they each react with the same quantity of "dephlogisticated air" (oxygen):

Inflammable air: $H_2 + \frac{1}{2}O_2 \rightarrow H_2O$

Nitrous air: $NO + \frac{1}{2}O_2 \rightarrow NO_2$ (reddish gas)

The question Harrington poses is why fire and light are generated with "inflammable air" but not with "nitrous air" even as the two release the same quantity of phlogiston.

And here is another inconsistency—we "know" that exhaled air is somewhat "injured" since it is has absorbed phlogiston and is not as breathable as the air freshly inhaled moments earlier. "Clearly," phlogiston is being captured in the lungs. A snake eats phlogiston-rich animals (i.e., containing fats).[11] Its body is cool, and so the lungs should be huge in order to rapidly expel large quantities of phlogiston extracted from the meat. In fact, the lungs in a snake are small—why does the snake not incinerate from the trapped heat? In contrast, herbivores like cows eat phlogiston-poor (fat-free) food, are warm, and have large lungs to rapidly expel phlogiston.[11] Why do cows not starve? Frankly, Harrington does not have any good answers himself, but that is not his mission.

Views of a Libertarian Chemist: The Nefarious Smithsonian Institution and Other Plots

Toward the end of the nineteenth century Gustavus Detlef Hinrichs, M.D., LL.D., Professor of Chemistry, St. Louis College of Pharmacy, had embarked on a desperate crusade to save America's chemical industry and its educational establishment from the incorrect atomic weights forced on our unsuspecting nation by its nefarious government.[12] The source of all evil was one Frank Wigglesworth Clarke, Chief Chemist of the Geological Survey, whose

> recalculations have been formally endorsed by the Secretary of the Smithsonian Institution and published officially at the expense of the Smithson

> Fund; it has, finally, been sent out under the official frank as registered matter. The deficiency of the postal service—partly so resulting—is made up by Congressional appropriations.
>
> The same author Clarke is also habitually sent by authority of the National Government and at public expense, as delegate to the Congresses of Chemists, and put in charge of National Exhibits at home and abroad. This highest possible official consideration has enabled him to exercise a ruling influence in the American Chemical Society.[13]

Sounds pretty alarming—"the new world order of atomic weights." So what's the "big hoo-hah" here? Seems like Professor Hinrichs had adopted Prout's hypothesis[14–16] as a strict fundamentalist.

In 1815 and 1816 William Prout (1785–1850)[14] published two papers in which he asserted that the densities of gases were simple whole-number multiples of hydrogen.[14–16] This was only a decade after Dalton first postulated atomic weights, and there were considerable uncertainties in experiments and formulas. Nonetheless, the concept was an attractive one since it implied the possibility of a simplest "primary material" out of which all other atoms were composed. Prout postulated the existence of this "protyle" from which all other atoms would be composed. It had an almost religious simplicity. In 1819, Berzelius published a complete series of relative masses of elements and compounds, which included many fractional atomic weights (relative to $H = 1$) that would not agree with Prout's hypothesis. However, there were considerable uncertainties in formulas and atomic weights. In 1825, Thomas Thomson published his table of atomic weights: all, including chlorine (36), for example, were whole-number multiples of hydrogen (1). Berzelius said of Thomson's atomic weights:[14]

> Much of the experimental part, even of the fundamental experiments, appears to have been made at the writing desk; and the greatest civility which his contemporaries can show its author, is to forget that it was ever published.

Berzelius soon regretted his accusation, and Partington notes that Thomson was scrupulously honest.[14] Attempts to salvage Prout's hypothesis included suggestions that hydrogen might contain exactly two or perhaps four "protyles."[17] Other attempts included suggestions that, statistically speaking, while most atoms of chlorine, for example, might have an atomic weight of 36, a few might be 35 and 37; even fewer, 34 and 38; and so on. Partington even notes ideas reminiscent of Newton's "worn atoms."[18]

However, the coffin of Prout's hypothesis was nailed shut around 1865 by the careful analytical studies of Jean Servais Stas. Figure 166 depicts a magnificent apparatus for total analysis of silver iodate ($AgIO_3$).[19] How could one possibly argue with a "Rube Goldberg–looking" apparatus like that? The gasometer (A) on the left delivered a steady stream of nitrogen gas purified by a *kaliapparat* (B) filled with concentrated sulfuric acid, followed by drying tubes (C) containing anhydrous calcium chloride, and then a gas furnace *D* containing a fused-glass tube filled with finely divided copper. Nitrogen, for flushing the system, could be made via combustion of ammonia. Its impurities would include unreacted ammonia (trapped in *B*), water (removed by C) and residual oxides of nitrogen (reduced to nitrogen in *D*). The balloon flask (*H*), containing silver iodate,

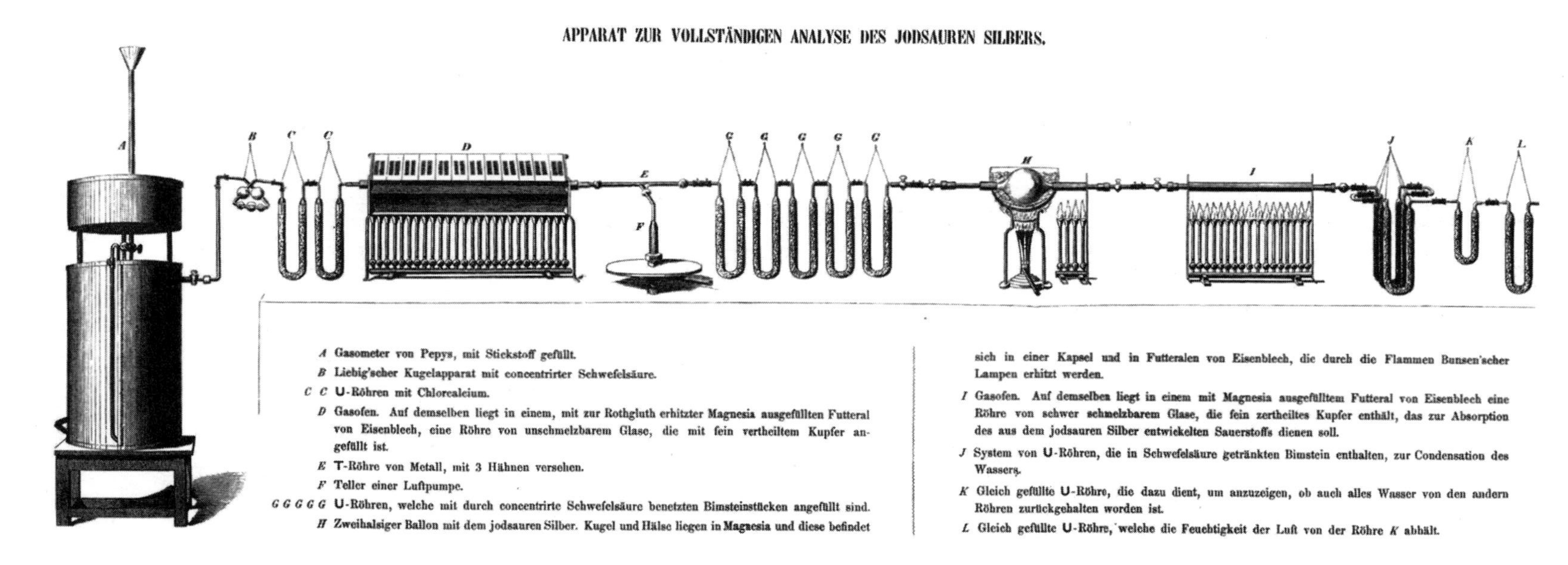

FIGURE 166. ▪ Apparatus used by Jean Servais Stas for total analysis of silver iodate. Stas' exacting and precise work on atomic weights laid a firm basis for development of the periodic table. However, Gustavus Detlef Hinrichs referred to Stas as "The Greatest False Scientist." Late in his career, Hinrichs became a certifiable scientific crank who perceived a vast governmental conspiracy to hide the true atomic weights of the elements. (From Stas, *Untersuchungen Über Die-Gesetze Der Chemischen Proportionen*, 1867).

is heated slowly over a bunsen burner and the oxygen generated (leaving molten AgI) is captured in gas furnace *I* containing finely divided copper.

Unimpressed by Stas and his apparatus, Hinrichs was nothing if not straightforward:[20]

> Ever since I understood the conditions of the chemical elements in reference to a single, primitive substance, (that is, since 1855), I have most faithfully labored in this field.

His idea was almost a half-century old when his book *Absolute Atomic Weights*[12] was published in 1901. His opinion of Jean Stas was summarized in a paragraph titled:[21] "The Greatest False Scientist."

Hinrichs diagnosed his contemporary John William Mallet, whose atomic weights of gold, lithium, and aluminum were not to his liking, as suffering from *Morbus Stasii* complicated by "the incipient stages of '*Furor Clarkii*'."[21] The work of Edgar Fahs Smith's student W.L. Hardin "represent nothing but his own imagination."[21] Under Henri Moissan "good French laboratory work is spoilt or falsified, by reducing it by German atomic weights."[21] Berzelius' problems are summarized as: "Great Chemist, Poor Balance."[21] William Ramsay's studies of atomic weights were infected by his use of Clarke's tables. And Hinrichs calculates using his particular statistical techniques that[21] "*Berzelius* was in 1826, a 10,000,000,000 times better chemist as Ramsey in 1893.

Thirty years before the aforementioned atomic weights book was published, Hinrichs was professor of physical science at the State University of Iowa. Although he did not accept the concept of periodicity, Hinrichs' "chart of the elements" has been credited as the first spiral classification of the elements.[22] In his 1871[23] text he listed atomic weights (pre-Stas) that included fractional weights such as aluminum (27.4), chlorine (35.5), copper (63.4), platinum (197.4), selenium (79.5), strontium (87.6), and zinc (65.2). The remaining 37 elements in his table (a total of 63 were known by this time) were pleasing whole-number multiples of hydrogen (=1). There is no explicit hint that Hinrichs was concerned about these departures from Prout's hypothesis in 1871. Perhaps he did not want to confuse his students. However, it is certainly clear that the Clarke-adopted 35.45 (rather than 35.5) for chlorine, for example, would have "bugged" Hinrichs, as did Darwin's theory of evolution.[20]

Hinrichs may have also been a bit "grouchy" in addition to being "certifiably cranky." Here is the first of his five "laboratory rules:"[24]

1. "BE QUIET—Talk not to your fellow students, and only in low whispers to your teacher. Walk to and from the balance so that your steps are not heard. Early thus learn to show reverence for truth and its investigation; the laboratory should be a temple of science."

Or, perhaps a monastery. Still, quiet whispers and silent footsteps in the teaching lab seem more attractive to me the older and grouchier I get.

---

1. A. Greenberg, *A Chemical History Tour*, John Wiley and Sons, New York, 2000, pp. 257–259.
2. A. Lightman and O. Gingerich, *Science*, Vol. 255, pp. 690–695, 1991. I am grateful to Dr. Joel F. Liebman and Dr. Larry Dingman for helpful discussions on this topic.

3. Scientific "anomalies" appear to me to be very closely related to Stent's "premature discoveries"; see G. Stent, *Scientific American*, Vol. 227, No. 6, pp. 84–93, 1972.
4. J.R. Partington, *A History of Chemistry*, Macmillan and Co. Ltd., London, 1962, Vol. 3, pp. 237–271.
5. Partington, op. cit., p. 200.
6. T.S. Kuhn, *The Structure of Scientific Revolutions*, second edition, University of Chicago Press, Chicago, 1970.
7. E.R. Atkinson, *Journal of Chemical Education*, Vol. 20, pp. 244–245, 1943.
8. Partington, op. cit., 1964, Vol. 4, p. 597.
9. Partington (1962), op. cit., p. 490.
10. L. Stephen and S. Lee, *The Dictionary of National Biography*, Vol. VIII, Oxford University Press, London, 1921–1922 (reprinted 1963–1964), pp. 1320–1321.
11. (R. Harrington), *Thoughts on the Properties and Formation of the Different Kinds of Air; with Remarks on Vegetation, Phosphori, Heat, Caustic Salts, Mercury, and on the Different Theories upon Air*, R. Faulder, J. Murray, and R. Cust, London, 1785, pp. 278–285.
12. G.D. Hinrichs, *The Absolute Atomic Weights of the Chemical Elements*, C.G. Hinrichs, Publisher, St. Louis, 1901.
13. Hinrichs, op. cit., p. iv.
14. Partington (1964), op. cit., pp. 222–226.
15. A.J. Ihde, *The Development of Modern Chemistry*, Harper & Row, New York, 1964, pp. 154–155.
16. Partington (1962), op. cit., pp. 713–714.
17. Partington (1964), op. cit., pp. 875; p. 886.
18. Partington (1964), op. cit., p. 882.
19. J.S. Stas, *Untersuchungen Über Die-Gesetze Der Chemischen Proportionen Über Die Atomgewichte Und Ihre Gegenseitigen Verhältnisse*, Verlag Von Quandt & Händel, Leipzig, 1867, pp. 187–200; see folding plate at the end of the book. This is the first German edition. The first French edition was published in 1865.
20. Hinrichs, op. cit., pp. 292–295.
21. Hinrichs, op. cit., pp. 20–35.
22. G.N. Quam and M.B. Quam, *Journal of Chemical Education*, Vol. 11, p. 288, 1934.
23. G. Hinrichs, *The Elements of Chemistry and Mineralogy*, Griggs, Watson, & Day, Davenport, 1871, p. 101.
24. Hinrichs (1871), op. cit., p. (161)—second leaf following p. 158.

WHY IS PROUT'S HYPOTHESIS STILL IN MODERN TEXTBOOKS?

Prout observed in 1815 that gas densities were whole-number multiples of the density of hydrogen gas. This led to his idea that all atomic weights are whole-number multiples of the atomic weight of hydrogen and that hydrogen might well be the "primary substance" from which all other elements are made. However, subsequent observations such as the atomic weight of chlorine (ca. 35.5) led to "protyles" having half the weight of the hydrogen atom. Chemical analyses and formulas at the time of Prout and for the next 50 or so years contained sufficient errors to cast doubts on published decimal-place accuracies of atomic weights. However, as analytical chemistry improved, it became abundantly clear toward the end of the nineteenth century that the deviations from unity, $\frac{1}{2}$ or even $\frac{1}{4}$, were real and significant and our "cranky" friend Hinrichs (previous essay) should have accepted this. As so often happens in science, these tiny discrepancies were hinting at things much more profound—the subatomic structure of matter.[1–3]

When chemistry is first taught to a student, the first exam may have "fill-in-the blanks questions" such as

The atomic number is the—number of protons

The atomic mass number is the—number of protons plus neutrons

The number of protons equals the number of—electrons

These simplified concepts tend to exemplify the apparent utility of Prout's hypothesis as an organizing principle. The aomic mass number, often incorrectly truncated to "atomic mass," treats protons and neutrons as equals. Taken too literally, the mass of uranium–238 would appear to be roughly equal to that of 238 hydrogens (protium or hydrogen–1 atoms) with some tiny discrepancy understood as arising from the 0.1% difference in mass between protons and neutrons. In fact, if we take the masses of 92 protons, 146 neutrons, and 92 electrons, the total mass is 240.0 amu.

The subatomic structure of the atom began to emerge in the late nineteenth and early twentieth centuries, thanks to the development of the Crookes [after Sir William Crookes (1832–1919)] tube and the work of J.J. Thomson. Isotopes, atoms of the same element (atomic number) having different atomic masses, emerged independently from two lines of investigation.[1] The newly discovered radioactive elements often included species that were chemically identical but had distinct radioactive decay properties.[1] Thus, two distinct species of uranium were discovered, each having a unique decay pattern. B.B. Boltwood at Yale discovered a new element, "ionium," which was an intermediate in the decay of uranium-II, but not uranium-I.[1] However, he learned that "ionium" was chemically identical with thorium. Frederick Soddy coined the term "isotope" ("same place"—i.e., in the periodic table) in 1913.[1] In 1919, F.W. Aston modified a technique of Thomson's and discovered that ionized neon atoms produced two different ions, one of mass 20, the other of mass 22.[1] It was the difference in atomic mass that accounted for isotopes. Not long afterward, chlorine was found to be a mixture of two isotopes (mass numbers 35 and 37) with a statistical average atomic mass of 35.45 amu.

Following the discovery of the long-suspected neutron in 1932, almost 25 years after Einstein's theory of relativity, here is where things stood with regard to Prout's hypothesis:

1. The simplest hydrogen isotope is protium. It has a proton (relative charge +1) in its nucleus and electron (relative charge –1) outside the nucleus.
2. The masses and (relative) charges of the three major subatomic particles are as follows (with the mass of the carbon-12 atom for reference):[3]

Particle	Charge	Mass (amu Relative to Carbon 12)
Proton	+1	1.0073
Neutron	0	1.0087
Electron	–1	0.0005486
Carbon-12 atom	0	12.0000000 (assumed)

The fact that the proton is 0.1% lighter than the neutron is not consistent with the literal interpretation of Prout's hypothesis.

3. Since we now know that a neutron free of the nucleus has a half-life of 17 minutes as it decays to a proton, an electron and an antineutrino of negligible mass, it might be tempting to think of the source of Prout's hypothesis as effectively the total mass of the proton and one electron. The total for the two isolated particles, 1.0078, is still less than that of the neutron. Nonetheless, it might also be tempting to consider the "protyle" to have the mass of the neutron.

4. Clearly, isotopes are the major source of the discrepancy with Prout's hypothesis. Hydrogen is 99.986% protium, the lightest isotope (1.0078 amu = mass of proton plus electron) and only 0.014% deuterium (2.0141 amu).[2] The amount of tritium is ultratrace. This coincidence is why hydrogen so often "works" as the apparent "protyle" in Prout's hypothesis. If, for example, protium were 80% and deuterium 20% in naturally occurring hydrogen, Prout's hypothesis would never have existed. For chlorine the two natural isotopes occur in significant amounts: chlorine–35, 75.53%; chlorine–37, 24.47%. The observed mass in naturally occurring chlorine is the weighted average: 35.45 amu—impossible to rationalize using Prout's hypothesis.

5. Another major discrepancy is the packing effect of nuclear particles. Thus, if we sum up the masses of all four nuclear particles in helium–4 (two protons + two neutrons = 4.03190 amu) and compare the sum to the observed mass (i.e., minus the two electrons) of the helium nucleus (4.00150 amu), the discrepancy (0.03040 amu) furnishes the energy (strong force, $\Delta E = \Delta mc^2$) that binds the nucleus.[3] This energy is about a million times more powerful than the chemical forces released by explosion of dynamite or TNT. The "mass defect" in uranium-238 (92 protons, 146 neutrons) is an amazing 1.9353 amu. Luckily, the calculated "excess" (1.9356 amu) from the sum of the masses of the nuclear particles (239.9356 amu), reduced by the binding energy equivalent mass (1.9353 amu), leaves us blissfully happy that the nuclear mass is 238.0003 amu (virtually identical to the atomic mass number). Now if we add those 92 electrons, another 0.050 amu is added, still too little to shake our blissful, sloppy complacency in using the atomic mass number to specify atomic weight.

As stated earlier, one might try imagining the neutron to be the "protyle" of matter or even the "primary material." However, a physicist would respond today that quarks are the primary material. The mass of the neutron is said to comprise these quarks plus their energy. It is a scary thing for a chemist to learn, however, that physicists admit that they do not yet really understand the fundamental nature of mass.

---

1. J.R. Partington, *A History of Chemistry*, Macmillan and Co. Ltd., London, 1964, Vol. 4, pp. 929–947.
2. *The New Encyclopedia Britannica*, Encyclopedia Britannica, Inc., Chicago, 1986, Vol. 14, pp. 343–348.
3. T.L. Brown, H.E. LeMay, Jr., and B.E. Bursten, *Chemistry The Central Science*, seventh edition, Prentice-Hall, Upper Saddle River, NJ, 1997, pp. 43–46; 771–791.

SECTION VIII
SOME FUN

CLAIRVOYANT PICTURES OF ATOMS—A STRANGE CHYMICAL NARRATIVE

> I remember the occasion vividly. Mr. Leadbeater was then staying at my house, and his clairvoyant faculties were frequently exercised for the benefit of myself, my wife and the theosophical friends around us. I had discovered that these faculties, exercised in the appropriate direction, were ultramicroscopic in their power. It occurred to me once to ask Mr. Leadbeater if he thought he could actually *see* a molecule of physical matter. He was quite willing to try, and I suggested a molecule of gold as one which he might try to observe. He made the appropriate effort, and emerged from it saying the molecule in question was far too elaborate a structure to be described. It evidently consisted of an enormous number of some smaller atoms, quite too many to count; quite too complicated in their arrangement to be comprehended. It struck me at once that this might be due to the fact that gold was a heavy metal of high atomic weight, and that observation might be more successful if directed to a body of low atomic weight, so I suggested an atom of hydrogen as possibly more manageable. Mr. Leadbeater accepted the suggestion and tried again. This time he found the atom of hydrogen to be far simpler than the other, so that the minor atoms constituting the hydrogen atom were countable. They were arranged on a definite plan, which will be rendered intelligible by diagrams later on, and were eighteen in number.[1]

This narrative appears early in the second edition (published in 1919) of the strange but fascinating book *Occult Chemistry—Clairvoyant Observations on the Chemical Elements*, authored by Annie Besant and Charles W. Leadbeater and first published in 1908.[2] The fact that a deluxe edition of the book was published as late as 1951[3] is evidence of the enduring allure of this richly illustrated text. Well, what are we to make of its contents? Figure 167 depicts the structure of a "chemical atom" of sodium. *The method of examination employed was that of clairvoyance.*[4] This chemical atom consists of upper and lower parts (each composed of a globe and 12 funnels) and a connecting rod. The parts inside the funnels, globes, and rod and counted below are the "smaller atoms" referred to above:

> We counted the number in the upper part: globe–10; the number in two or three of the funnels—each 16; the number of funnels—12; the same for the lower part; in the connecting rod—14; Mr. Jinarajadasa reckoned: 10 + (16 × 12) = 202; hence 202 + 202 + 14 = 418: divided by 18 = 23.22 recurring. By this method we guarded our counting from any pre-possession, as it was impossible for us to know how the various numbers would result on addition, multiplication and division, and the exciting moment came when we waited to see if our results endorsed or approached any accepted weight.[5]

PLATE I.

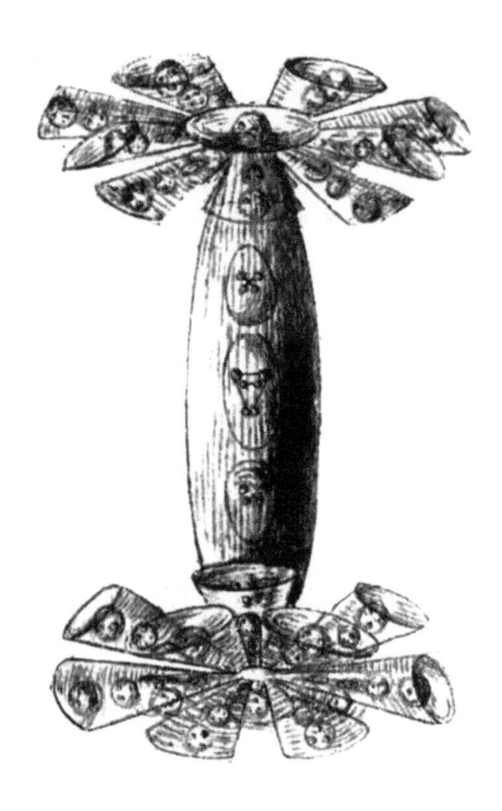

SODIUM.

FIGURE 167. ■ The structure of the sodium atom obtained through clairvoyance and published in the classic book (*Occult Chemistry*, 1908). There are a total of 418 smaller "physical atoms" in the sodium "chemical atom." Clairvoyance also showed that a hydrogen atom consisted of 18 smaller physical atoms. So . . . 418 divided by 18 = 23.22, in pretty darned good agreement with 22.99! (Q.E.D.)

Et, voilá! The accepted atomic weight of sodium is 23.0—in pretty darned good agreement with 23.2, it seems. In total, 57 of the 78 recognized elements were examined as well as one previously unknown, "occultum," a "chemical waif" tucked between hydrogen and helium. Also, six new "varieties" of known elements were reported—not a bad day's haul. The agreement between accepted atomic weights and the clairvoyant count of "smaller atoms" is impressive, and this is one major component of the scientific validation of the atoms derived through clairvoyance.

The clairvoyant uncertainty in obtaining an observation is noteworthy. Note that only two or three funnels are sampled for counting, and uncertainties in counting of 1 or 2 "smaller atoms" are admitted by the investigators. Pretty difficult to keep this vision from flickering in and out without getting a severe headache. So the limitations here might illustrate a kind of "clairvoyance uncertainty principle."[6]

But what of the constituent "smaller atoms?" They are shown in Figure 168—these fundamental ("smaller" or "ultimate physical") atoms are found to be male and female. For the male atoms, force whorls in from fourth-dimensional space (the astral plane) and *out* into the physical world. The female atoms take force *in* from the physical world and whorl it, in the opposite screw sense, back into the astral plane. *Hmm.* The relationship to the male (sulfur) and female (mercury) imagery of early alchemy[7] is pretty apparent—Sol and Luna, the "atoms family."

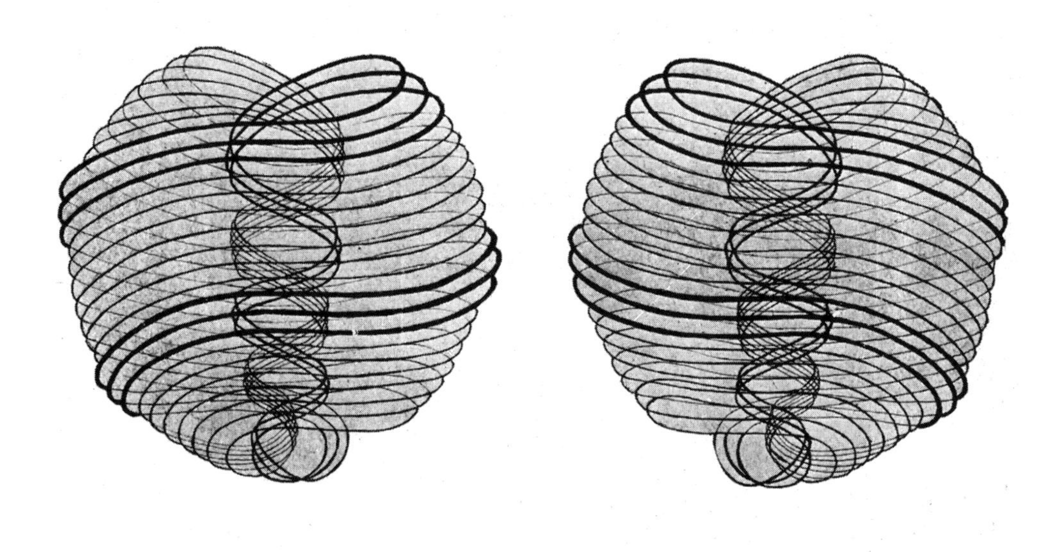

FIGURE 168. ▪ The smaller physical atoms are, by the way, male and female. We have returned to Sol and Luna, sulfur and mercury. Thus, the hydrogen atom has nine female smaller physical atoms and nine male physical atoms. (From *Occult Chemistry*, 1908.)

The other powerful scientific validation of the chemical atoms derived by clairvoyance is their seeming consistency in explaining the chemical and physical properties so neatly organized by the periodic law. Figures 169 and 170 depict family types of chemical atoms, not protozoans as they might appear to the uncritical eye (or even *d* and *f* orbitals to a wishful-thinking chemist!). The point can be illustrated succinctly—the structural type I in Figure 169 is classified as the "dumbbell" class and includes copper, silver, and gold, three coinage metals found in group 11 of the periodic table. However, Besant and Leadbeater also place sodium in this class—the structure of the sodium atom (see Figure 167 again) makes it one of the dumbbell class. However, I suggest, gentle reader, that you avoid accepting a sodium penny unless you love to burn money (and your hand as well). The confusion is clarified by the errors incurred by the authors' use of the periodic roller coaster supplied to them by the truly eminent scientist Sir William Crookes (1832–1919), with whom they had maintained some correspondence (see Figure 171). (In a later book[8] by the aforementioned Jinarajadasa another version of Crookes' periodic table appears, but this one specifically includes "elements discovered first by clairvoyant investigation"—talk about chutzpah!) Crookes, the inventor of the vacuum tube that led to the discovery of electrons as well as X rays, developed an interest in spiritual and highly speculative ideas and what William Brock terms "metachemistry"[9]—clearly an appropriate scientific correspondent for the authors of *Occult Chemistry*.

In summary, it appears that even in the early twenty-first century, the theory may require some further study and some modifications. Although the Bohr atom and the Lewis–Kossel–Langmuir explanation of the octet rule came after the first edition and before the second edition of *Occult Chemistry*, the latter was little modified. There is also no evidence that Besant and Leadbeater saw any need to include the concept of atomic numbers developed by Moseley during this same period. Models that can survive the assault of more modern theories, including quantum mechanics and their supporting experimental data, must be powerful indeed. I am thus recommending grant support and imagining the nature of a research budget:

National Séance Foundation
Proposal Title: *Atoms and Astral Plane Interactions*
Budget

I. Personnel:
   Clairvoyant (100% academic load)
   Artist/Recorder (50% academic load)
   Physical atom counter/mathematician (50% academic load)
II. Equipment
   Clairvoyaniscope (one)
   Astral plane direction detector (clairvoyaniscope option kit Y2K) (one)
   Calculator (one)
III. Supplies
   Aspirins (10 gross)
   Herbal teas (10 gross)
IV. Facilities and administration (85% of basic budget)

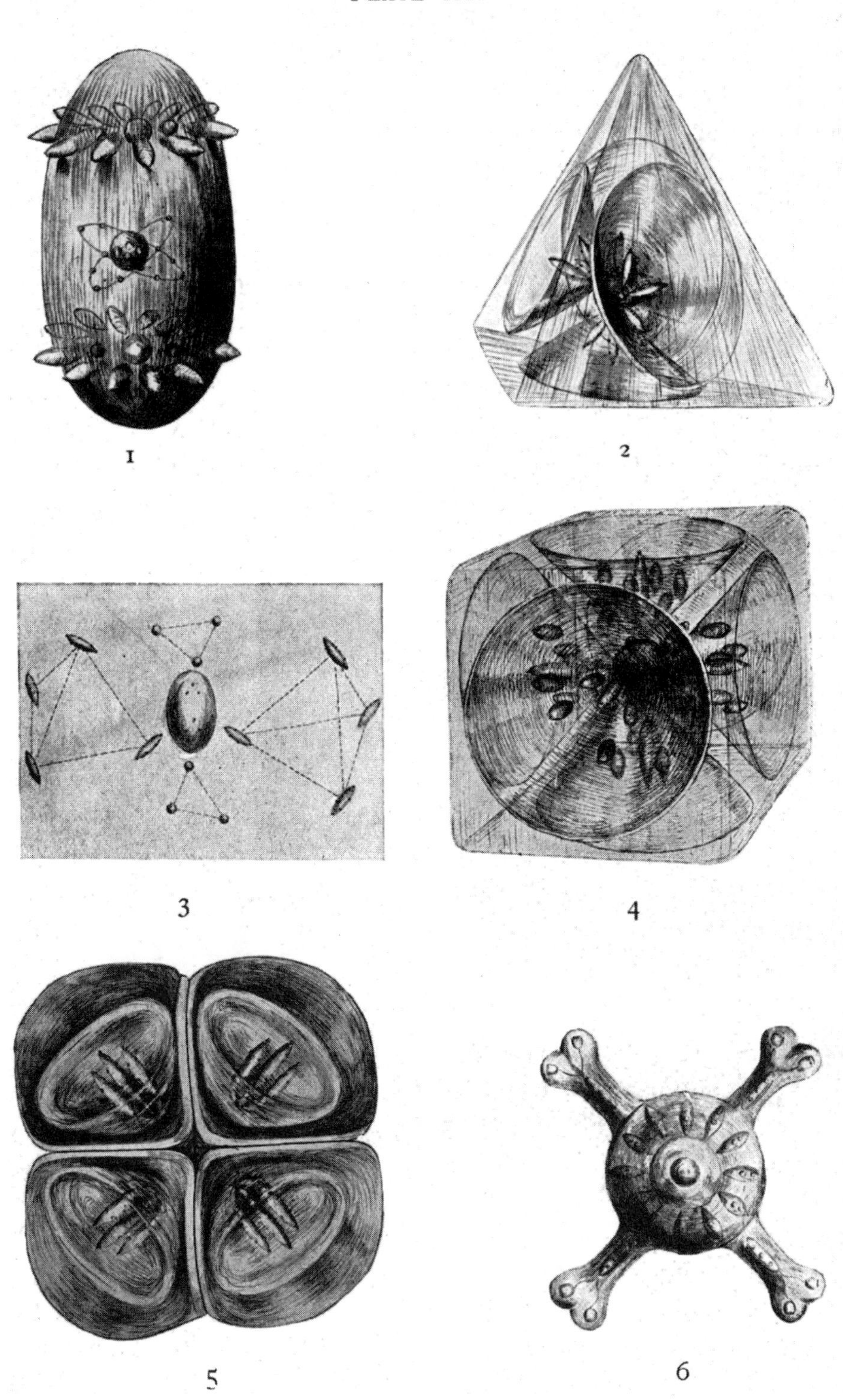

FIGURE 169. ▪ Here is why the Periodic Law works according to Besant and Leadbeater—atoms in the same families share similar shapes. For example, look at structure 1 in this figure—representing the structure of copper, silver and gold atoms. Since some early classifications actually grouped sodium with copper, silver, and gold (see Figures 154 and 171), it would appear perfectly sensible that these four atoms would bear some familial resemblance (see Figure 167). Fortunately, we do not make coins of sodium, so our change does not burn holes in our pockets. (Figures from *Occult Chemistry,* 1908.)

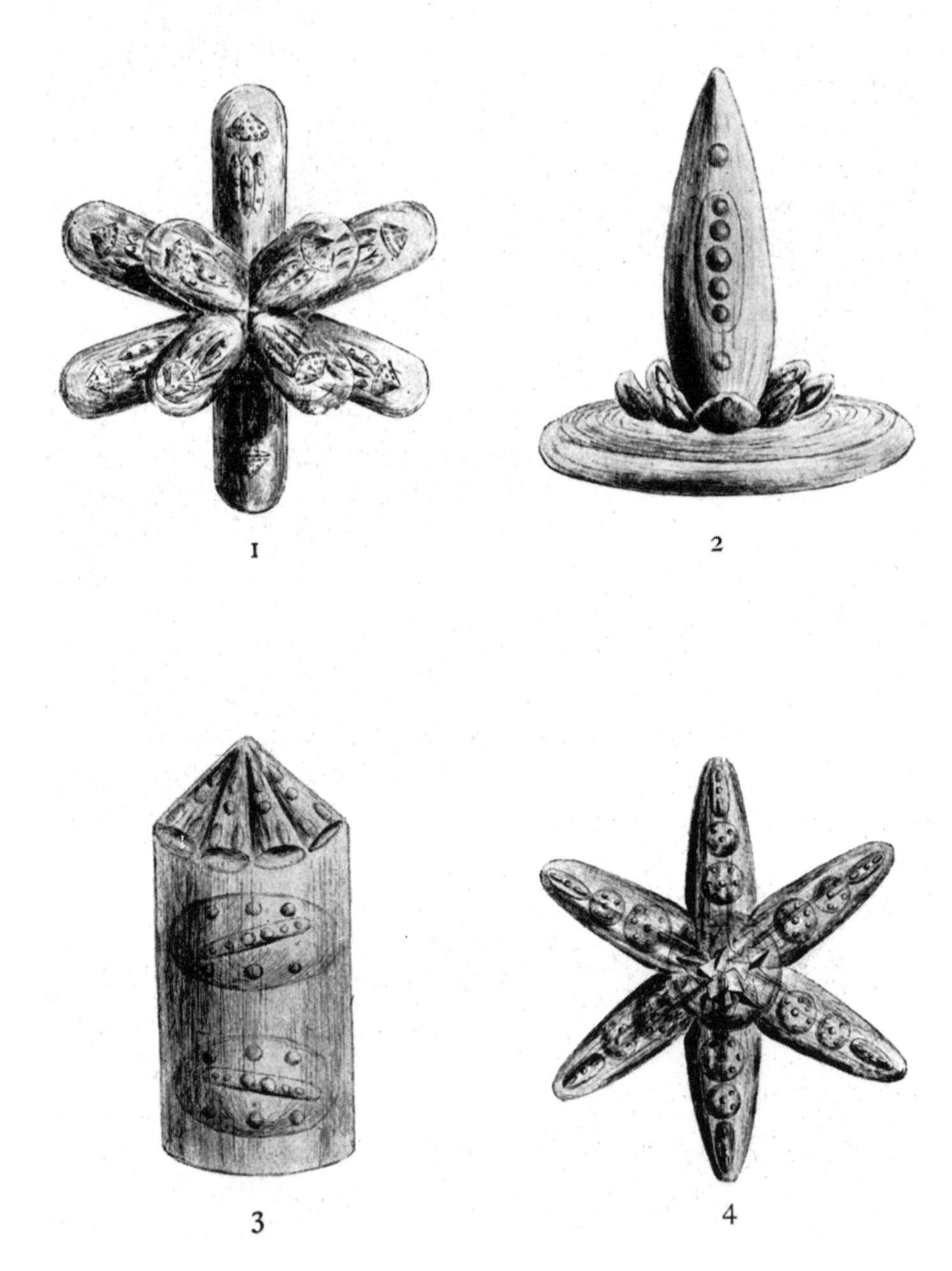

FIGURE 170. ▪ More chemical atom structural types (from *Occult Chemistry*, 1908). Professor Pierre Laszlo commented on their uncanny resemblance to atomic orbitals.

I confess that I am having difficulty in mechanically interpreting the manner in which clairvoyant pictures of atoms are obtained. But I have always had similar problems trying to fathom how the Philosopher's Stone changes lead to the "purer" metal gold through a mysterious process termed "projection." Although Robert Boyle was credulous about alchemy and wrote "a strange chymical narrative" describing a reverse transmutation, he probably would not have been en-

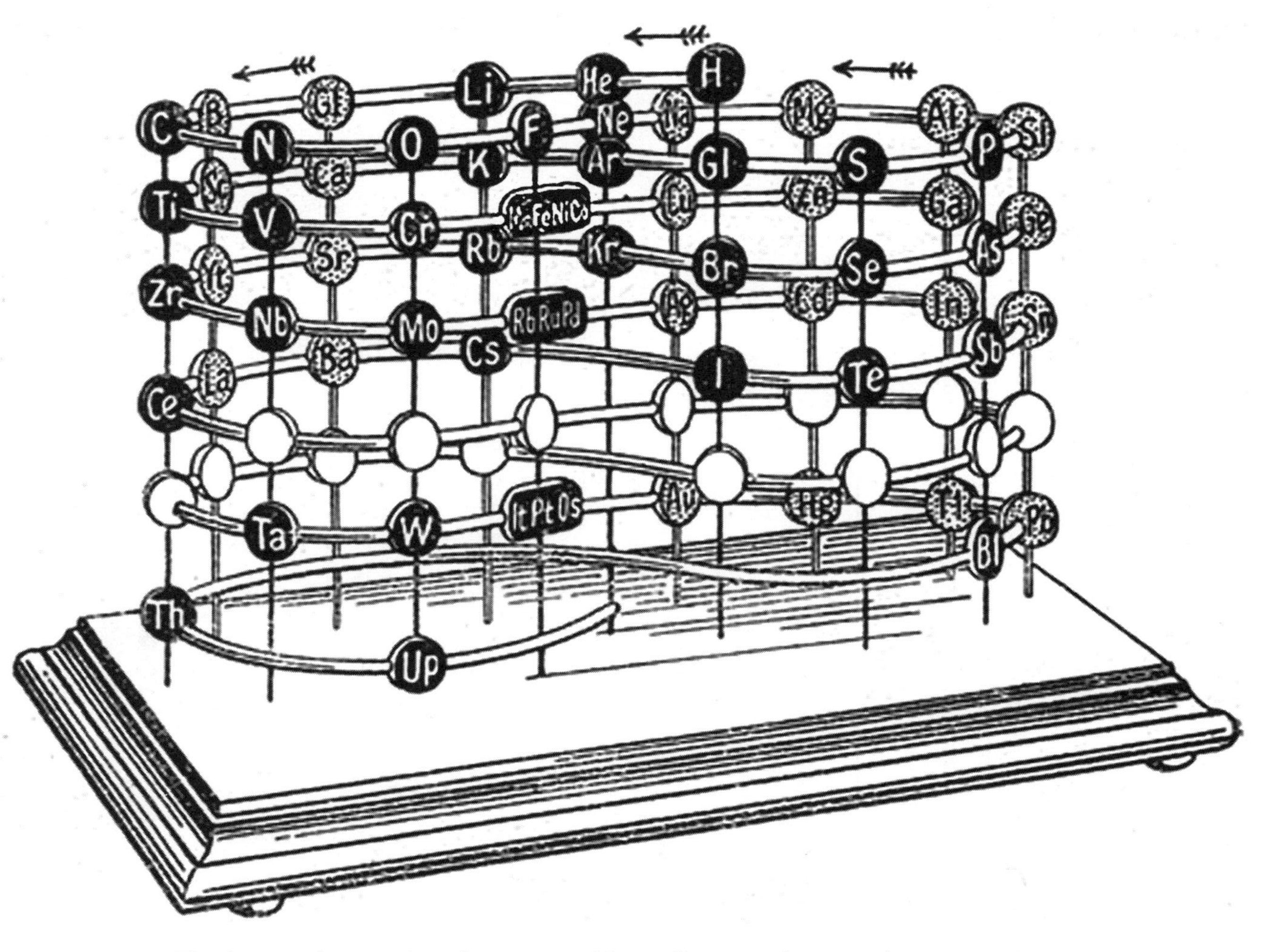

FIGURE 171. ■ The figure-eight periodic roller coaster of Sir William Crookes. A truly eminent physicist whose vacuum tube helped to establish the properties of the electron in the late nineteenth century, Crookes was also sympathetic to occult beliefs. (From *Occult Chemistry*, 1908.)

thused about clairvoyance as an experimental technique.[10] Well, methinks Mr. Leadbeater has a wonderfully appropriate name, and it would not surprise me to learn that *he*, at least, understood both clairvoyance *and* projection.

Actually, the more interesting co-author is Mrs. Annie Besant (1847–1933), who Emsley describes as "a fiery social reformer with socialist tendencies and boundless energy."[11] Mrs. Besant, originally married to a vicar but eventually separated, secretly published a pamphlet questioning the divinity of Jesus Christ and later an article on birth control. She helped organize a strike by the "poorest of the poor and the lowest of the low" (mostly women and children) against a London manufacturer of matches in 1888 and scored a smashing victory for the rights of the workers.[11] In 1889, she converted to the doctrines of the Theosophical Society, which emphasized human service and spiritualism and served as its president from 1907 to 1933, living in its home city in Madras, India.[11] Mrs. Besant was an early advocate of India's independence and formed the Indian Home Rule League in 1916.[12] In sum: a totally "difficult" and wonderful woman.

1. A. Besant and C.W. Leadbeater, *Occult Chemistry—Clairvoyant Observations on the Chemical Elements*, revised edition, Theosophical Publishing House, London, 1919, pp. 1–2.
2. A. Besant and C.W. Leadbeater, *Occult Chemistry—a Series of Clairvoyant Observations on the Chemical Elements*, Theosophist Office, Adyas (Madras), 1908.
3. A. Besant and C.W. Leadbeater, *Occult Chemistry—Clairvoyant Observations on the Chemical Elements*, third edition, Theosophical Publishing House, Adyas (Madras), 1951.
4. Besant and Leadbeater (1908), op. cit., p. 2.
5. Besant and Leadbeater (1908), op. cit., p. 3.
6. Unpublished discussions with Professor Joel F. Liebman, who places occult chemistry in the realm of "arts and séances" and further suggests the National Séance Foundation as a potential research funding source.
7. A. Greenberg, *A Chemical History Tour*, John Wiley, New York, 2000, p. 36.
8. C. Jinarajadasa, *First Principles of Theosophy*, third edition, Theosophical Publishing House, Adyar (Madras), 1923, pp. 156–181.
9. W.H. Brock, *The Norton History of Chemistry*, W.W. Norton & Co., New York and London, 1993, pp. 454–459.
10. R. Boyle, *An Historical Account of a Degradation of Gold, Made by an Anti-Elixir: A Strange Chymical Narrative*, R. Montagu, London, 1739 (the original 1678 edition was anonymous); see A. Greenberg, *A Chemical History Tour*, John Wiley, New York, 2000, pp. 92–94.
11. J. Emsley, *The Thirteenth Element; The Sordid Tale of Murder, Fire and Phosphorus*, John Wiley & Sons, Inc., New York, 2000, pp. 89–96.
12. *Encyclopedia Britannica*, Vol. 2, Encyclopedia Britannica, Inc., Chicago, 1986, p. 165.

THE PERIODIC HELIX OF THE ELEMENTS

In 1869, Mendeleev ordered the 63 then-known elements according to increasing atomic mass and placed them in rows having related chemical properties. This original vertical periodic table (Figure 153) was soon replaced by the horizontal form familiar to us today.[1] Other representations were also feasible, and these included spirals and helices[2] (see Crooke's 1898 figure-eight spiral in Figure

171). In 1916 W.D. Harkins and R.E. Hall published a wondrous periodic helix of the elements (Figure 172).[3] A month after Harkins and Hall submitted their paper to the *Journal of the American Chemical Society,* Gilbert N. Lewis submitted his paper "The Atom and the Molecule" to the same journal. His simple electron-dot formulas allowed researchers to merely glance at the periodic table and predict the nature of bonding (single, double, or triple bonds) between atoms of the main-group elements.[4]

Three years earlier, Henry G.J. Moseley (1887–1915) discovered that the square root of the frequencies of X rays emitted from different metallic cathodes was directly proportional to simple cardinal numbers that he termed "atomic numbers."[5] The atomic number—the integer number of positive charges in an atom's nucleus, and not the atomic weight—is the true determinant of an element's identity. It is the atomic number that provides the continuous one-by-one "roll call"[6] of elements that underlies periodicity. The eighth element after lithium (#3) is sodium (#11), and eight elements later comes potassium (#19). All three share very similar chemical properties. If the first 19 elements had been placed in strict order of atomic weights, #19 would have been argon (atomic weight = 39.95) and potassium would have been #18 (atomic weight = 39.10). Chemically, of course, this would have been nonsense. It was never a problem for Mendeleev back in 1869 because, happily for him, the inert gases had not yet been discovered.

Moseley's stepwise "walk" through the periodic table clearly indicated missing members of a larger and more complex family of 85 then-known elements. Descend Harkins' "staircase" (Figure 172) to the very bottom, and you finally arrive at the heaviest element known to Moseley—uranium (atomic # 92). (Uranium had first been isolated as an oxide in 1789; the pure metal was reported in 1841.[7]) This is the source of the "magical number" 92, part of our "Chemical Kabbala"—the number of "naturally occurring" elements fixed in our minds by Moseley. The reality is much more complex. Note in the foreground of Figure 172 a vacancy, corresponding to element #87 just below cesium (Cs), two vacancies below manganese (Mn) for #43 and #75, and one below iodine (I) corresponding to element #85. If we descend into the dark, dingy, and dangerous "basement," we discover, upon passing thallium (Tl), clutter and confusion. Radiation, first discovered in 1898 by Henri Becquerel,[8] was a by-product of naturally occurring transmutations of elements exchanging identities before our very eyes. There are six different lead (Pb) isotopes in Figure 172. Below xenon (Xe) we see mysterious #86, an emission from thorium (Th Em) and also from Marie Curie's radium[8] (Ra Em), the latter briefly named "Nitonium" (Nt).

Let us escape the radioactive basement and ascend the staircase. Just above tantalum (Ta, #73) there is a break and we must "scuttle up" a "rope ladder" of 15 elements. The topmost of these, lanthanum (La, #57), is connected by a strange loop to #58 (cerium, Ce). We have encountered the "rare earth" elements that are today recognized to include lanthanum, the 14 "lanthanides" (#58–71), as well as the lighter elements yttrium (Y, #39) and scandium (Sc, #21). When Harkins and Hall first published their helical representation, it was assumed that elements #57 through #72 were all rare earths.

The marked differences in chemical reactivities between adjacent elements (e.g., sulfur versus chlorine) that guided Mendeleev were largely absent in the 17

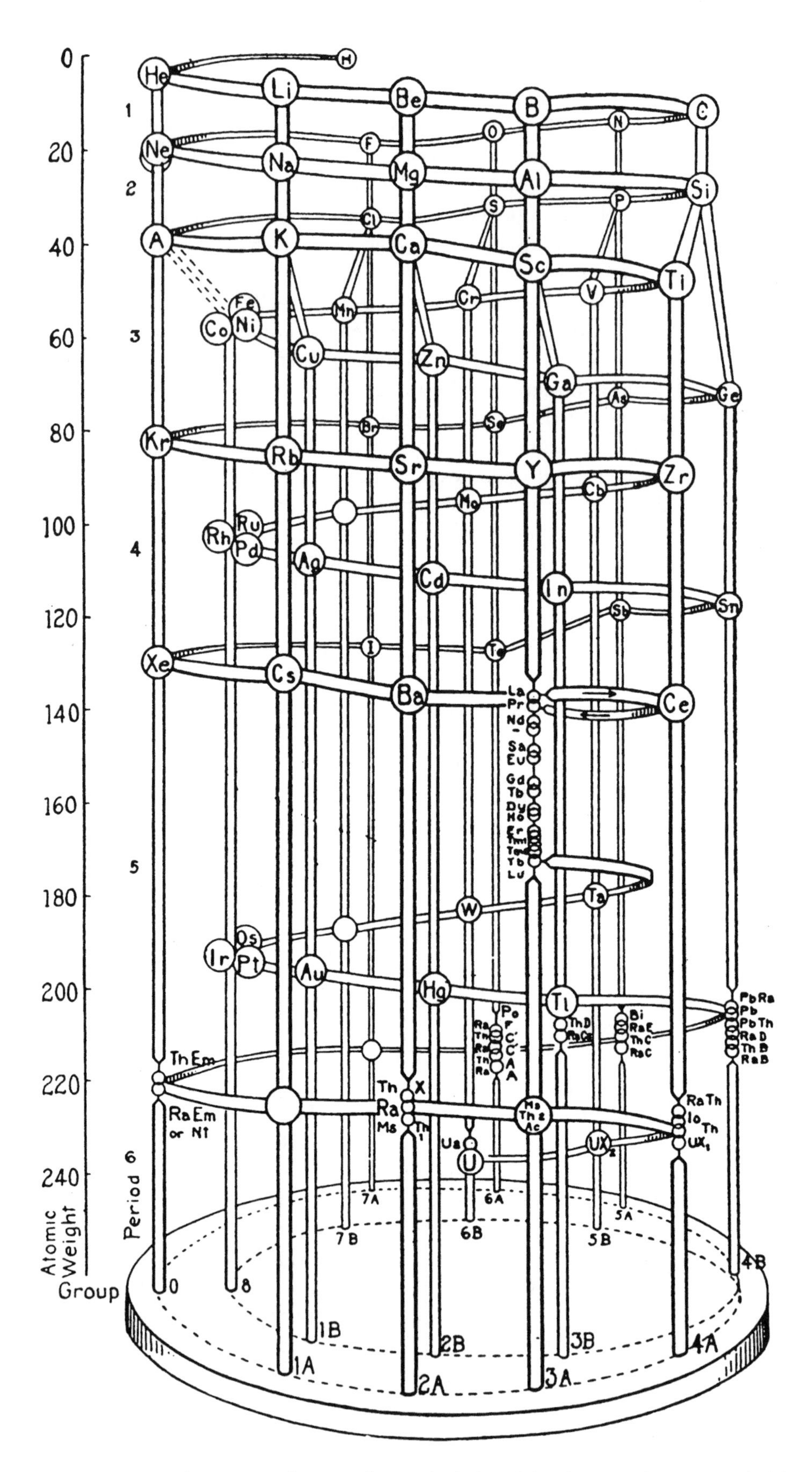

FIGURE 172. ■ The 1916 Harkins–Hall periodic helix of the elements (from *Journal of the American Chemical Society*, 1916, with permission of The American Chemical Society).

rare earths. Their chemistry was so similar (all commonly formed valence 3, MX_3-type compounds) that they were exceedingly difficult to separate. This is the source of a 150-year-long saga in the history of chemistry.[9,10] In 1794 John Gadolin obtained an unknown "earth" (a now-extinct term for oxides) from a black ore called *ytterbite* in the Swedish village of Ytterby and discovered the metal yttrium. In 1803, Jons Jacob Berzelius and Wilhelm Hisinger (Sweden) and Martin Klaproth (Germany) isolated and announced almost simultaneously another new element, cerium, from the mineral cerite. Credit for the discovery is given to Klaproth because he was the first to *seemingly* purify it. In fact, improvements in separation techniques, the development of new techniques, and the use of the spectroscope developed by Gustav Kirchhoff and Robert Wilhelm Bunsen guided the isolation of all 17 rare earths (Figure 173) from the original "two." In 1907, Georges Urbain (France), Carl Auer von Welsbach (Austria), and Charles James (United States) almost simultaneously announced separation of the final rare earth. This, too, was not without adventure and controversy, but ultimately Urbain was credited with the discovery of lutetium (Lu, #71).[9,10] Moseley confirmed the placement of these rare earths in the family of elements and noted one missing element (atomic number 61) and predicted its existence. In Figure 172, the mysterious element 61 is a blank space between neodymium (Nd) and samarium ("Sa"). Another missing element, #72, assumed by Urbain and Moseley and others to be a rare earth, was sought in vain from ore samples that had yielded the tight-knit family of 17.

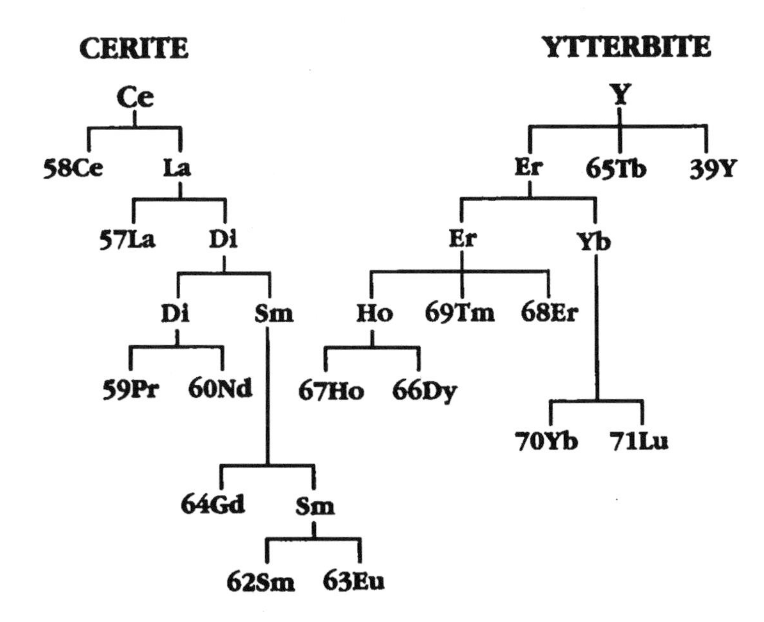

FIGURE 173. ▪ Reprinted with permission from "A Natural Historical Chemical Landmark: Separation of Rare Earth Elements, University of New Hampshire, October 29, 1999." Copyright 1999 by the American Chemical Society.

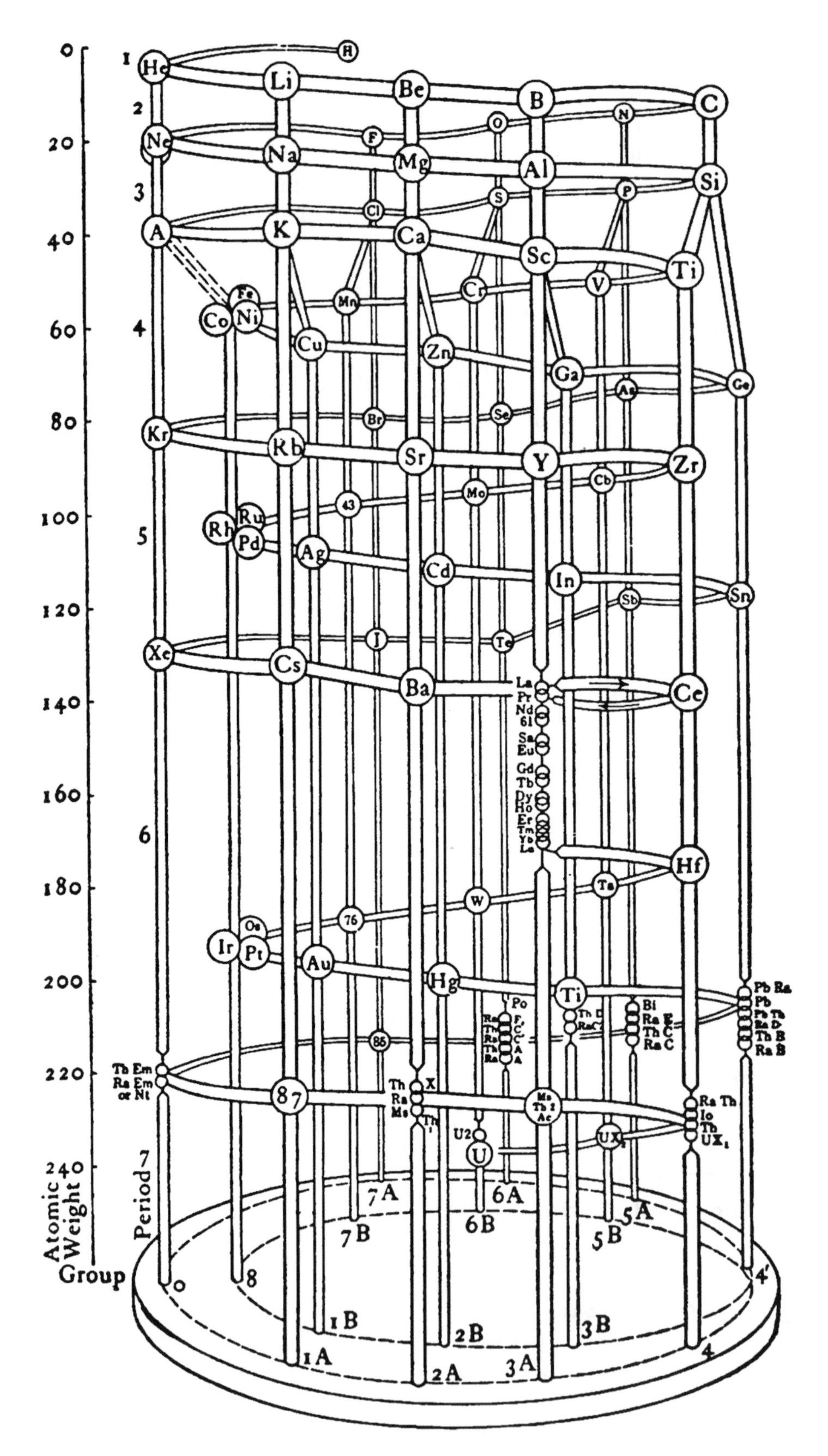

FIGURE 174. ▪ Later version of the Harkins–Hall periodic helix including the 1923 discovery of halfnium (Hf, #72—*not* a rare earth as some originally thought) and the isolation of rhenium (Re, #75) in 1925, although it is not explicitly named here. (Used with permission from *Journal of Chemical Education*, 1934.)

Figure 174 is an updated Harkin's helix published in 1934.[1] The truly distinctive aspect of science is its ability to make predictions and test them—the more daring, the better. The blank spaces in Mendeleev's original periodic table (Figure 153) were audacious predictions of new elements such as gallium (*eka*-aluminum) and germanium (*eka*-silicon). All of the new elements predicted by Moseley were found within the next 24 years. The mystery of hafnium (Hf, #72) was solved when Niels Bohr applied his version of quantum numbers and realized that #72 was not similar to #57–#71 but should be quite similar in chemical behavior to zirconium (Zr), which is commonly tetravalent. Once this was realized, the new element was found in tiny quantities in zirconium ores, laboriously separated and reported in 1923.[10] Similar considerations led to the isolation of rhenium (Re) in 1925 (although it is still listed as #75 in Figure 172).[10] The era of nuclear chemistry began in earnest in the late 1930s. Technetium (Tc, #43), francium (Fr, #87), astatine (As, #85), and the final rare earth, promethium (Pm, #61—see Figure 174) were isolated over the course of a decade. Moseley would have been 60 years old had he lived to witness the discovery of promethium in 1947. Sadly, he was drafted during World War I and was shot and killed in the battle of Gallipoli.

Transuranium elements (with atomic numbers greater than 92) were first discovered in 1940.[11] The first two, neptunium (Np, #93) and plutonium (Pu, #94) actually do occur naturally in ultratrace amounts. However, during the 1940s significant quantities were made using nuclear bombardment. A series of increasingly heavy, highly unstable synthetic new elements, reported over the course of the next five decades, provided a very extended periodic table.[11] If we continue to use the spiral staircase analogy, then nuclear chemists have been digging below the radioactive basement into an even stranger subbasement. However, there is no reason to assume that the helix must start lightest at the top and descend with ascending atomic number. As an organic chemist, the arrangement in Figure 174 is just fine—my elements, hydrogen, carbon, nitrogen, and oxygen are closest to heaven. However, one person's heaven is another person's hell. I suspect that nuclear chemists would better appreciate an ascent with increasing atomic number. Their goal has long been an "island of stability" (perhaps a "cloud of stability"?) of superheavy elements.[11] An organic chemist such as I might view their goal to be "Atlantis."[11] In 1999, it appeared that the promised land (or cloud) had been reached with the nuclear synthesis of element #118.[11] However, the claim was subsequently withdrawn, and the goal remains elusive.[12]

1. A. Greenberg, *A Chemical History Tour*, John Wiley and Sons, New York, 2000, pp. 214–216.
2. G.N. Quam and M.B. Quam, *Journal of Chemical Education*, Vol. 11, pp. 27–32, 217–223, 288–297 (1934).
3. W.D. Harkins and R.E. Hall, *Journal of the American Chemical Society*, Vol. 38, pp. 169–221 (1916).
4. Greenberg, op. cit., pp. 266–268.
5. Greenberg, op. cit., pp. 271–274.
6. J.R. Partington, *A History of Chemistry*, Macmillan and Co. Ltd., London, 1964, pp. 950–951.
7. A.J. Ihde, *The Development of Modern Chemistry*, Harper & Row, New York, 1964, pp. 747–749.
8. Greenberg, op. cit., pp. 263–266.
9. W.H. Brock, *The Norton History of Chemistry*, W.W. Norton & Co., New York, 1993, pp. 327–330.
10. American Chemical Society, *A National Historic Chemical Landmark—Separation of Rare Earth*

Elements, University of New Hampshire, Durham, New Hampshire, October 29, 1999, Division of the History of Chemistry, American Chemical Society, Washington, DC, 1999.
11. Greenberg, op. cit., pp. 282–285.
12. R. Monastersky, *The Chronicle of Higher Education*, August 16, 2002, pp. A16–A21.

WHITE LIGHTNING IN AN ATOM, A KISS OR A STAR

Chemist Primo Levi's powerful book, *The Periodic Table*,[1,2] employed 21 elements as chapter titles, to explore symbolically his experiences, memories, and dreams as an Italian-born Jew working in World War II Turin. For example, in the opening chapter, Argon, Levi likens his Italian renaissance ancestors and their heirs to the inert gases:[3]

> The little that I know about my ancestors presents many similarities to these gases. Not all of them were materially inert, for that was not granted them. On the contrary, they were—or had to be—quite active, in order to earn a living and because of a reigning morality that held that "he who does not work shall not eat." But there is no doubt that they were inert in their inner spirits, inclined to disinterested speculation, witty discourses, elegant, sophisticated, and gratuitous discussion . . . . Noble, inert, and rare: their history is quite poor when compared to that of other illustrious Jewish communities in Italy and Europe.

(Incidentally, Levi mistakenly assumed that Professor Neil Bartlett was awarded a Nobel Prize in Chemistry for discovering in 1962 that the inert gas xenon reacts to form chemical compounds.[3] Although Levi's history is wrong, I think his judgment is sound.)

Fifty years before publication of *The Periodic Table*, Edwin Herbert Lewis, "an eccentric English Professor at Chicago's Lewis Institute,"[4] authored *White Lightning*[5] (Figure 175), a 354-page novel divided into 92 chapters each named for elements in order of (the recently discovered) atomic number. It is a Mendeleevianly confident book—Chapters 43, 61, 75, 85, and 87 are unnamed and are particularly mysterious.[6] The novel relates the coming of age of a Marvin Mahan, his tempering through the bombs and gas of World War I, and his emergence in the 1920s as a brilliant young radiochemist. "White lightning" is employed as a metaphor throughout the book for the energy hidden in matter—Marvin is "this imp of bottled lightning";[7] earth viewed from Venus is "a steady point of white lightning."[8] To let the metaphor explode from the bottle, Marvin ponders:[9]

> the cheek of a girl, which feels so smooth to the lips, is really a starry sky full of electric suns and moons. The tension between each sun and its moons is all that keeps the cheek from exploding when you kiss it. And here he had been calling them all "darlin'"! Well, he might have known that girls were composed of electricity. He had often felt it thrilling up his arm.

More ominously, Lewis predicts the use of nuclear weapons:[10] "Nothing but subatomic lightning will teach the Germans anything." Ironically, the author also

FIGURE 175. ■ The front cover and spine section for the 1923 novel *White Lightning* by the mildly "eccentric" Edwin Herbert Lewis. The author was quite knowledgeable and current about chemistry (particularly nuclear chemistry), and his 354-page novel consisted of 92 chapters starting with hydrogen and ending with uranium. Chapters 43, 61, 75, 85, and 87 were, of course, untitled, but Chapter 72 was titled "Hafnium" since it was reported in 1923, the year of publication. Lewis' two most lasting gifts to posterity were the University of Chicago Alma Mater and his daughter, the distinguished poet and novelist Janet Lewis, who died in 1998 at the age of 99.

predicts inevitable war with Japan over natural resources and colonization in Asia. Marvin's left hand is also a powerful symbol since it contains, receives, and releases lightning (Figure 175). A bomb blows it off in the war, and, thus maimed, he has lost youthful perfection and innocence.

While I am not a licensed "lit-crit" (literary critic), I think that a 1916 reviewer of another Lewis' book had him properly pegged as a novelist: "The plot moves swiftly with the help of incredible coincidences and improbable romances."[11] Nevertheless, let us give the author very considerable credit as a knowledgeable and sophisticated observer of chemistry. He was amazingly well informed and current about the complex revolution in the understanding of the structure of the atomic nucleus that was very much in motion as he wrote *White Lightning*. Marvin reads of Henry G.J. Moseley's discovery of atomic numbers in 1914:[12] "This unknown Moseley had found it—a sure way to determine the amount of electricity concealed in the heart of an atom . . . Think of it—an atom of lead is a small universe of compressed lightning carrying eighty-two electric moons in its sky. . . . If a gram of radium emits enough energy to lift five hundred tons a mile high, a gram of disintegrated lead ought to turn every wheel in a great factory!" (And when Moseley is shot and killed at the age of twenty-seven at the battle of Gallipoli—"Lead driven through the one brain that really understood lead"[13]). Marvin attends Yale and works under the supervision of (the very real) Professor Bertram Borden Boltwood, discoverer of "ionium" (soon identified as a thorium isotope from the radioactive decay of uranium).[14] Indeed, Lewis cites the work of Soddy and Aston and their discoveries of isotopes, excitedly relates Rutherford's nuclear transmutation of nitrogen and provides the contemporary understanding of isotopes that rationalizes extra unit masses as due to nuclear protons neutralized by nuclear electrons. The discovery of the neutron by Chadwick occurred in 1932, almost a decade after *White Lightning* was published.

Occasionally, Lewis *does* "hit the mark." Marvin's first two "Darlin's" are Cynthia and Gratia. Jean, the woman who becomes his wife in Chapter 92, witnesses her mother's early death from a sudden stroke on the same day she learns that her brother was killed in the war. Although that chapter (10) is highly contrived, it likens neon to a cold, indifferent and amoral universe—"But all the time the noble gas called neon remained unmoved. Like some quiet-eyed chemist looking down the future, it heard no explosion."[15] And Jean, emotionally numbed, vows chastity in Chapter 18 (Argon) and will allow no man to woo her—"She would be ready for them, as inert as a nun."[16]

But our author cannot resist a periodic stretch and so in Chapter 31 (Gallium, an element predicted by Mendeleev to fill a gap in his periodic table), we have Marvin lightning-struck by his future wife:[17]

> Just as Mendeleyeff had prophesied an element like boron and an element like aluminum, so had he unconsciously known that there must be a girl as impassioned as Cynthia and as exquisitely self-contained as Gratia.

And it gets much worse:

Chapter 25	"Jimmy's face grew much pinker than manganese salts." (Ouch!)
Chapter 27	Somehow I knew just as I started to read Chapter 27 that the Laurentians *had* to be cobalt blue. (What else?)

Chapter 31 "She might not exactly melt in his hand as the metal gallium melts, but she would yield." (Help!)

Chapter 38 Begins: "The happy youth rowed off to his own hired island and for a time sat watching the port lights coming up the river, red as a nitrate of the thirty-eighth element." (I'd prefer—"It was a dark and stormy night . . .")

Chapter 50 (Tin, if you are still paying attention): "I like canned milk first rate" (Of course you would).

Chapter 59 Praseodymium. How *do* you capture the reader's interest with this one?

Well, it all ends happily. Argon *can* form a compound.[18] And (I nearly forgot) Jean initially spurns Marvin's proposal and invites him to return and visit her three years to the day with his wife. During this three-year purgatory, Marvin begins to make his mark, Jean develops an interest in chemistry and sets up a simple laboratory, and her gifted intellect leads her to admire Marie Curie and discover, on her own, some of the fundamental chemical questions of the day. Marvin returns, learns that he will occupy an endowed chair in chemistry in Palo Alto (i.e., Stanford) and is finally accepted by Jean. Although Pierre and Marie Curie are perhaps suggested here, Antoine and Marie Anne Pierrette Lavoisier are probably more apropos.

And what of our idiosyncratic author Mr. Lewis? His most important literary contribution was arguably his daughter—Janet Lewis (1899–1998).[4,19] She was a renowned poet, playwright, and novelist whose most famous work remains *The Wife of Martin Guerre*. She wrote the libretto for William Bergsma's opera of the same title, and her work might be reasonably counted as one of the sources for the French Film *The Return of Martin Guerre*. It is a wonderful thing to imagine a father–daughter dialog that included a mutual interest in Native Americans—they are ubiquitous in *White Lightning* and are the subject of Ms Lewis' first book of poetry (*The Indians In The Woods*, 1922). And part of their loving conversation might have included a weaving together of science and poetry as she has done so beautifully in this brief work:[20]

Early Morning

The path
The spider makes through the air,
Invisible,
Until the light touches it.
The path
The light makes through the air,
Invisible,
Until it finds the spider's web.

---

1. P. Levi, *The Periodic Table*, Schocken Books, Inc., New York, 1984 (original Italian edition published in 1975).
2. A. Greenberg, *A Chemical History Tour*, John Wiley & Sons, New York, 2000, pp. 10,12.

3. Levi, op. cit., pp. 3–4. For a discussion of Neil Bartlett and the discovery of xenon compounds, see P. Laszlo and G.J. Schrobilger, *Angewandte Chemie, International Edition in English*, Vol. 27, pp. 479–489, 1988.
4. *Los Altos Town Crier*, Dec. 9, 1998. See also The University of Chicago Library Catalog Webpage for Edwin Herbert Lewis (1866–1938), writer and rhetorician, University of Chicago alumnus, and faculty member from 1896 through 1934 at the Lewis Institute in Chicago, now part of Illinois Institute of Technology. His most lasting work is the words to the University of Chicago "Alma Mater." See *http://webpac.lib.uchicago.edu/webpac-bin*.
5. E.H. Lewis, White Lightning, Covici-McGee, Chicago, 1923. Herein also lies a brief story. Pascal Covici, was a relative of my wife Susan (née Covici). He owned a bookstore in Chicago and started to publish books in 1922 (Covici-McGee; Pascal Covici; Covici-Friede). Although *White Lightning* was obscure, Covici became widely respected for the quality of books and their artwork. He is quite fairly said to be the discoverer of John Steinbeck, whose first successful novels were published by Covici-Friede in the 1930s (see T. Fensch, *Steinbeck and Covici: The Story of a Friendship*, Paul S. Eriksson, Burlington, 1979). I, too, have a famous relative—my father's cousin whose biography is also in print: T. Carpenter, *Mob Girl—a Woman's Life in the Underworld*, Simon & Schuster, New York, 1992. But the less said about that, the better.
6. Element 43: technetium (Tc, discovered 1939); 61, promethium (Pm, 1945); 75, rhenium (Re, 1925); 85, astatine (At, 1940); 87, francium (Fr, 1939). Lewis' book was quite up-to-date—Hafnium (Hf) was discovered in 1923, the year *White Lightning* was published, and one can imagine the author happily updating the title of Chapter 72 in the galley proofs. Chapter 86 is titled "Niton" (now Radon); Chapter 91 is titled "Brevium" (now Protactinium). For a brief table on the discovery of the chemical elements, see A.J. Ihde, *The Development of Modern Chemistry*, Harper & Row, New York, 1964, pp. 747–749.
7. Lewis, op. cit., p. 4.
8. Lewis, op. cit., p. 32.
9. Lewis, op. cit., p. 9.
10. Lewis, op. cit., p. 79.
11. *The Book Digest*, 1916, p. 337.
12. Lewis, op. cit., p. 16.
13. Lewis, op. cit., p. 38.
14. J.R. Partington, *A History of Chemistry*, Vol. 4, Macmillan & Co., Ltd., London, 1964, pp. 944, 946.
15. Lewis, op. cit., p. 54.
16. Lewis, op. cit., p. 75.
17. Lewis, op. cit., p. 132.
18. The HArF molecule, is a ground-state molecule observable only at very low temperatures in a solid matrix. Its decomposition to HF and Ar is hugely favored thermodynamically but a tiny (8 kcal/mol) activation barrier allows its covalently held atoms to "shake, rattle and roll" (i.e., vibrate) under these unearthly conditions (see K.O. Christie for a brief discussion of "A Renaissance in Noble Gas Chemistry" in *Angewandte Chemie, International Edition*, Vol. 40, pp. 1419–1421 (2001). One can only hope that Marvin and Jean have greater affinity.
19. See Stanford University American Literary Studies homepage: *http://www-sul.stanford.edu/depts/hasrg/ablit/amerlit/lewis.html*. Note that Ms Lewis taught creative writing and literature at Stanford and co-founded with her husband, author Yvor Winters, a professor at Stanford, a literary journal *Gyroscope*. All of this seems to have occurred three or four years after the fictional Marvin Mahan accepted the endowed chair at Stanford. 'Tis a mystery.
20. J. Lewis, From *The Selected Poems of Janet Lewis*, edited by R.L. Barth, p. 91. Reprinted with the permission of Swallow Press/Ohio University Press, Athens, Ohio. It is most interesting that Janet Lewis was a writer-in-residence at the Djerassi Resident Artists Program. (See C. Djerassi, *This Man's Pill: Reflections on the 50th Birthday of the Pill*, Oxford University Press, Oxford, 2001, p. 239). Professor Djerassi co-authored with Professor Roald Hoffmann the play *Oxygen* cited elsewhere in the present text (*The Art of Chemistry*). Djerassi coined the term "science-in-fiction" (see pp. 151–167 in *This Man's Pill*) and Edwin Herbert Lewis' book *White Lightning* was perhaps, something of an early contribution to this genre.

"TRADE YA BABE RUTH FOR ANTOINE LAVOISIER!"

Babe Ruth is the "Father of Modern Baseball" because his home-run hitting revolutionized the game. In 1918, none of the 16 Major League *teams* hit more than 27 home runs.[1] That year, the 23-year-old Ruth pitched the Boston Red Sox to the World Championship by winning two of their four World Series victories *and* tied for the League lead in home runs by an individual (11).[2] In 1919, his final year with the Red Sox, Ruth hit 29 of his entire team's season total of 33 home runs. The cash-poor Red Sox promptly sold Ruth to the wealthy New York Yankees and have not won the World Series since 1918 (the infamous "Curse of Fenway"). As a Yankee in 1920, Ruth hit 54 home runs—greater than the season totals of 14 of the 15 other Major League teams. He hit 59 home runs in 1921 and established, in 1927, the modern record of 60 that held for 35 years. In 1930 Ruth signed for a salary of $80,000 per year and, when told that he was making more than the president of the United States, was said to have responded: "Well, I had a better year than he did." History suggests that Ruth was probably right. Antoine Laurent Lavoisier, the "Father of Modern Chemistry," totally redefined the field of chemistry and revolutionized it. One of the 40 "Farmers" in the Ferme Générale, Lavoisier might conceivably have had a higher salary in 1789 than did Louis XVI, although I doubt it, but he certainly would not have dared to brag about it. It is also fair to say that Lavoisier had a better year in 1789 than the king did. It is thus eminently fair to call Babe Ruth "The Antoine Lavoisier of Baseball." One might even consider calling Lavoisier "The Babe Ruth of Chemistry."

Trading a Babe Ruth baseball card for a Lavoisier card would seem like "a steal" to *me*. However, the grim reality is that there is today a very active investor's market in baseball trading cards, while chemistry trading cards, "hot" in Belgium and Holland over sixty years ago, are not exactly "selling for a premium." And since I know of no chemists signing 10-year $252 million contracts[3] with no-cut clauses, there will probably be no renaissance in chemistry cards in the near future.[4]

The handsome portraits in Figure 176 are from tobacco trading cards issued by *La Cigarette Oriental de Belgique* in 1929 or 1930 (the narrations on the backs of the cards are in French and Flemish). Figure 176a depicts Carl Wilhelm Scheele, a brilliant Swedish apothecary of very modest means who made the original discovery of oxygen but failed to understand its role in combustion. Figure 176b is a portrait of the aristocratic Antoine Laurent Lavoisier, the "Father of Modern Chemistry." Figure 176c shows the boyishly handsome Humphry Davy whose chemical demonstrations entranced women as well as men at Royal Institution "chemistry nights" starting in 1801. Figure 176d depicts Claude Berthollet. His discovery of the "mass law effect" raised questions that would vex Dalton's atomic theory until the inconsistencies were fully understood. In contrast, the law of combining gas volumes of Gay-Lussac (Figure 176e) strongly supported the atomic theory of the modestly attired Quaker, John Dalton (Figure 176f). Justus Liebig (Figure 176g) was one of the founders of "animal chemistry" (i.e., biochemistry). His work in analytical chemistry

FIGURE 176. ▪ Collectors' cards portraying famous chemists issued in 1938 for *La Cigarette Oriental de Belgique*. See color plates. Although Topps issued bubblegum trading cards in the early 1950s that included Marie Curie and Louis Pasteur, there seems to be no current market for a "Stars of Chemistry" bubblegum trading card series. *Quel domage*! (I am grateful to Jamie and Steve Berman for this information.)

helped tame the "primeval forest" of organic chemistry. Robert Bunsen (Figure 176h), German chemist and physician, set about making a spectrosope for analysis of trace metals, but the light source is now the Bunsen burner known to everybody who has taken high school chemistry. Figure 176i is a portrait of Alfred Nobel, whose wealth was derived from the manufacture of explosives and who willed his fortune to establish the world-renowned prizes bearing his name that include the Nobel Peace Prize.

Figure 177 displays six laboratory scenes. The Chocolat Poulain card depicting Gay-Lussac on the top right of Figure 177 raised the art of the trading card to a new level. The middle right card, dating from the 1930s, is a German rendition of perhaps a turn-of-the-century laboratory, and the card at the bottom right shows Louis Pasteur in his laboratory. The three colorful cards at the left of Figure 177 are from a 1930s series published by an Italian company advertising Liebig's Meat Extract. The top card paints an imagined scene in the laboratory of the legendary eighth-century Arab physician and alchemist Geber (Jabir ibn Hayyan).

The second card at the left of Figure 177 depicts Lavoisier[5] and Berthollet at the Sorbonne in Paris, although neither held an appointment there. Berthollet defended phlogiston theory between 1780 and 1783. During 1783 the true nature of water was discovered by Cavendish and the compound carefully synthesized from the elements and decomposed back into the elements by Lavoisier. Water was found to be a compound consisting of precisely eight parts by weight of oxygen and one part hydrogen. It was not, after all, "dephlogisticated phlogiston." In April 1785, Berthollet became the first French chemist of prominence to support Lavoisier's new theory of oxidation.[6] He remained a friend of Lavoisier, survived the French Revolution while maintaining his integrity, and accompanied Napoleon on his military and scientific expedition to Egypt in 1798. It was during this expedition that Berthollet made the curious discovery of deposits of soda (sodium carbonate) on the shores of salt lakes that led to his formulation of the mass law theory.[7] With his characteristic integrity as a senator, Berthollet voted to depose his friend in 1814 in order to end the disastrous war led by Napoleon.[6]

The scene depicted at the bottom left of Figure 177 is that of Justus Liebig's laboratory in Giessen, and the original is in the Museum of the University of Giessen. Most of the figures have been identified.[8] The figure seated in the front center, dreamily applying the mortar and pestle, is Liebig's student Adolph Friedrich Ludwig Strecker, my great-great-great-great-great-grandfather, chemically speaking (yes, I am also a chemical descendant of Liebig—see the Epilogue in this book). Just to the left of Strecker is Heinrich Will, who will soon succeed Liebig as Director at Giessen. At the rightmost part of this picture is August Wilhelm Hofmann, Liebig's greatest student, who accepted the position of Professor at the Royal College of Chemistry after it was declined by his mentor.[9]

Now what do we make of the six trading cards in Figure 178 that advertise Justus Liebig food products with a facsimile of his autograph (just like those on baseball cards)? Liebig had been writing scientific books about food chemistry since the 1840s. He held a theory about the vital importance of "meat juice" for diet and health.[10] He prepared a "chicken tea" by allowing minced chicken to sit

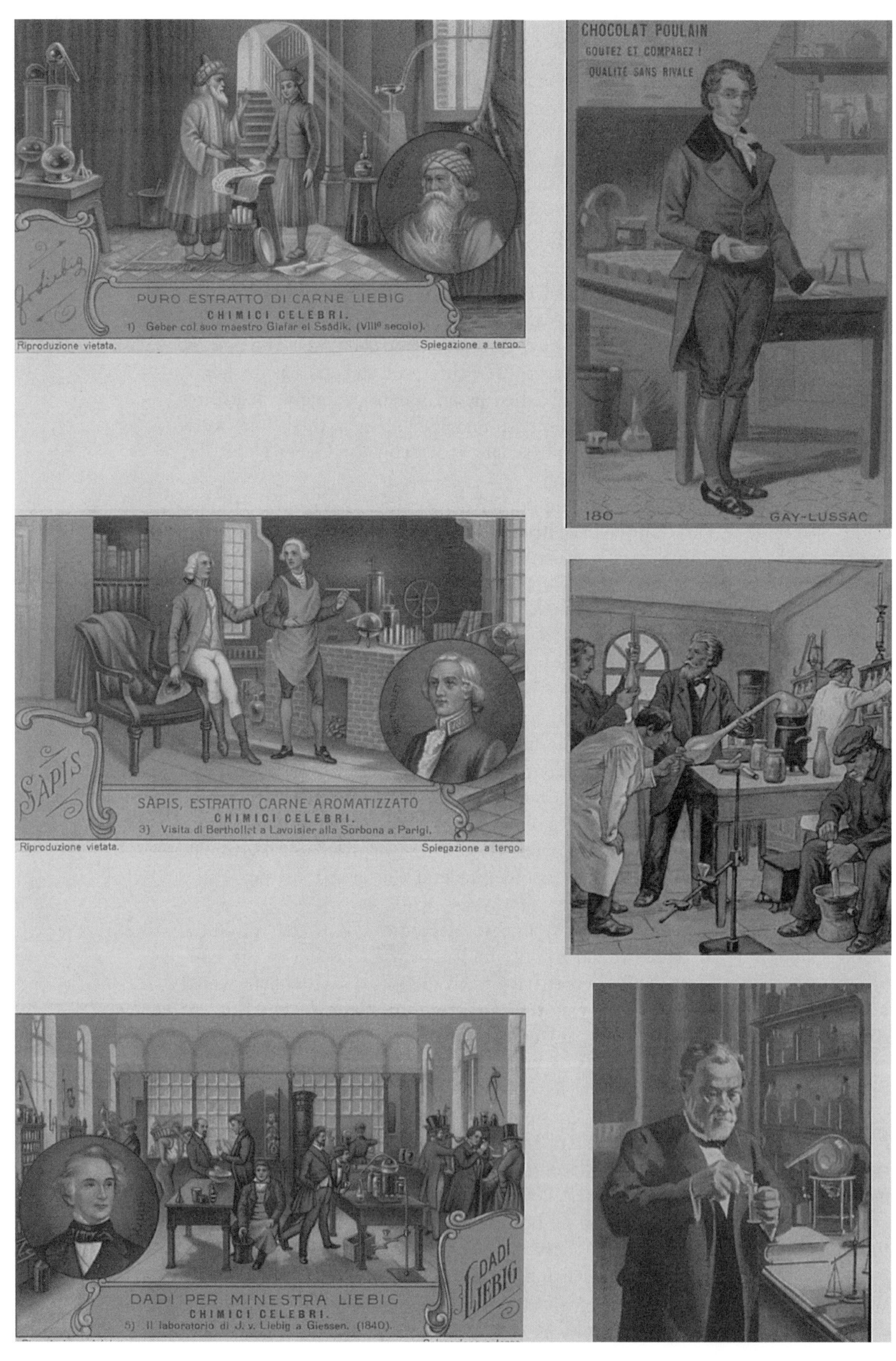

FIGURE 177. ■ Collectors' cards issued during the 1920s and 1930s depicting chemistry laboratories and famous chemists. See color plates. The figure of Gay-Lussac published for *Chocolat Poulain* ("Taste and Com-

FIGURE 178. ▪ Justus Liebig was one of the fundamental pioneers in biochemistry (animal, plant, and food chemistry). He held strong views about the value of "meat juice" in the diet and lent his name to commercial endeavors. (Today there is a company that sells Linus Pauling vitamin C tablets.) The Justus Liebig Company sponsored these Belgian cards printed during the 1930s that tout their line of food items. See color plates. I am grateful to Jamie and Steve Berman for this information.

---

pare! Quality without Rival") is particularly well done. The card at the lower left is a version of the famed drawing of Justus Liebig's laboratory housed at the University of Giessen. The chemist in the front–center, dreamily applying the mortar and pestle, is my *chemical* great-great-great-great-great-grandfather, Adolph Strecker.

in cold water for hours with a few drops of hydrochloric acid added to soften the meat. Frequent drinks were shown to cure all manner of illness. However, his most popular preparation was his meat extract. In 1865, a German railway engineer named Georg Christian Giebert hired Liebig as a director of the newly created Liebig Extract of Meat Company. Shares were sold on the London Stock Exchange.[10] To this day, Liebig Extract of Meat is still sold in Germany. The collectors cards in Figure 178, narrated in Dutch, were issued in the 1930s. They are attractive and on the backs have quite informative histories of chemistry.[11] Card 1 ("Liebig Blocks, Give Strength and Taste") tells of "The Sacred Art in Ancient Egypt" in a succinct but quite informative manner. Card 2 ("Liebig's Ravioli with Egg: The Finest Italian Dish") describes the history of panaceas beginning in Arab lands, including the work of the eighth-century alchemist Geber. Card 3 ("Clear Liebig Bouillon (Cubes): Perfect Chicken") describes the thirteenth-century Catalan mystic Ramond Lully (Ramon Llull) to whom numerous alchemical books have been (fictionally) attributed. Card 4 ("Liebig Tomato Concentrate: Intense Taste") provides an informative glance at the life of Paracelsus, who made and prescribed synthetic medicines consisting of metallic compounds rather than the traditional extracts and distillates derived from plants and animals. Card 5 ("Liebig Meat Extract: The Friend of the Connoiseur") provides an interesting discussion of a late-eighteenth-century English physician and member of the Royal Society, James Higginbotham. Higginbotham claimed to have found the "Philosopher's Stone." When the Royal Society took him to court to substantiate his claims, he poisoned himself in front of his colleagues. Card 6 ("Liebig Aroma: Seasons Food") tells of one Joseph Balsamo, also known as Count Alexander of Cagliostro. He created a stir at the Court of Louis XVI as a miracle worker and "gold-maker." One could only imagine Lavoisier "making minced meat out of him," but the opportunity apparently never came. Marked by numerous scandals, Balsamo moved to Rome, where he was seized during the Inquisition and died in captivity in 1795.

The backs of the portrait cards in Figure 176 describe details of the various chemists' lives, but we admire the statistical line format on the backs of baseball cards that also usually include some catchy little career summary lines. Let us try our hand for an "improved" presentation for the back of the Justus Liebig card (Figure 176g):

No. 57

BARON JUSTUS "THE GATEKEEPER" VON LIEBIG

b. 1803 in Darmstadt, Germany Height: 5' 9"* Weight: 145 lb*

Writes blackboard: Right-handed* Erases blackboard: Left-handed*

Justus, who prefers to be called "Herr Professor Doktor" or sometimes just plain "Baron" to his friends virtually invented precise analysis of organic substances, laid the foundation for understanding organic chemistry, and is one of the fathers of biochemistry. His exacting standards as Editor of the *Annalen der Chemie und Pharmazie* earned him the nickname of "The Gatekeeper." He was the first Major League chemist to sign a lucrative product endorsement contract. His hobby is "xtreme" chemical debate in which he proudly notes "I take no prisoners."

CAREER STATISTICS*

Lectures Started	Lectures Completed	Students Influenced	Analyses Completed	Debating Penalties (in minutes)	Journal Started
3251	3251	705	2348	3655	1

*These are fictional except for the number of students influenced and journal started.

Chemical historian William H. Brock points out that Liebig was a very public scientist who took strong positions on issues of great interest to the public such as farming, nutrition, and public health.[12] Brock compares him,[12] in this regard, to Linus Pauling, who dominated twentieth-century chemistry and also played a public role in the debate on atmospheric testing of nuclear weapons, the environment, vitamin C, peace, and public health.[13] Pauling was also no stranger to controversy, and the back of his (hypothetical) collectors' card would include students influenced (tens of thousands), Nobel Prizes (2), and orthodox Communists and McCarthyites offended (all).

1. D.S. Neft and R.M. Cohen (eds.), *The Sports Encyclopedia: Baseball*, St. Martin's Press, New York, 1989.
2. For those amiable readers who are not baseball *cogniscenti*, pitchers are notoriously weak hitters, and to have one lead the League in home runs borders on the outrageous. Had Ruth been a weak hitter, he probably would have been elected to the Hall of Fame on the basis of his pitching alone.
3. This is an actual contract, signed in December 2000 by baseball player Alex Rodriguez and the Texas Rangers. Incidentally, the 2001 Texas Rangers finished last in their division. We are heading for the billion-dollar sports contract. Perhaps universities should be applying to professional athletes for grants to support academic infrastructure and research.
4. A tobacco company in the Canary Islands, *Obsequio De La Fabrica De Cigarillos*, published a series of collectors' cards of Nobel Prize winners, including chemists, in 1952. Closer to home, the Topps Company printed collectors' cards in 1952 of famous people, including Marie Curie that closely resembled the company's wonderful baseball cards.
5. What did Lavoisier really look like? There are, of course, no photographs, but see M. Beretta, *Imaging a Career in Science—the Iconography of Antoine Laurent Lavoisier*, Science History Publications/USA, Canton, 2001.
6. J.R. Partington, *A History of Chemistry*, Macmillan and Co. Ltd., London, 1962, Vol. 3, pp. 496–516.
7. A. Greenberg, *A Chemical History Tour*, John Wiley and Sons, New York, 2000, pp. 168–170.
8. A.J. Ihde, *The Development of Modern Chemistry*, Harper & Row, New York, 1964, p. 263.
9. W.H. Brock, *Justus Von Liebig—the Chemical Gatekeeper*, Cambridge University Press, Cambridge, UK, 1997, pp. 112–114.
10. Brock, op. cit., pp. 216–233.
11. I am grateful to my former chemistry professor Dr. Arno Liberles for obtaining this translation.
12. Brock, op. cit., pp. viii-ix.
13. L. Pauling, *How to Live Longer and Feel Better*, W.H. Freeman and Co., New York, 1986.

THE SECRET LIFE OF WANDA WITTY[1]

This essay is dedicated to the countless chemistry students who occasionally (very occasionally, mind you) allow their minds to drift off during class as they doodle and dream—for example, the young lady who owned Cooley's *Chemistry* (Figure 179)[2] over a century ago and speaks to us, on a sultry southern day in late May, perhaps, through her drawings.

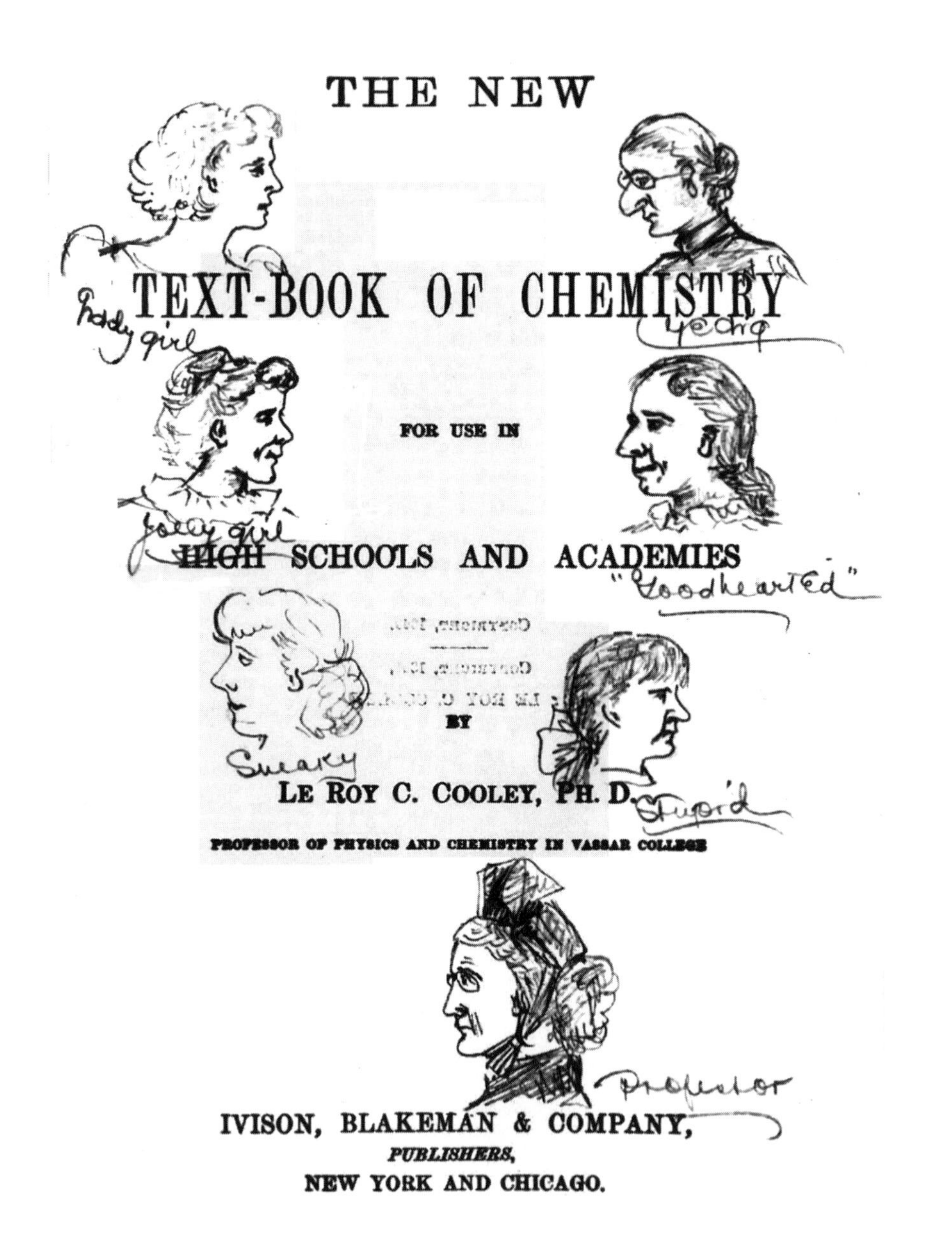

THE NEW

TEXT-BOOK OF CHEMISTRY

FOR USE IN

~~HIGH~~ SCHOOLS AND ACADEMIES

BY

LE ROY C. COOLEY, PH. D.

PROFESSOR OF PHYSICS AND CHEMISTRY IN VASSAR COLLEGE

IVISON, BLAKEMAN & COMPANY,

PUBLISHERS,

NEW YORK AND CHICAGO.

FIGURE 179. ▪ What became of the young lady who doodled so imaginatively on her high school chemistry textbook well over a century ago? What did she daydream about in class? Read the accompanying essay and discover her "secret life."

> So class, we have already learned how to obtain saltpetre (potassium nitrate, you know) from barns and charcoal by burning wood under oxygen-poor conditions. Now we discuss the final component of gunpowder—sulfur. Sulfur is an amorphous yellow solid that is commonly obtained from pyrites by heating in a closed environment and watching the vapors drift lazily to the upper sides of the vessel and

"She is so smart yet so severe in her manner," thought Wanda. "I wonder if she ever had more choices than teaching in this tiny old high school. We call her 'Professor' but I think she could have been an army general and I her adjutant."

"Colonel Witty, we are low on ammunition, short on food and bandages, there are no explosives and a Yankee regiment is advancing in the valley below!" The steam engine that powered the regiment's locomotive could be heard very softly in the distance (*pocketa-pocketa-pocketa*) as advance troops used mortars and sniper fire to clear out Rebel resistance near the dairy farm. "We must stop that ammo train," said the General with grim determination. "I know, m'am, but we are under fire and . . . My God! Goodheart Ed has been wounded! The men are so brave, but they are squeamish." And with that Colonel Wanda Witty ripped the sleeve off of her shirt, grabbed a bottle of whiskey from one of the sergeants, ran with her rifle in the other hand to Ed, drank a swig, poured the remaining contents over the wound, and dressed it with a patch of her uniform. It would still be another two hours before the train would pass by and then Salisbury would be lost and all of those soldiers released from the prison to rejoin their compatriots. *No explosives—a hopeless situation*. And then Colonel Witty was remembering the pyrite formation she had casually observed near the barn a few days earlier. Suddenly, she gave orders to one of the men to collect a few pounds of pyrite, place it into a copper still, and heat it over the fire. Another soldier was ordered to burn a few pounds of wood completely in a vessel having only a small opening for air. She then commandeered three more men to collect the oldest dung from shady moist parts of the barn, expose it to the air for 15 minutes, and then place the mass in boiling water. She then collected campfire ashes (rich in pearl ash or potassium carbonate) and added them to the cooling pot. A mass of white solid appeared, and the solution cooled and was poured through mosquito mesh. *Pocketa-pocketa-pocketa*—but louder now. Every man in the outfit was ordered by Wanda to pour some of the solution into his mess kit and boil off the water until dryness. In this way, saltpetre magically appeared as white crystals in every mess kit were promptly scraped out and collected. Powdery yellow sulfur was scraped from the top of the pyrite heating kettle and the charcoal remains from the wood collected in a vat. *Pocketa-pocketa-pocketa*. "What was that formula that you taught us last week General?!" screamed Wanda. Armed with the formula she mixed the gunpowder, ran for the tracks, felt the wind of a sniper's bullet as it barely grazed her, and placed the gunpowder in a can on the tracks. *Pocketa-pocketa-pocketa!!* The wires were speedily connected to the plunger and that was immediately pushed just in the nick of time—*POCKETA-POCKETA*. . . . Bang! . . . "Bang!?" . . . Not . . . *KABOOOM!!?*

Bang! "Professor" again rapped the ruler on Wanda's desk as the class giggled. "And what was my last sentence?" she asked. "But General . . ." the giggles became gales of laughter. Wanda then noticed that her shirt was fully intact and no cow dung was in sight. "And how many times must I tell you 'Never let your mind wanda' . . . witty, eh?" and with that, "the General" victoriously dismissed the class for the day. And so, on that warm spring day over a century ago, Wanda sat pondering the fate of a S/hero born 100 years too soon and started to dream. "The General's been kidnapped by Prussians!" shouted "Giddy Girl" as she ran into the classroom toward Wanda. Standing up, touching the hilt of her sword, and flattening an errant hair curl with her fingers, she again prepared for the rescue—Wanda Witty—the indomitable, undefeated to the last.

---

1. This is written in homage to humorist James Thurber, author of *The Secret Life of Walter Mitty*, and with happy recollections of reading Thurber's works with my daughter Rachel.
2. L.C. Cooley, *The New Text-Book of Chemistry for Use in High Schools and Academies*, Ivison, Blakeman & Co., New York & Chicago, 1881. The drawings on the title page were drawn by a student, who I think is female by the nature of these drawings and written names of friends; the year 1891 is also written in the same hand.

PAULING'S CARTOON CARNIVAL

> When Linus did a traditional demonstration, dropping bits of sodium into a bowl of water and igniting the hydrogen formed, he added an instructive twist. He became extremely excited, in imitation of a stereotypical mad chemist. He would shout and jump about, run to the other end of the lecture table pour gasoline into a bowl, leap back, and throw in chunks of sodium. Amazed at the lack of an explosion, or any reaction, his frightened students had an unforgettable lesson. Pauling also often posed questions to the class, and the first student to answer was rewarded by receiving a candy bar tossed forth by Linus with gusto.[1,2]

Linus Pauling (1901–1994) was the first to effectively bring the modern quantum mechanics of Schrödinger, Heisenberg, and Pauli in the 1920s to the bench chemists of the 1930s and the university students of the 1940s.[3] The mathematical arguments of the quantum mechanics were beyond the reach of the vast majority of the contemporary chemists of the day. Furthermore, the "elementary" problems tackled by the physicists (H atom, He atom, H_2 molecule, He_2^+ ion–molecule) lacked practical chemical utility and interest. One aspect of Pauling's genius lay in his ability to develop simple conceptual methods for learning—models useful to chemists who lacked the full theoretical foundation. So many of these heuristic concepts and models still grace our twenty-first-century textbooks almost unchanged over three score years. Among these concepts are

1. Electronegativity
2. Hybridization
3. Resonance

(a)

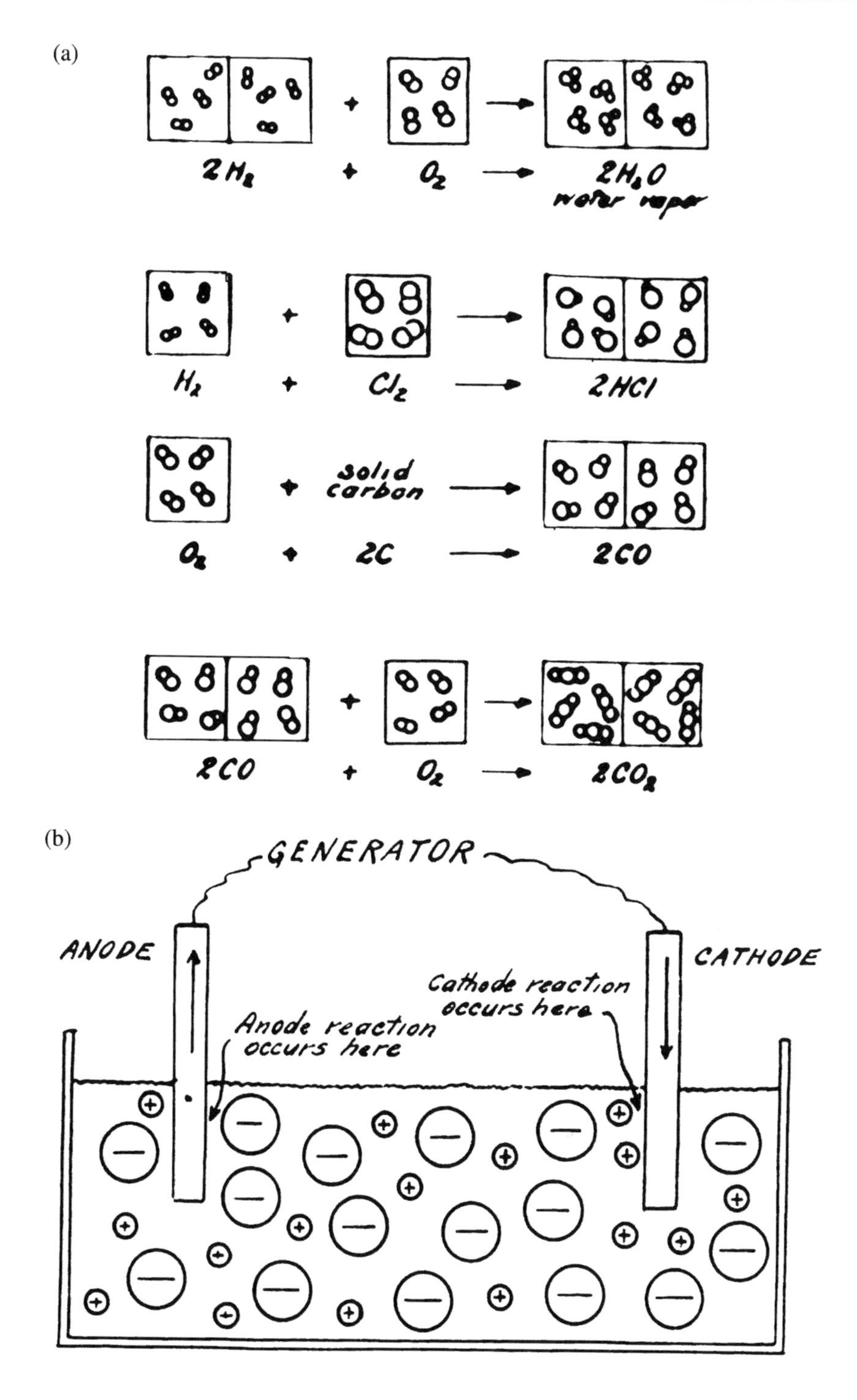

FIGURE 180. ■ Linus Pauling's *General Chemistry*, first edition published in 1947, was a landmark in the teaching of chemistry. The two figures (a and b) shown here are from his self-published draft, printed in 1944, of his famous future textbook [Pauling, *General Chemistry* (privately printed, 1944, courtesy of the family of Linus Pauling)].

Pauling's classic text, *The Nature of the Chemical Bond,*[4] first edition in 1939, third edition in 1960, has a title that is equally magisterial and mysterious. It evokes Lucretius' ancient classic *De Rerum Natura* (*On the Nature of Things*). Indeed, his goal was perhaps no less than leading a diving expedition into the depths of chemical bonds to sample their electronic contours, waves, and currents.

Pauling's 1947 college textbook, *General Chemistry,*[5] was arguably even more influential than *The Nature of the Chemical Bond.* Professors will often employ a draft of a forthcoming textbook in the course they are teaching to test its effectiveness. In a 1944 draft of his future text[6] we see two rather basic, if pedestrian, schematic diagrams likely drafted by the author—one illustrating Avogadro's law of combining volumes of gases (Figure 180a) and the other, an electrochemical cell (Figure 180b). And here we might try to imagine Pauling, as always, impatiently reaching beyond the possible. Crystallography and electron diffraction had provided data on the spatial arrangements of bonded atoms, allowing the modeling of nearly static molecules. Pauling's heuristic models allowed one to plumb these molecules' electronic depths and contours.

So why not go beyond the static models of molecules and picture their birth and death stories—their dynamics—the collisions and rearrangements occurring at dimensions 10,000 times smaller than microscopic, on a timescale of femtoseconds (0.000000000000001 second) at air speeds approaching a mile per second? Impossible during the 1940s (and for decades afterward)! But why not imagine such events—say, the change of electron-density contours during the slow-motion collision of chlorine molecules with the surface atoms of metallic sodium en route to forming crystalline table salt (Figure 181a). In this endeavor Pauling was assisted by the artist-architect-engineer Roger Hayward. And note Hayward's rendition of the electrolysis of water (Figure 181b) and compare it with Pauling's earlier skeletal schematic (Figure 180b).

Figure 182 is from the 1964 book *The Architecture of Molecules,*[7] co-authored by Pauling and Hayward, a coffee-table art book for nonscientists and scientists alike. At the time this book was printed, computer-generated molecular graphics was nothing but a gleam in the eyes of computer scientists. But the paintings by Hayward often included visions of the molecular boundaries and contours derived from the best theoretical studies of the day. Figure 182, fittingly enough, depicts the heme molecule. There are four such molecules embedded within four protein chains in the huge hemoglobin supermolecule: two α chains of 141 amino acid residues each and two β chains of 146 amino acid residues each. Pauling and his co-workers discovered that sickle-cell anemia is an inherited disease characterized by replacement of a single polar amino acid in the β chain, glutamic acid, by a nonpolar amino acid, valine. Pauling's discovery of the α helix structure of proteins, based upon the principles of structural chemistry he helped to pioneer, and the discovery of the absolute molecular basis of sickle-cell anemia were astounding capstones to a career that included winning the 1954 Nobel Prize in Chemistry and the 1962 Nobel Peace Prize.(received in 1963).[3]

(a)

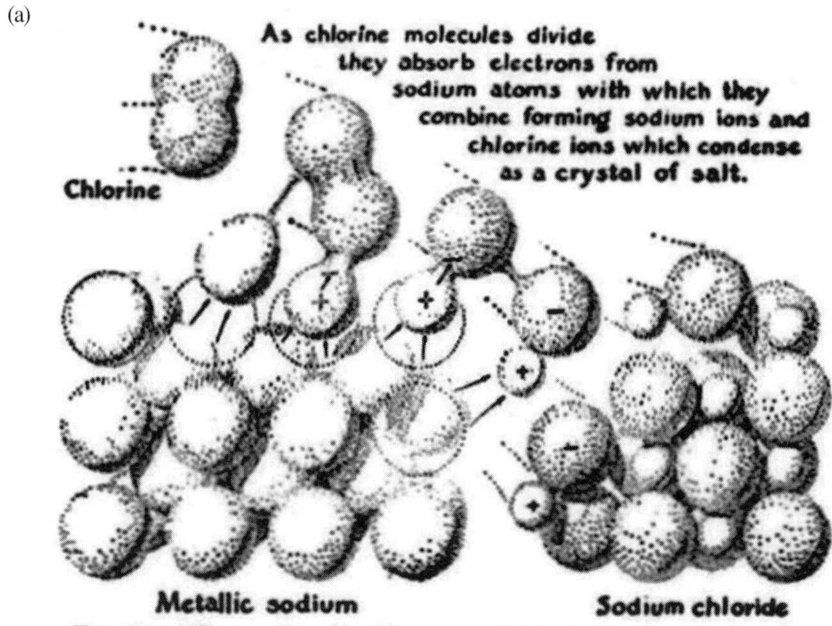

(b)

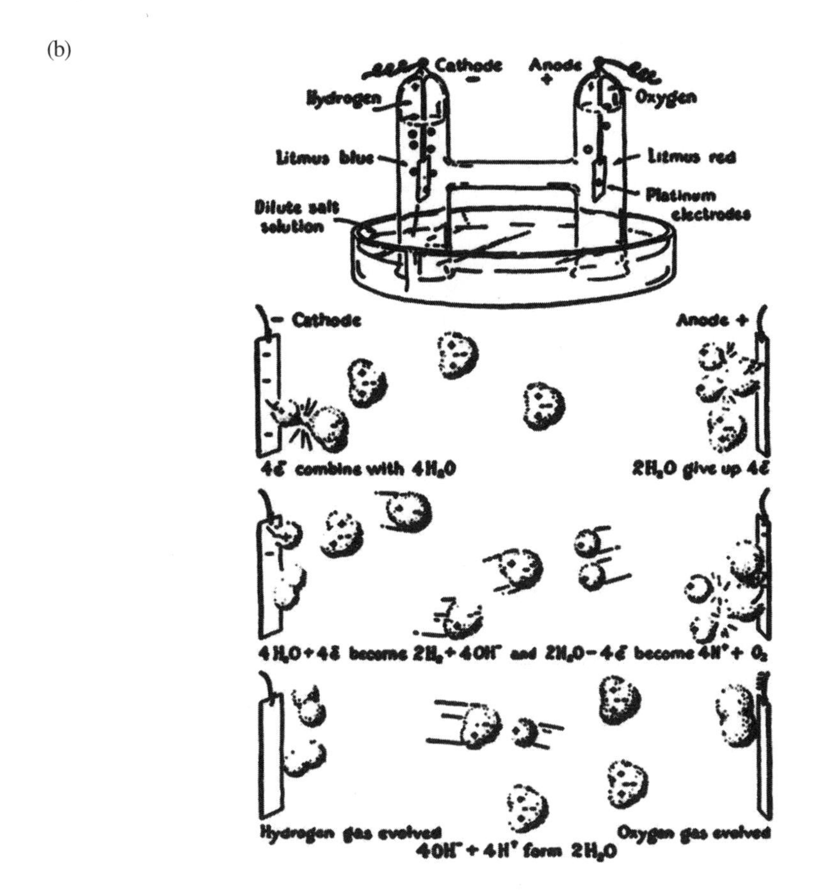

FIGURE 181. ■ The first edition of Pauling's *General Chemistry* united Pauling with artist Roger Hayward. *General Chemistry* by Linus Pauling © 1947 by Linus Pauling. Used with permission of W.H. Freeman and Company. The simple elegance and the dynamics of Hayward's illustrations depicted here certainly contrast with the static drawings in Figure 180. Pauling, as usual, was reaching beyond conventional representations to try to imagine the very rearrangements of nuclei and electron clouds that we understand today occur in femtoseconds (quadrillionths of a second).

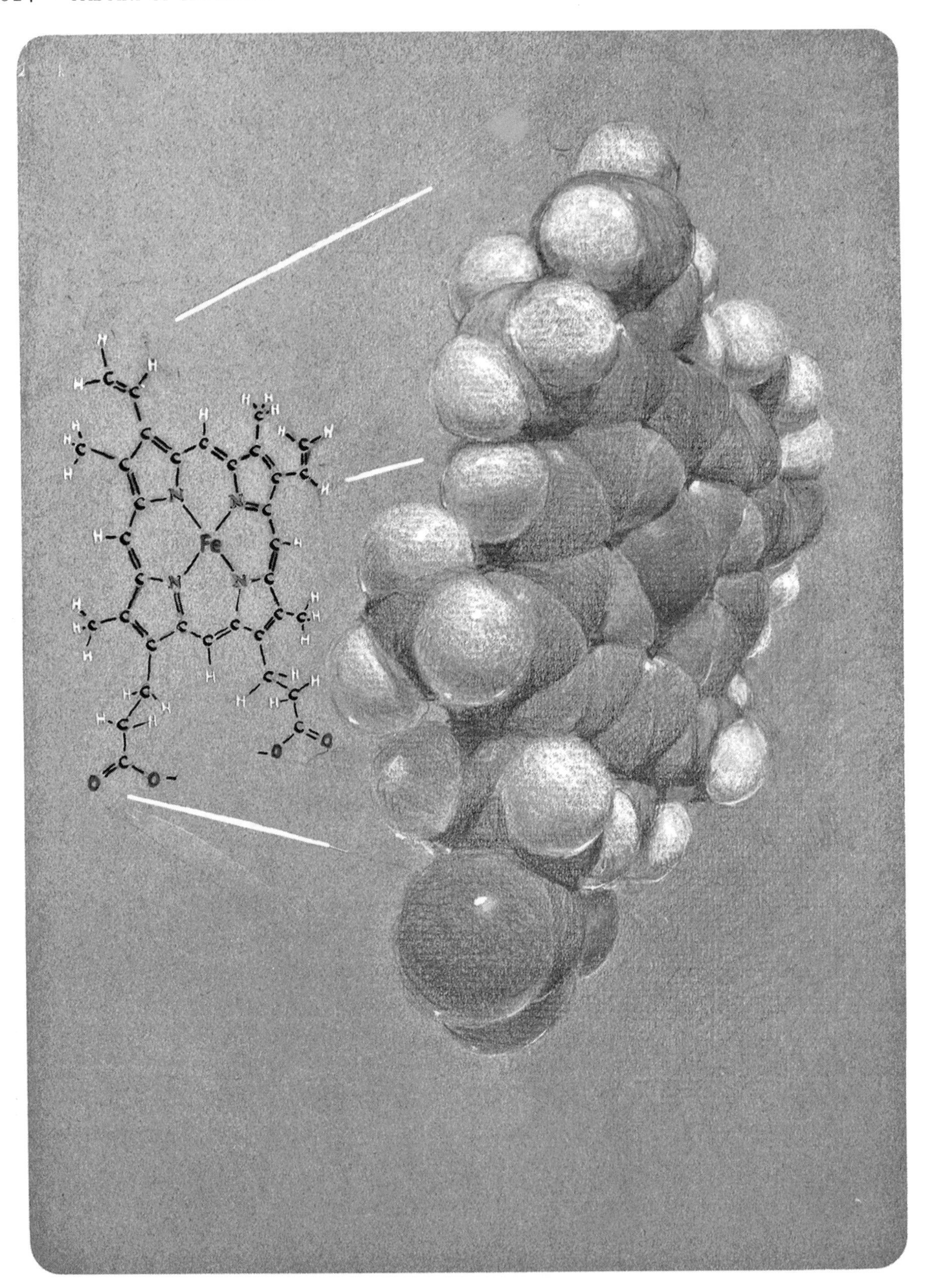

FIGURE 182. ▪ Linus Pauling continued to be "way ahead of the curve" as he co-authored with long-time artist friend Roger Hayward *The Architecture of Molecules* in 1964. Before the era of molecular graphics, the depiction of the heme molecule, one of 57 color drawings in this lovely book, exemplified their enlightened attempt to convey the beauty of chemistry to the public. See color plates. (From *The Architecture of Molecules* by Linus Pauling and Roger Hayward © 1964 W.H. Freeman and Company. Used with permission.)

1. This is a description of Pauling's teaching style furnished second hand by Professor Dudley Herschbach, a pioneer in molecular dynamics and 1986 Nobel Laureate in Chemistry; see Z.B. Maksić and W.J. Orville-Thomas (eds.), *Pauling's Legacy—Modern Modeling of the Chemical Bond*, Elsevier Press, Zurich, 1999, p. 750.
2. For more jubilant dancing, see Edmund Davy's description of brother Humphry's discovery of potassium metal (A. Greenberg, *A Chemical History Tour*, John Wiley & Sons, New York, 2000, p. 182).
3. T. Hager, *Force of Nature*, Simon & Schuster, New York, 1995.
4. L. Pauling, *The Nature of the Chemical Bond*, Cornell University Press, Ithaca, 1939.
5. L. Pauling, *General Chemistry*, W.H. Freeman and Co., San Francisco, 1947.
6. L. Pauling, *General Chemistry* (privately printed in Pasadena, CA, lithoprinted by Edwards Brothers, Inc., Ann Arbor), 1944.
7. L. Pauling and R. Hayward, *The Architecture of Molecules*, W.H. Freeman and Co., San Francisco and London, 1964.

HERE'S TO LONG LIFE (L'CHAIM)!

Linus Pauling (1901–1994) wrote extensively about maintaining good health, and he lived his life accordingly.[1,2] I recall attending his lecture commemorating the Seeley Mudd Chemistry Building at Vassar College in 1987. At the age of 86 he presented an energetic, stimulating hour-long lecture that included a number of excursions, all of which neatly reconnected with the main theme of his talk about diet and health.

As admirable as Pauling's life and health were, at least two famous chemists lived to be centenarians: Michel Eugène Chevreul (1786–1889) and Joel H. Hildebrand (1881–1983). The two overlapped for eight years, and thus the lives of these two chemists combined to span the entire period of 1786–1983. In his 1964 monograph, Partington ponders and fantasizes a bit about the Frenchman Chevreul: "He died in my lifetime and he could have spoken to Lavoisier."[3] Chevreul began his chemical studies under the eye of Nicolas Vauquelin in 1803 at the *Muséum d'Histoire Naturelle* in Paris and retained his association with the museum for almost 90 years.[3,4] Chevreul's first publication appeared in 1806 when he was 20 and treated the analysis of bones. This was two years before Dalton published his atomic theory. He worked at the dawn of organic chemistry and was a pioneer in the daunting world of animal chemistry. In a decade starting around 1813, Chevreul discovered the true nature of the biblical art of soap-making. Saponification, the reaction of lard with lye, yielded fatty acids as well as glycerol. By examining numerous animal fats, he collected a "library" (in modern combinatorial terms) of fatty acids. Chevreul established the test of an unchanging melting point as the measure of purity for his new substances. As early as 1818, Chevreul effectively anticipated Berzelius' definition of isomers, 12 years hence, when he "defined a 'chemical species' as formed from the same elements in the same proportions and in the same arrangements."[3]

Chevreul's most profound impact derived from his studies of colors and dyes. The French established dominance in the textile dyeing industry during the reign of Louis XIV in the late seventeenth century. In 1691 the factory works of

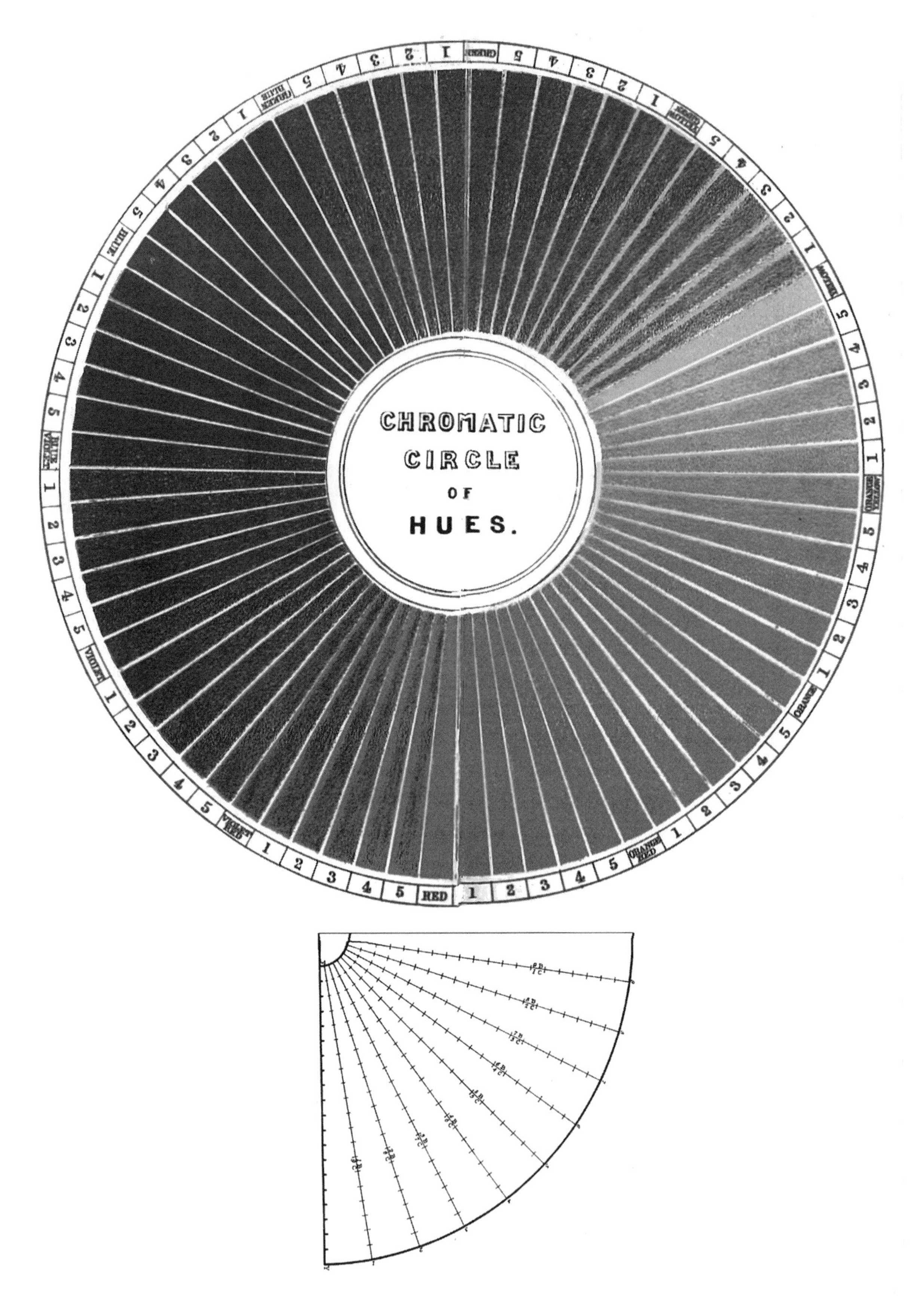

FIGURE 183. ▪ This is the color wheel, in black and white, pioneered during the nineteenth century by the famous early organic chemist Michel Eugène Chevreul (1786–1889), who first published in 1806, and later published his final paper at the age of 97 and sent his last scientific communication to his beloved *Muséum d'Histoire Naturelle* at the age of 102.

the Gobelins family became the official location of the government's dye industry and dominated the European industry for well over a century. In 1824 Chevreul succeeded Claude Berthollet as Director of dyeing at the *Manufactures Royales des Gobelins*. He developed a color wheel in which a third dimension was introduced where white forms the base and black the apex. The wheel was divided into 72 sectors, and the arc connecting the periphery of the circle to the apex was divided into 10 sectors. (Figure 183 shows Chevreul's color circle, in black and white, along with the "vertical arc."[6]) Taken together, the two form a hemisphere with all possible color and shade combinations. His work on the juxtaposition of colors had a profound impact on neo-Impressionists such as George Serat.[4] Chevreul's final paper (concerning vision) was published in 1883, at the age of 97, and his last scientific communication with his beloved *Muséum d'Histoire Naturelle* was presented at the age of 102.[3,4]

Joel H. Hildebrand[7] published his first paper, derived from his doctoral dissertation at the University of Pennsylvania (1906), in the *Journal of the American Chemical Society* in 1907.[8] It was abstracted in Volume 1 of *Chemical Abstracts*.[8] Following a period as an instructor at "Penn," Hildebrand was recruited to Berkeley by Gilbert N. Lewis in 1913—the beginning of a 70-year association with that campus. His lifetime of research work focused in part on electrolytes, their ionic nature only first disclosed by Arrhenius in 1884. However, his most profound impact was as a chemical educator. In the first edition of his influential *Principles of Chemistry*, published in 1918, Hildebrand was the first to include the more recently published (1916) Lewis dot structures in a textbook. In Figures 184a–184c we see drawings from G.N. Lewis' 1923 text,[9] including a page from Lewis' 1902 notebook. Hildebrand's lucid work was published in seven editions, with the final one appearing in 1964—a 46-year run! Hildebrand himself loved running and skiing. He coached the U.S. Ski Team at the 1936 Olympics in Berlin. He celebrated his 77th birthday with a rapid half-mile swim. Hildebrand was president of the American Chemical Society in 1955. His final published work was a history of electrolytes published in 1981,[10] the year of his 100th birthday. In 1982, the Berkeley Chemistry building housing his office was named Hildebrand Hall and the 101-year-old Professor Emeritus commented "The regents got tired of waiting for me to die before naming it."[7]

---

1. L. Pauling, *Vitamin C and the Common Cold*, W.H. Freeman and Co., San Francisco, 1970.
2. L. Pauling, *How to Live Longer and Feel Better*, W.H. Freeman and Co., New York, 1986.
3. J.R. Partington, *A History of Chemistry*, Macmillan and Co. Ltd., London, Vol. 4, 1964, pp. 246–249.
4. C.C. Gillispie, *Dictionary of Scientific Biography*, Charles Scribners Sons, New York, 1971, Vol. III, pp. 240–244.
5. C. Singer, *The Earliest Chemical Industry*, The Folio Society, London, 1948, pp. 255–260.
6. M.E. Chevreul, *The Principles of Harmony and Contrast of Colours and Their Applications to the Arts* (transl. C. Mantel), third edition, George Bell and Sons, London, 1899, pp. 56–57.
7. W.D. Miles and R.F. Gould (eds.), *American Chemists and Chemical Engineers*, Vol. 2, Gould Books, Guilford, 1994, pp. 128–130.
8. J.H. Hildebrand, *Journal of the American Chemical Society*, Vol. 29, pp. 447–455, 1907 [*Chemical Abstracts*, Vol. 1, No. 1832 (1907)].
9. G.N. Lewis, *Valence and the Structure of Atoms and Molecules*, The Chemical Catalog Co., Inc., New York, 1923, pp. 29, 82, 86.
10. J.H. Hildebrand, *Annual Reviews of Physical Chemistry*, Vol. 32, pp. 1–23, 1981.

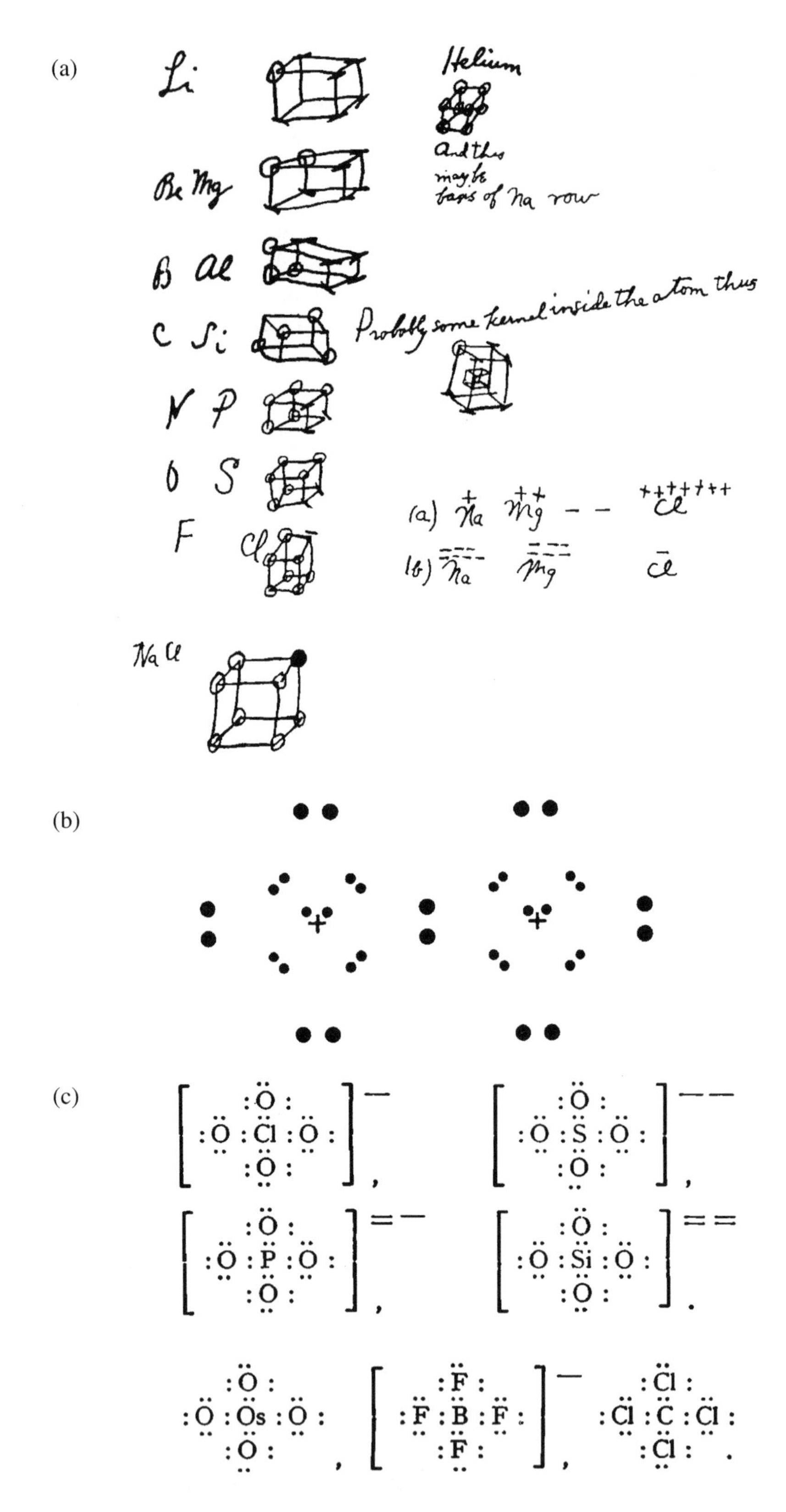

FIGURE 184. ▪ Joel H. Hildebrand (1881–1983) became a faculty member under Gilbert N. Lewis at the University of California, Berkeley in 1913. In his influential *Principles of Chemistry*, first published in 1918, Hildebrand was the first to use the 1916 Lewis structures in a textbook (figures from Lewis' 1923 book are shown here). Hildebrand's seventh and final edition was published in 1964, and his last chemistry paper was published during the year of his hundredth birthday.

SO YOU *WEREN'T* JOKING, MR. FEYNMAN![1]

Way back in the middle of the last century (1959, to be more specific), Nobel Laureate physicist Richard P. Feynman gave an after-dinner talk ("There's Plenty of Room at the Bottom") that challenged scientists to explore the uncharted realms of nanotechnology.[2,3] He observed that the limits to miniaturization were truly reached only at the level of molecules and atoms. Among his wonderfully prescient, and typically bold, speculations was the following thought:[3]

> But I am not afraid to consider the final question as to whether, ultimately . . . in the great future . . . we can arrange the *atoms* the way we want, the very atoms, all the way down! What would happen if we could arrange the atoms one by one the way we want them (within reason, of course; you can't put them so that they are chemically unstable, for example).

At this atomic as well as slightly larger, "mesoscale,"[4] levels, the physical laws would be a tantalizing and unpredictable mixture of classical physics and quantum mechanics.[3]

Two years before Professor Feynman gave his talk, Russia had successfully launched the 184-pound space satellite Sputnik into earth orbit. This small instrument package, launched atop a huge multistage rocket, presaged the incredible advances in miniaturization during the following decade that would culminate in Americans walking on the moon's surface. Miniaturization produced, almost as a by-product, advances in technology that have placed high-powered computers in most modern homes. Indeed, it is now the computer industry that is perhaps the primary driver for nanotechnology, although biomedical science will surely drive this revolution during the twenty-first century. Profound questions on the limits of storage capacity and speed of communication have, in turn, raised the most fundamental questions about matter. For example, the hydrogen atom nucleus has the property of magnetic spin—hydrogens may be spin $\frac{1}{2}$ or $-\frac{1}{2}$, a binary choice of virtually equal probability. Could individual hydrogen atoms attached to molecules be the basis for molecular computers? What about DNA?[5]

Nanotechnology has, until recently, been dominated by a "top–down approach."[2,6] For example, a bulk material such as silicon may be etched to form a complex microchip, by using ultraviolet photolithography, to a level of 100 nanometers (nm).[5] Dimensions smaller than that are extremely expensive to achieve.[6] Feynman's dream of moving atoms one by one was achieved some two decades later by Heinrich Rohrer and Gerd K. Binnig of IBM Zurich who subsequently shared the 1986 Nobel Prize in Physics. The atomic force microscope (AFM), a modification of the scanning tunneling microscope (STM), was successfully employed to physically rearrange matter by pushing atoms one by one. The STM image of the "quantum corral," formed by using an AFM to place 48 iron atoms one at a time into a circle, has become an icon of modern science.[7] Indeed, visionaries have imagined nanoassemblers ("nanobots") capable of assembling nanostructures from atoms or molecules. Here, of course, Avogadro's number is not a friend. As Nobel Laureate Richard E. Smalley—a co-discoverer of C_{60} ("buckyball")—has noted, assemblage of one mole of chemical bonds [the number in just 9 mL (milliliters) of water] would take a harried "nanobot," working at a billion new bonds per second, about 19 million years,[8] although I expect that productivity incentives and overtime pay could reduce this by 10%. Smalley notes that if "nanobots" could be designed to self-replicate prodigiously, then this problem might be overcome. However, this could introduce some potentially dangerous problems—it brings to my cartoonish mind an image of *The Sorcerer's Apprentice in Fantasia:* Mickey Mouse threatened by, say, a trillion dumb, frenetic, and possibly even malevolent brooms. There are other interesting problems as well. Smalley refers to the "fat fingers" and "sticky fingers" problems.[8] He observes that, typically, a new chemical bond will be influenced by about 5–15 atoms near the reaction site. A "nanobot's arms," themselves made of atoms, would have to get close to the bond-making site, thus forcing aside the neighboring atoms needed to determine the bond's fate. Also, one would expect that the atoms to be moved are likely to "stick" to the arms—surface effects are much more significant at nanoscale than at typical macroscopic levels. And finally, Smalley[8] notes the "love" problem. One might try to force atoms together but, as Feynman also remarked, in essence, that the rules of bonding and thermodynamics will ultimately determine whether the atoms "marry" or "cohabitate" as "significant others."

The answer to these limitations might well lie in using the "bottom–up approach"—the self-organization of molecules. A self-help program might paraphrase this approach thus: "Make Avogadro's mumber work for *you*!" And indeed countless trillions of molecules lining up repetitively in harmony with Nature's rules of chemical bonding may be an effective strategy for mass assembly of nanoscopic motors in chemical beakers. Let us examine one elementary example of the bottom–up approach—the molecular switch[9,10] depicted in Figure 185.[9] This molecule is an example of a simple catenane—in this case, two cyclic molecules interlooped much as two links in a chain. Each loop is a separate molecule fully capable of an independent existence. Now this particular molecule is designed so that there may be attraction (or repulsion) between an inner segment of one loop and an inner segment of the other loop. When the molecule in conformation [A°] ("switch open") is deliberately oxidized to [A^+], the center segment of one loop loses an electron and assumes a positive charge, is repelled by the four sets of positive charges in the interior of the other ring and circumrotates to conformation [B^+] in which electrostatic repulsion is reduced. Perhaps a bit surprisingly, when [B^+] gains its electron, by reduction to near-zero bias, to form [B°] ("switched closed"), it does not immediately circumrotate to the original [A°].[9] Both "switch positions" are thus stable and further, controlled, reduction of [B°] returns it to [A°].

There is an interesting point to be made about catenanes. The first catenane was made in 1960, as deliberately as then possible, by Edel Wasserman.[11] He cleverly closed a 34-carbon open-chain molecule, with reactive ester linkages at the two termini, in the presence of an equimolar amount of the corresponding 34-membered cyclic alkane. It is hard to form large rings and the overall yield of the freshly closed 34-membered ring was typically under 20%. But of this 20%, about 1% of the new rings closed while threaded in the "partner" cycloalkane, thus forming a loop about the other ring resulting in a catenane. This extremely low yield reflected the very low probability of catching a long chain through the

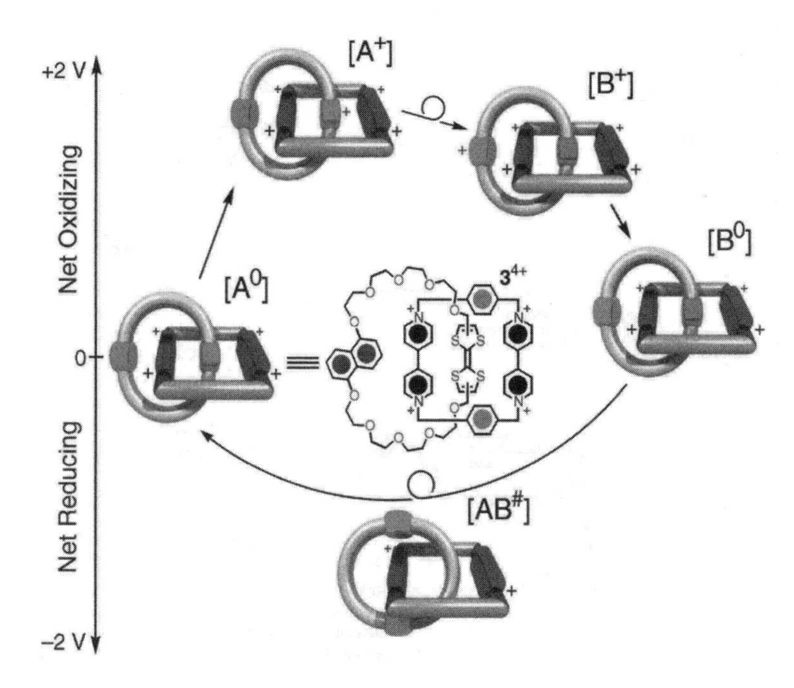

FIGURE 185. ■ A catenane molecule developed by Professor Stoddart and colleagues to function as a controllable (bistable) nanoswitch (see text). [Reprinted with permission from *Accounts of Chemical Research 1*(2001), copyright (2001) American Chemical Society.]

center of a ring and then closing the threaded chain into its own cycle. Such catenanes[12] were barely more than interesting curiosities for years. They raised questions that seemed to be mere chemical semantics—are two neutral unreactive catenated cyclo-$C_{34}H_{68}$ really two molecules physically concatenated, or does this truly represent a new species? Wasserman noted different chromatographic behavior for the catenane relative to its "topological isomer" (the two separated rings taken together). It certainly was a definition for "isomer" unimaginable to Berzelius 130 years earlier. The application of esoteric species such as catenanes to serious problems in technology is a wonderful example of the benefits of pure research not immediately imaginable to practical "administrative types." One of my favorite illustrations of this principle derives from the search for the exotic, short-lived, "administratively-uninteresting" 1,4-benzenediyl by Jones and Bergman in 1972.[13] Fifteen years later, a new class of natural and synthetic anticancer agents were discovered with the reactive 1,4-benzenediyl nucleus at their core.[14]

The modern catenane in Figure 185 was designed so that the loops in the separated rings interact strongly. Furthermore, an aspect of this strong interaction "demands" that the appropriate long-chain precursor "recognize" and move into the interior of the complementary ring compound (see Figure 186).[10] This self-organization precedes the final synthetic step that closes the long chain into a ring and forms catenanes in an incredible 70% yield. Indeed, this is much the same way nature "thwarts" entropy by preorganization via molecular recognition prior to a catalyzed reaction. Figure 186 also illustrates a similar self-organization pathway for forming another exotic topological isomer—a rotaxane in which a long chain threading a cyclic molecule (in high yield due to molecular recognition) is then chemically capped with large terminal groups.

Nanotechnology raises some very interesting questions. For starters, is one gold atom truly gold? That is, is it metallic? The answer must be no since "metallicity" requires total delocalization of electrons over many (hundreds?) of metal atoms. So, when does a cluster of gold atoms begin to resemble gold? Here is another thought–when we think of a machine that stamps out a gear from a metal plate, we would not think of the stamping machine as a "catalyst." Clearly, the machine causes change and is itself unchanged after the operation except for slight wear. However, in real life enzymes eventually "wear down" and lose their potency. But of course the stamping machine is not a "catalyst" because it is causing mechanical, not chemical, change. However, suppose that a nanomachine or "nanobot" somehow facilitates very rapid assembly (i.e., chemical ring closure following self-organization of the two components) of the catenane in Figure 185 from its precursors. This would clearly be a chemical change and the nanomachine, nanoassembler, or "nanobot" would be both chemical catalyst and machine.[15]

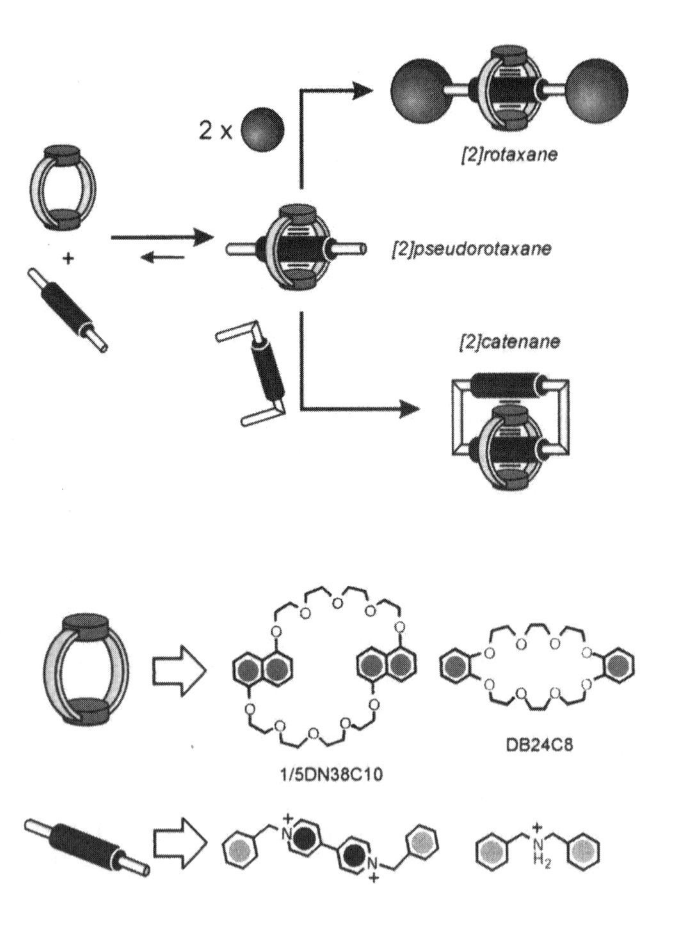

FIGURE 186. ■ The use of molecular recognition and self-organization to form catenanes and rotaxanes capable of performing as controllable nanoswitches (see text). [Reprinted with permission from *Accounts of Chemical Research* (2001), Copyright (2001) American Chemical Society].

But why should this seem so strange? It has been done for billions of years.[15] Molecular recognition drives the organization of the double helix, its replication as well as its transcription to produce specific proteins. Indeed, it is molecular recognition that forms the complex of protein and RNA in the ribosome. It is molecular recognition that attracts transfer RNA (t-RNA) and their specific passenger proteins to the surface of the ribosome. The result is a supermolecular complex that links single amino acids into chains to form specific proteins. Clearly, the ribosome may simultaneously be regarded as a machine and a catalyst.[15] And what about assembling "nanobots" by these same self-organizational principles? Where will the resulting "gray goo" (novelist K. Eric Drexler's term[16]) stop? Will they continue to serve us or mutate into supercolonies having their own agendas and TV programs?

---

1. This is, of course, derived from the title of the autobiographical book by the late Professor Richard P. Feynman: R.P. Feynman, *Surely You're Joking, Mr. Feynman!—Adventures of a Curious Character,* W.W. Norton & Co., New York, 1985.
2. G. Stix, *Scientific American,* Vol. 285, No. 3, pp. 32–37, Sept. 2001.
3. Feynman's 1959 talk "There's Plenty of Room at the Bottom" can be found at *www.its.caltech.edu/~feynman.*
4. The definition of "nanotechnology" offered by Mihail C. Roco of the National Science Foundation indicates, among other things, that materials and systems must "have at least one dimension of about one to 100 nanometers" (see Stix, op. cit.). One nanometer (nm) is, of course, one-billionth of a meter (1×10^{-9} m). The sizes of atoms are typically presented in most textbooks in Ångstroms (Å). An ångstrom (1×10^{-10} m or 1×10^{-8} cm) is one-tenth the size of a nanometer. The diameter of an iron atom (in the metal) is roughly 2.5 Å or 0.25 nm. Thus, 100 nm would correspond to about 400 iron atoms in a line. A white blood cell is about 10 micrometers (10 μm) in diameter (see P. Morrison, P. Morrison, and the Office of Charles and Ray Eames, *Powers of Ten,* Scientific American Books, Inc., New York, 1982. This also corresponds to 10,000 nm or 40,000 iron atoms in a straight line. Objects on the order of 1–100 nm could be termed "mesoscale" if we consider subatomic particles to be at the lower end of the scale.
5. C.M. Lieber, *Scientific American,* Vol. 285, No. 3, pp. 59–64, Sept. 2001.
6. G.M. Whitesides and J.C. Love, *Scientific American,* Vol. 285, No. 3, pp. 39–47 (Sept. 2001).
7. A. Greenberg, *A Chemical History Tour,* John Wiley and Sons, New York, 2000, pp. 295–298.
8. R.E. Smalley, *Scientific American,* Vol. 285, No. 3, pp. 76–77, Sept. 2001.
9. A.R. Pease, J.O. Jeppesen, J. Fraser Stoddart, Y. Luo, C.P. Collier, and J.R. Heath, *Accounts of Chemical Research,* Vol. 34, pp. 433–444, 2001.
10. R. Ballardini, V. Balzani, A. Credi, M.T. Gandolfi, and M. Venturi, *Accounts of Chemical Research,* Vol. 34, pp. 445–455, 2001.
11. E. Wasserman, *Journal of the American Chemical Society,* Vol. 82, pp. 4433–4434, 1982.
12. G. Schill, *Catenanes, Rotaxanes and Knots,* Academic Press, New York, 1971.
13. R.R. Jones and R.G. Bergman, *Journal of the American Chemical Society,* Vol. 94, p. 660, 1972.
14. M.D. Lee, T.S. Dunne, M.M. Siegel, C.C. Chang, G.O. Morton, and D.B. Borders, *Journal of the American Chemical Society,* Vol. 109, pp. 3464–3465, 1987.
15. G.M. Whitesides, *Scientific American,* Vol. 285, No. 3, pp. 78–83, Sept. 2001.
16. K.E. Drexler, *Scientific American,* Vol. 285, No. 3, pp. 74–75, Sept. 2001.

EPILOGUE

This book concludes with two essays that are somewhat personal in nature. Although appropriate to themes developed earlier, they do not fit smoothly into the historical flow of the book. Their placement at the end might at first appear to be exercises in self-indulgence and self-aggrandizement. In fact, although the second essay describes my own chemical genealogy, I am not a significant player in the history of our field. The real points of the genealogy essay are the flow of chemical history, the evolution of education, the fact that at some level these connections matter, and, finally, the sheer delight of discovery. The first essay describes my memories of a young genius, Robert E. Silberglied, during our early and middle teenage years. It is relevant to this book at two levels. Chemistry describes hidden reality. Robert's studies of the communications between butterflies uncovered the hidden reality of sexual selection in the ultraviolet, a light range invisible to us, but apparently like neon signs to them. The chemical pigments and material structures of the wings govern these behaviors. But the true *raison d'être* of this essay is the fun in imagining the youths of some of the geniuses visited so briefly here.

A NATURAL SCIENTIST

Do we recognize truly creative scientific talent when we witness it at an early age? Should we nurture it or just get out of the way and let it develop on its own?

Robert E. Silberglied died on January 13, 1982 in the crash of Air Florida flight 90 in Washington, DC at the age of 36.[1] I last saw him when he was 16, yet he remains for me a vital force—the combination of quirkiness and creativity so typical of a natural-born scientist.[2]

We first met as 12-year-olds at a Brooklyn junior high school. I had then a fairly high opinion of my own scientific abilities: I read natural history, insect, and dinosaur books as well as books by Isaac Asimov, and walked around with complex books on nuclear physics, very prominently displayed, that I hoped would enlighten me if I carried them long enough. At least, they might impress girls! I drew designs of impossible-to-build rockets. My "specialty" was liquid fuels, and I assumed that some day I'd get a hold of hydrazine and liquid oxygen or red fuming nitric acid. My valve designs consisted of multiple layers of cardboard—load the liquids (how?), then run like hell. Fortunately, I was a rocket-design "theoretician" in contrast to other enthusiasts who used available solid fuels and sometimes injured themselves performing real experiments.[3] Around this time, I first heard rumors of a kid who, according to our crowd, was a "scientific brain," and I had to "check him out."

Robert was short, wore plain glasses, and was hopeless in gym class. His best defenses in the Brooklyn schoolyard were his wit and the fact that there was no glory in beating him up. Early in our friendship he took me up to his room in an aged and very modest apartment house and showed me his insect collection. Unlike my own random walks through insect collecting, Robert had a systematic laboratory with homemade nets, insect-killing jars, killing fluid (actually lighter fluid—more on that later), relaxing fluid, and mounting boards and pins. His insects were scrupulously mounted with proper pins sticking through labels bearing their scientific names and places of capture, written in a very tiny but neat hand (more on that, too). Clearly, Robert was doing science on a much higher plane than I was. We hunted insects at the Brooklyn Botanical Gardens, and he taught me how to make a sweep net (metal clothing hanger plus a fine-mesh curtain). He would sweep the opening of the net back and forth along the side of a bush and then place it opening-down on the grass. The results were thrilling: an entomological grab bag—literally hundreds of beetles, bugs, aphids, leaf hoppers, flies, wasps, and ants to harvest at our leisure.

Robert was also known among our junior high group for his famous "scroll wristwatch." He had removed the works from an old wristwatch and had scrolled "gyp sheets," written in a tiny yet neat entomologist's hand, onto the watch's rollers. Why didn't we report him? I guess we were taken with his ingenuity; we probably received a vicarious thrill from this bold, if secret, challenge to the authorities who ran the school, and, in any case, his grades were barely Bs. Frankly, I suspect he never used the watch "in battle" but kept it for security liked a nuclear-tipped ICBM.

Here are some other highlights of Robert's activities; for instance, he would carry (insect-killing) lighter fluid around on rainy days and spray some onto puddles hoping to see an old, religious lady discard a lit cigarette butt into the puddle, and witness "a miracle." He obtained a catalog from a Florida reptile supply house and fantasized about releasing 750 chameleons (only $15.00!) into the junior high. In fact, he did "collect" (I suspect purchase) a load of praying mantis cocoons and did place them in hidden spots throughout the school and was generous enough to give me a dozen. They never did hatch in the malevolent junior high school environment. However, during an abnormally warm February they did hatch in my bedroom, covering the walls with hundreds of minimantids and forever traumatizing my baby sister Roberta.

During our senior year at Erasmus Hall High School I saw little of Robert. I was totally "into" sports, and he had come under the sway of a gifted zoology teacher. Through the grapevine I learned that he had retired the infamous wristwatch and was making As and applying to Cornell's "Ag School." I moved to Englewood, New Jersey during the last half of my senior year and, having read that fossils could be found along the Navesink River, invited him to cross the Hudson and take a bus to Red Bank. Our fossil hunt was unsuccessful. However, with his ever-present gear, Robert caught a bumblebee, meticulously removed the stinger, tied one end of a thread to the abdomen of the disarmed bee and the other to his shirt button, and rode victoriously home, "bee-buzzing" all the way. This occurred in Spring, 1963 and it was the last time I saw Robert.

Some eighteen years later my two children, David and Rachel, who were eight and six years old, respectively, gave me the excuse to reexperience insect collecting. I purchased for them the insect collecting paraphernalia that *I* had always wanted. I thought of Robert and, on a hunch, looked him up in *American Men and Women in Science*. There he was—B.S., Cornell; Ph.D., Harvard; he was now a faculty member at Harvard and a curator of the university's butterfly collection. I wrote to him and reminded him of our earlier experiences together in minute (if painful) detail. He wrote back, congratulated me on my memory, and invited us all to visit him in Cambridge. He signed off now as "Bob." My family was too busy at the time to accept his offer. About five years later I again looked him up and found the word "deceased." Only years later did I learn that he died in the icy waters of the Potomac.

As a teenager, Robert had won a science fair with a study of variations in butterfly markings in different parts of New York City. I suspect that the recognition was incidental to his scientific interests, and I am certain that he had no assistance from his parents or an established scientific laboratory. The aforementioned Smithsonian Institution Website informed me that Robert rose to the rank of Associate Professor and Associate Curator of Lepidoptera, Museum of Comparative Zoology at Harvard. He first obtained an appointment at the Smithsonian Tropical Research Institute in 1976 and finished his career there. He was also an environmental activist who devoted himself to the protection and management of Lignum vitae Key in Florida. Among many scientific accomplishments, Robert was particularly recognized for his studies of the importance of ultraviolet light in the mating habits of butterflies.

The two plates A and B in Figure 187 are from his last paper, "Visual Communication and Sexual Selection among Butterflies," completed a few days before he died and published posthumously.[4,5] Silberglied noted that butterflies have the widest spectral range of vision among animals—the full human visible region as well as ultraviolet down to 300 nm. The male *Colias eurytheme* was found to reflect ultraviolet light (invisible to humans) from the dorsal (back or top) surface of its wings, and this reflection, not its color, attracted females. In the top plate (A), we see a control male of *Colias eurytheme* (top left) dyed yellow by magic marker on the underside of its wing. Although its color and appearance, even on the dorsal surface, have been altered, it still reflects UV light from the dorsal surface and mates successfully. The other five males in A have been dyed various colors on the dorsal surfaces of the wings, both changing visible colors and suppressing UV reflection. These five males were rejected by females. In plate B are shown males of *Colias philodice*, which are known to absorb UV light on the dorsal surfaces of their wings rather than reflect it. Yellow dye was applied to the surface of the control male (upper left) and the same group of colored dyes as in plate A applied dorsally. Since all six dorsal surfaces absorbed UV light, all six males mated successfully.

In his autobiographical book *Naturalist*, the renowned Professor Edward O. Wilson, Robert's Ph.D. mentor, called him "a gifted naturalist and a polymath taxonomist."[6] In a posthumous dedication to the symposium book containing his final written paper, the editors note in part:[7] "Bob Silberglied captivated all who met him with his infectious enthusiasm and boundless energy. This was never more true than at the Symposium meeting, when he was in great form, buzzing with ideas, information and humour. His terrible death, in the Washington air disaster of 13th January 1982, not only robbed biology of a considerable talent but also took from us a delightful friend." Amen to that!

FIGURE 187. ▪ Butterflies painted to prove the role of ultraviolet light in the mating behavior of butterflies. See color plates. This plate is from the final paper presented by Professor Robert E. Silberglied, a boyhood friend of the author and a "most unforgettable character." Reprinted from *The Biology of Butterflies, Symposium of the Royal Entomological Society of London,* No. 11, R. I. Vane-Wright and P. R. Ackery (Editors), pp. 207–223 (1984) by permission of the publisher Academic Press (an imprint of Elsevier Science).

I was most fortunate to be touched by this embryonic genius at such an early stage of my own life, and I have been trying throughout my adult teaching life to find him again.

---

1. The Smithsonian Institution Archives retain Robert's papers (1960–1982) and related materials (to 1984) in a collection (Record Unit 7316) described by Rebecca V. Schoemaker (see *http://www.si.edu/archives/faru7316.htm*).
2. Here I happily acknowledge my friendship with Joel F. Liebman dating back to Fall 1967, when we met as first-year graduate students in chemistry at Princeton University. Although Joel and I were 20 when we met, we were "young adults," scientifically speaking, and thus mostly "formed." It would have been fun to have known Joel at, what entomologist Silberglied might have referred to as, his pupal stage. Joel, too, won science fairs as an early teenager without help of parents or any scientific establishment. Finally, Joel's continued scientific creativity, his "manic" punning, his appreciation for the absurd, and his innate kindness perhaps allow me to imagine Robert as an adult.
3. In 1997 I had the good fortune to meet Dr. Slayton A. Evans, Jr., Professor of Chemistry at the University of North Carolina at Chapel Hill. Slayton, an African-American, grew up in rural Alabama, and like so many other boys in the "Sputnik generation," also developed an interest in rockets. He related to me how one of his (solid-fuel) rockets blew up near the house, leaving a small crater in the yard. Alarmed by the explosion, his mother called out to her precocious son "What happened?" To which he replied "Nothing, Mom." To which she replied "Well, make sure that 'nothing' never happens again." Professor Evans died on March 24, 2001 following a prolonged illness. On the university Webpage the following words appear: "His grace and dignity touched everyone he met in a unique way." Girls were not particularly encouraged into the sciences and engineering during those days. But many were not to be denied and I think of my contemporary Dr. Marye Anne Fox, Chancellor of North Carolina State University and a member of the National Academy of Sciences, as one formidable example. Dr. Joan Valentine, the first female graduate student in chemistry at Princeton, Professor of Chemistry at UCLA, and journal editor, is another. Indeed, my wife Sue would, as a young girl, unfashionably watch and assist her dad, Wilbert Covici, in his building and repair activities. As a result, her mechanical abilities put mine to shame. Similarly, my sister Ilene ("Dee Dee") Franklin was not encouraged to imagine a career as a scientist, but she became a gifted chemistry teacher and is now stimulating her second generation of high school students. My other sister, Roberta, a talented dancer, might have become a business executive.
4. R.E. Silberglied, in *The Biology of Butterflies. Symposium of the Royal Entomological Society of London*, No. 11, R.I. Vane-Wright and P.R. Ackery, Academic Press, London, 1984, pp. 207–223.
5. See Plate 4 and its description in the book cited in Siberglied, op. cit.
6. E.O. Wilson, *Naturalist*, Island Press/Shearwater Books, Washington, DC, 1994, pp. 276–279.
7. R.I. Vane-Wright and P.R. Ackery, op. cit., p. xxi. After the present essay was written, and the manuscript sent to the publisher, an article on the aftermath of Air Florida Flight 90 appeared (*The New York Times Magazine*, August 4, 2002, pp. 36–41). I am grateful for the details about Robert's life that it provides, including the fact that he proposed marriage (joyfully accepted) on the morning of his ill-fated flight.

DESCENDED FROM FALLOPIAN TEST TUBES?

A Long Overdue Question

I am embarrassed to admit that I was 55 years old when I finally researched my chemical genealogy. It is not, however, completely my fault. My "chemical father," Pierre Laszlo, who directed and signed my Ph.D. thesis at Princeton University, was educated in France, had a couple of slightly unconventional twists in his postgraduate education and never really talked with our research group about his chemical lineage, nor were we curious enough to ask. But Pierre's visit to New Hampshire in October 2001, provided some relaxing moments, and I finally popped the question: "Who was my 'chemical grandfather'?" He informed me that it was Edgar Lederer, at the Sorbonne in Paris, an organic chemist and pioneer in chromatography. Pierre also supplied the identity of Lederer's advisor—Richard Kuhn at Heidelberg.

That night, I leaped into the World Wide Web and within minutes discovered that Kuhn, who was a Nobel Laureate, completed his doctorate under Richard Willstätter, who was awarded the Nobel Prize in 1915, that Willstätter studied with Adolph von Baeyer, the 1905 Nobel Laureate, who studied with August Kekulé. I was overjoyed to find this distinguished "family" history and proud to discover familial traits in myself. Kekulé was one of the fathers of structural organic chemistry. He first realized that carbon forms four bonds and that benzene, the fundamental aromatic molecule, is composed of hexagonal rings. I have long considered myself a structural organic chemist and published articles about aromatic compounds and "aromaticity." I have even had the occasional "snake dream," although never so useful as Kekulé's. Among Baeyer's numerous accomplishments was the development of the first theory explaining the high reactivity of angle-strained organic molecules such as cyclopropane. I co-authored, with long-time friend Joel F. Liebman, the book *Strained Organic Molecules* in 1978 and published other papers in this research area. So, this rapid discovery of distinguished chemical lineage back into the mid-nineteenth century (with more to come), was a particular thrill to one whose real family heritage prior to the twentieth century was quite literally demolished, buried, and paved over in some shtetels in eastern and central Europe.

From Germany to France to America—a Twentieth-Century Odyssey

The twentieth century history of my chemical forefathers is fascinating and was dramatically impacted by events in Germany between 1920 and 1950. Following his 1915 Nobel Prize, for purifying chlorophyll and laying the basis for determining its structure, my "great-great grandfather" Richard Willstätter, a Jew, was appointed Professor of Chemistry at the University of Munich.[1,2] Willstätter was a close friend of Fritz Haber, another Jew, who performed the chemical miracle of "fixing" atmospheric nitrogen to form fertilizers (and explosives) for which he won the 1918 Nobel Prize.[2] Haber was a devoted patriot and developed poison gas as a weapon to save the Fatherland during World War I. During the 1920s, he tried to develop a method to extract gold from seawater to allow Germany to pay its war reparations.[3,4] When the Nazis assumed power in 1933, "Jew Haber"[4] was forced to quit his Directorship of the Kaiser Wilhelm Institute for Physical Chemistry in Berlin. He accepted a position in Cambridge, England but died of a heart attack the following year.

In 1924, at the age of 53, Willstätter quit his professor's chair to protest rising anti-Semitism that he felt was the reason for denying an appointment at Munich to Dr. V.M. Goldschmidt.[1,2] His career, abruptly truncated, became increasingly difficult. His life endangered after the Nazis' anti-Semitic campaign of 1938, Willstätter was forced to flee, settling in Switzerland in 1939, but only after considerable difficulty. The principled and gentle Willstätter described his eventful life in an excellent work of scientific autobiography, *Aus meinem Leben* (*From my Life*),[5] published in 1948, some six years after his death.

Richard Kuhn completed his Ph.D. under Willstätter's direction at Munich in 1922, performing early exploratory work on enzymes and carotene.[6,7] Following a period in Munich, then Zurich, Kuhn returned to Germany as a Professor and a Director in the Kaiser Wilhelm Institute (later the Max Planck Institute) for Medical Research in Heidelberg. His work on carotene and vitamin A earned him the 1938 Nobel Prize. However, he had to refuse the award. Hitler had banned acceptance of Nobel Prizes by Germans after the 1935 Nobel Peace Prize was awarded posthumously to a concentration camp prisoner, German pacifist Carl von Ossietzky, who had died of tuberculosis.[8] In 1949, Kuhn finally accepted his medal and certificate in Stockholm. Kuhn's student Lederer developed chromatography in order to separate different isomers of carotene. Lederer was Jewish and fled Germany in March 1933, just four days ahead of the Gestapo's visit to the Kaiser Wilhelm Institute.[9] He was aided by Kuhn's assistant André Lwoff[9] and emigrated to France, and that is how my genealogy became a *généalogie*. Lwoff shared the 1965 Nobel Prize in Physiology.

Confusions and Conceits

And now came the enjoyable task of tracing my distinguished "family" history from Baeyer to Kekulé to . . . but wait! Closer reading of historical sources disclosed that when Willstätter tried to join the inspirational Baeyer's group at Munich, the great master steered him to his department colleague Alfred Einhorn with whom he then completed his doctoral dissertation.[3–5] Einhorn had deduced a structure for cocaine. Willstätter suspected that it was incorrect and asked his director if he could work on the problem. Einhorn refused and gave him an unrelated research problem. Willstätter consulted his true inspirational advisor Baeyer, who finessed this problem in academic politics. Baeyer suspected that tropine was very similar in structure to cocaine and suggested its investigation to Willstätter. After hours, so to speak, Willstätter solved the tropine structure that eventually led to the correct structure of cocaine. Furious, Einhorn refused to speak with Willstätter for years, that is, until his former student became his department director, at which time his views moderated.[3,4] Einhorn gets only the briefest possible mention in Partington's comprehensive history of chemistry.[10] However, he did invent novocaine,[5] and this legacy at least reduces some of the pain from having the steady Einhorn rather than the brilliant Baeyer in my lineage.

In the Willstätter–Einhorn relationship we have an example of a problem that arises occasionally in these genealogical searches—who "gets credit" for the famous scientist—the dissertation advisor or the intellect of true influence? This gets even trickier as we delve deeper into the past. And what about postdoctoral supervisors? Why are they not considered "parents"? Perhaps it is fair to say that it is the dissertation advisor who first spots the accidental puddles in the chemical toddler's lab and provides nurturing to chemical adulthood.

Three Hundred Years of German Chemical Heritage

The academicization of science and awarding of the Doctor of Philosophy (Ph.D.) degree was pioneered in nineteenth-century Germany.[11] The Doctor of Medicine degree is many centuries older. My search, at this point, led to a useful chemical genealogy page on the University of Illinois Website (*www.scs.uiuc.edu/ ~mainzv/Web_Genealogy*). Alfred Einhorn received his Ph.D. under Wilhelm Staedel at Tòbingen, and Staedel completed his doctoral dissertation with Adolph Friedrich Ludwig Strecker, also at Tòbingen. Strecker completed his Ph.D. at Giessen with Justus Liebig, arguably the most important organic chemist of the nineteenth century.

Justus Liebig developed the *kaliapparat* that revolutionized organic analysis. Only after precise analysis could complex formulas of compounds be obtained, leading ultimately to conclusions about valence and structure derived decades later.[12,13] In Figure 177 (bottom left), we see a rendering of the Liebig lab in Giessen. Do I spot great-great-great-great-great-grandfather Strecker? Yes, he's seated center front, mortar and pestle in hand.

Justus Liebig—the father of animal, vegetable, and food chemistry! "O frabjous day! Callooh, Callay!" I chortled in my joy.[14] Liebig was an academic failure in his early schooling and an overly passionate and emotional adult[15]—I recognized clear genetic similarities to myself. In Liebig's history, we encounter another "paternal" ambiguity, closely resembling that of Willstätter. Liebig himself recognized Joseph-Louis Gay-Lussac in Paris as his primary mentor. However, history informs us that Liebig completed his doctoral dissertation with Karl Friedrich Wilhelm Gottlob Kastner, with whom he was dissatisfied, at Erlangen.[12,13] And so we continue back in time:

Johann Friedrich August Göttling
↑
Johann Christian Wiegleb[16]
↑
Ernst Gottfried Baldinger
↑
Christian Andreas Mangold
↑
Georg Erhardt Hamberger
↑
Johann Adolph Wedel
↑
Georg Wolfgang Wedel[17]
↑
Werner Rolfinck[18]

Werner (or Guerner) Rolfinck! Rolfinck was born in 1599 in Hamburg, was educated in Wittenberg, Leiden, Oxford, Paris, and Padua, receiving his M.D. in 1625.[18] In 1638 he established the chemical laboratory in Jena and, in 1641, became the first chemistry professor at this university. Rolfinck's *Chemia in Artis Forman Redacta* (Figure 188), first published in 1661, opposed alchemy and presented medicinal treatments. Rolfinck achieved notoriety for his public dissec-

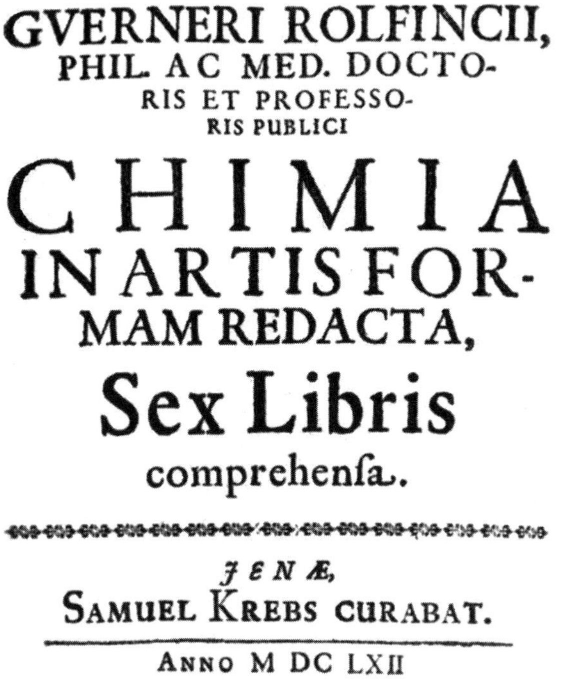

GVERNERI ROLFINCII,
PHIL. AC MED. DOCTO-
RIS ET PROFESSO-
RIS PUBLICI

CHIMIA
IN ARTIS FOR-
MAM REDACTA,
Sex Libris
comprehenſa.

JENÆ,
SAMUEL KREBS CURABAT.

ANNO M DC LXII

FIGURE 188. ■ Title page from the chemistry text published in 1662 by Dr. Guerner Rolfinck, a chemical forefather of the author. Rolfinck represented a critical transition from physicians educated in Padua to chemists teaching in Germany. He achieved some notoriety for his public dissections in Jena of executed criminals, for the purpose of medical instruction. For some time, human dissections were referred to as "rolfincking."

tions, in the anatomical theater he constructed at Jena, of executed criminals. For a period, such dissections were referred to as "rolfincking."[18]

Sixteenth-Century Venetian Anatomists

Rolfinck was the transitional figure in my "family" history. Born in Germany and enjoying a distinguished medical and scientific career in Germany, he had completed his M.D. in Padua under the supervision of Adriaan van den Spiegel (also Spieghel).[19] While Rolfinck could equally be thought of as iatrochemist and surgeon, van den Spiegel clearly belonged to the fields of anatomy and surgery. Born in Brussels in 1578, he was trained at Padua by Girolamo Fabrici and Giulio Casseri, and completed sometime between 1601 and 1604.

The University of Illinois genealogy Website continues to transport me back in time. Having reentered late-sixteenth-century Italy, I start to wonder where this will all end (or really, begin). Geographically closer to biblical lands, I begin to wonder whether Moses, a worker of gold, or Tubal-Cain,[20] the earliest metallurgist described in the Old Testament, were part of my chemical lineage, too.

In fact, my chemical genealogy runs through almost the entire sixteenth century in Padua and then fades. Why the transition from the northern Republic of Venice to Germany in the southern part of the Holy Roman Empire? The City of Venice and its Republic formed a thriving mercantile and cultural region during the fifteenth century. When the French invaded through the northwestern Republic of Milan at the end of that century and the Spanish were enlisted as allies, the two powers began an occupation of Italy that weakened the region. While it remained independent throughout the sixteenth century, the Venetian state was weakened economically and culturally and lost its place as Europe's intellectual center. Casseri[21] received his M.D. at Padua in 1580, and his teacher, Fabrici,[22] took his degree at Padua around 1569. As a faculty member at Padua, Fabrici shared all the traits of a modern university prima donna. In 1588 his students accused him publicly of neglecting his teaching in favor of his research. In 1611 he became notably embroiled with a colleague over the scheduling of courses. Nonetheless, his academic reputation and consulting practice thrived and he took Galileo on as a patient starting in 1606.[22]

Fabrici studied under Gabrielle Fallopio (1523?–1562)[23] at Padua. Fallopio became such a famous figure that many books were falsely attributed to him following his death. Only the *Observationes anatomicae* (1561) can be attributed to him with any certainty.[23] His pioneering studies in anatomy extended the work of an earlier Chair of Anatomy at Padua, Andreas Vesalius. Of course, we know Fallopio best as the discoverer of the fallopian tubes that deliver eggs into the uterus. Fallopio's contemporary, Realdo Colombo[24] (often incorrectly called Matteo Realdo Colombo), was born around 1510. Both he and Fallopio studied under Giovanni Antonio Lonigo, although there are many uncertainties here. Colombo succeeded Vesalius as Chair at Padua. Shortly thereafter, responding to rumors of his criticism by Colombo, Vesalius "denounced him as an ignoramus and a scoundrel."[24] In 1548, Colombo visited Rome and studied anatomy with Michelangelo.[24] and later described the role of the heart in the pulmonary system, thus predating William Harvey.[24] Fallopio succeeded Colombo as the Chair of Anatomy at Padua and also enjoyed angry relations with him just as Vesalius had.[24]

A modern novel, *The Anatomist*,[25] by Frederico Andahazi, recreates the pioneering surgeon Colombo amid the religious inhibitions and hypocrisies of sixteenth-century Padua. In the novel, he has discovered the true function of the *Amor Veneris*, further discussion of which is outside of the realm of my very proper and decent chemistry book. The fictional Colombo enrages the Church and meets a horrific end. In historical fact, Fallopio studied the *Amor Veneris*, and one wonders whether Andahazi invented a composite "Matteo" from Colombo and Fallopio.

Yet further back, before Lonigo, we find Antonio Musa Brasavola (1500–1555), a major contributor to Renaissance pharmacy and physician to Pope Paul III.[26] And while I still have another 2800 years or so to account for in my future efforts to connect Brasavola and Moses, it has, to date, been a thrilling journey.

1. C.C. Gillespie (ed.), *Dictionary of Scientific Biography*, Charles Scribner & Sons, New York, Vol. XIV, 1976, pp. 411–412.
2. B. Narins (ed.), *Notable Scientists From 1900 to the Present*, The Gale Group, Farmington Hills, Vol. 5, 2001, pp. 2438–2441.
3. J.R. Partington, *A History of Chemistry*, Macmillan & Co. Ltd., London, Vol. 4, 1964, pp. 636.
4. *The New Encyclopedia Britannica*, Encyclopedia Britannica, Inc., Chicago, 1986, Vol. 5, pp. 601–602.

5. R. Willstätter, *Aus mein Leben*, A. Stoll (ed.), Verlag Chemie, Weinheim, 1948; transl. L.S. Hornig as From My Life, W.A. Benjamin, New York, 1965.
6. Gillespie, op. cit., Vol. VII (1973), pp. 517–518.
7. Narins, op. cit., Vol. 3, pp. 1281–1282.
8. *The New Encyclopedia Britannica*, op. cit., Vol. 8, p. 1031.
9. See: *http://sun0.mpimf-heidelberg.mpg.de/History/Kuhn1.html* (obtained December 7, 2001).
10. Partington, op. cit., p. 860.
11. J. Ziman, *The Force of Knowledge—the Scientific Dimension of Society*, Cambridge University Press, Cambridge, 1976, pp. 57–62.
12. Partington, op. cit., pp. 294–334.
13. Gillespie, op. cit., Vol. VII, 1973, pp. 329–350.
14. With sincerest apologies to Lewis Carroll for bawdlerizing *Jabberwocky*.
15. A. Greenberg, *A Chemical History Tour*, John Wiley & Sons, New York, 2000, pp. 196–199.
16. Gillespie, op. cit., Vol. XIV, 1976, pp. 332–333.
17. Gillespie, op. cit., Vol. XIV, 1976, pp. 212–213.
18. Gillespie, op. cit., Vol. XI, 1975, p. 511.
19. Gillespie, op. cit., Vol. XII, 1975, p. 577.
20. J. Read, *Humour and Humanism in Chemistry*, G. Bell & Sons Ltd, London, 1947, p. 3.
21. Gillespie, op. cit., Vol. III, 1971, pp. 98–100.
22. Gillespie, op. cit., Vol. IV, 1971, pp. 507–512.
23. Gillespie, op. cit., Vol. IV, 1971, pp. 519–521.
24. Gillespie, op. cit., Vol. III, 1971, pp. 354–357.
25. F. Andahazi, *The Anatomist* (transl. by A. Manguel), Doubleday, New York, 1998.
26. J.R. Partington, *A History of Chemistry*, Macmillan & Co. Ltd., London, Vol. 2, 1961, p. 96.

INDEX